FUNDAMENTALS OF
GEOPHYSICAL
DATA PROCESSING
With Applications to
Petroleum Prospecting

McGraw-Hill
Book Company
New York
St. Louis
San Francisco
Auckland
Düsseldorf
Johannesburg
Kuala Lumpur
London
Mexico
Montreal
New Delhi
Panama
Paris
São Paulo
Singapore
Sydney
Tokyo
Toronto

JON F. CLAERBOUT
Department of Geophysics
Stanford University

Fundamentals of Geophysical Data Processing

WITH APPLICATIONS TO PETROLEUM PROSPECTING

This book was set in Times Roman.
The editors were Patrick A. Clifford and Madelaine Eichberg;
the production supervisor was Judi Frey.
The drawings were done by ECL Art Associates, Inc.
Kingsport Press, Inc., was printer and binder.

Library of Congress Cataloging in Publication Data

Claerbout, Jon F
    Fundamentals of geophysical data processing.

    (McGraw-Hill international series in the earth and
planetary sciences)
    Includes bibliographical references and index.
    1.   Prospecting—Geophysical methods—Data processing.
2.   Petroleum—Geology—Data processing.   I.   Title.
TN271.P4C6   622′.15′02854   75-22013
ISBN 0-07-011117-0

**FUNDAMENTALS OF
GEOPHYSICAL
DATA PROCESSING**
With Applications to Petroleum Prospecting

1 2 3 4 5 6 7 8 9 0 K P K P 7 9 8 7 6

CONTENTS

PREFACE

This book is based at the level of a bachelor's degree in physical science. Experience at Stanford indicates that a one-semester class in engineering systems theory provides helpful additional background. It will be readable to a general science and engineering audience and should be useful to anyone interested in computer modeling and data analysis in physical sciences. Inevitably, the book is strongly flavored by my own research interests which are presently mainly in exploration seismology. However, I have taken an interest in a good many of the data processing problems in general geophysics which have arisen in eight years of teaching graduate students and supervising research. This book is intended to be a textbook rather than a research monograph. The exercises are of a reasonable degree of difficulty for first-year graduate students, and most of them have been thoroughly tested.

I am indebted to a great many friends, associates, and former teachers for much of what I have learned. I have had many fruitful conversations with Steve Simpson, Enders Robinson, and John Burg about time series analysis. Ted Madden taught me much of what is written in this book on stratified media, but most importantly he infected me with the idea that the time had come to go beyond stratified media. John Sherwood and Francis Muir introduced me to reflection seismic prospecting and some unorthodox ways of thinking about it. Several generations of students were a great help in getting many of the "bugs" out of the text and the exercises. Phil Schultz, Don C. Riley and Steve Doherty prepared many of the figures in the final chapters. Mrs. Susana Erlin typed most of the manuscript and finally got the effort all together. My wife, Diane, inspired the continuing effort the project required.

Thanks for financial support over the past eight years is due mainly to Stanford University and the Chevron Oil Field Research Company, but also to the Petroleum Research Fund of the American Chemical Society, the National Science Foundation, and the Air Force Office of Scientific Research. Recent support from the sponsors of the Stanford Exploration Project (SEP) has enabled the rapid development of wave equation seismic data processing introduced in the last chapter. These sponsors are: Amoco, Arco, Chevron, Continental, Digicon, Dutch Shell, ELF-France, Exxon, GSI, INA-Yugoslavia, Mobil, Petrofina-Belgium, Petty Ray, Preussag-Germany, Seiscom Delta, Seismograph Service, Shell, Sun, Teledyne, Texaco, Total-France, Union, U.S. Geological Survey, United Geophysical, and Western Geophysical.

JON F. CLAERBOUT

INTRODUCTION

Geophysical data processing is the use of computers for the analysis of geophysical data. A major task in geophysics is to determine as much as possible about the constitution of the interior of the earth. Where direct penetration is impractical or impossible, seismological, electromagnetic, and gravity measurements are made and the task of making inferences from these measurements is begun. Through systematic application of the laws of physics and the principles of statistics, some of these interpretation tasks can be computerized. When the number of observations is small, it may be satisfactory to match them to the adjustable parameters in known analytic solutions to the equations of classical physics. Today, however, it is common to have massive numbers of observations which contain far more information about the earth than can be modeled by analytic solutions. A typical reflection seismic marine survey ship can collect about a trillion (10^{12}) bits of information per month.

Such massive amounts of data require both statistical reduction and the ability to compute theoretical solutions in many-parameter earth models. Use of digital computers to statistically analyze geophysical data began with the Geophysical Analysis Group (GAG), an industry-supported project at the Massachusetts Institute of Technology which ran from 1953 to 1957 [Ref. 1]. Theoretical geophysical calculations made a great step forward in 1954 when Norman Haskell [Ref. 2] published a famous paper in which he showed how seismic surface waves could be computed for an earth modeled by an arbitrary number of plane parallel layers, each with arbitrarily prescribed physical properties. This enabled utilization of the entire seismic waveform in fitting an arbitrarily stratified earth model. (By "stratified" it is meant that material properties are a function of one coordinate only, usually the depth or radius). Haskell's method has been intensively developed over the last twenty years to the point where we can now readily compute seismic and electromagnetic responses to arbitrary source distributions in any desired stratified model of the earth. Indeed, it seems that the stratified medium has nearly replaced the homogeneous medium as the most popular framework for publication in mathematical geophysics.

Seismograms often consist of hundreds of oscillations, most of which may be inexplicable. Elaborate methodologies have evolved for fitting seismograms to stratified media models with random variations on layer parameters and data. It is astonishing, however, to observe that explosion seismograms with all their complicated, inexplicable details are completely reproducible. Even earthquake seismograms will be reproducible when the source region is small. Thus, the introduction of random variables into data analysis often serves mainly to force fit the data to stratified models.

In contrast to our well developed stratified media tools, most of the questions presently being asked about the earth are really questions about its *departure* from the stratified model. Foremost are the matters of verifying the mechanics of continental drift, understanding earthquakes, and seeking to locate petroleum and minerals. Thus, today, the frontiers in geophysical data processing lie in the reconciliation of field data with two- and three-dimensionally inhomogeneous models of the earth. But before we start we need a good foundation in the traditional material.

Geophysical data processing begins with the study of the sampled time form of filter theory and spectral analysis. The mathematical restraints imposed by the principle of causality are very important. Even arbitrarily complex models of the earth are subject to this principle. Computational stability often hinges on perfectly strict adherence to it. Then basic concepts of resolving power, statistics, and matrices are reviewed preparing the reader for the general theory of least squares along with lots of examples. Least squares has been, of course, the principal vehicle for the reconciliation of data with theoretical models. While it remains in this prominent role, high-resolution techniques (maximum entropy) and robust techniques (the L_1 norm and linear programming) are challenging it.

Following this development of fundamental data processing ideas, the rest of the book is concerned with treating earth models of successively increasing complexity. First, we study multiple reflected plane waves in layered media from a base of only continuity, causality, and energy conservation (no more physics than that). Waves may be calculated from knowledge of the media and the media can be calculated from the waves. Then, the more general theory of mathematical-physical computations in stratified media is introduced and the essential features of finite-difference simulations of partial differential equations are surveyed.

The final chapters are devoted to wave extrapolation and data processing with partial differential equations. I developed this material with my graduate students over the past six years at Stanford University (see Refs. [3] to [8] and [36] and [37]). At present it is virtually unknown among scientists and engineers outside the world of geophysics. The basic objective is like that in holography. A wave field is observed on a plane (the surface of the earth) and the goal is to create a two- or three-dimensional model of the scattering objects to one side of (beneath) the plane. The main problems and the techniques used are quite different from holography. Velocity inhomogeneity, diffraction, interference, and multiple reflection are ubiquitous features of seismic propagation though they are rare in common visual experience. The eye is easily deceived in a house of mirrors or when looking into an aquarium.

The material on wave-equation data processing will soon need to be rewritten. The applicability of this new field to petroleum prospecting ensures further rapid developments. The Stanford Exploration Project (SEP) is now being supported by ten United States oil companies, eight exploration contractors, four European oil companies, and agencies of three different national governments. Many of these groups have their own research efforts. Consequently, in the final sections I have emphasized fundamentals at the expense of details.

We hope that we will someday be able to use earthquake waves for seismic imaging. Then we may see our first sharp pictures of the earth at depths below ten kilometers.

1

TRANSFORMS

The first step in data analysis is to learn how to represent and manipulate waveforms in a digital computer. Time and space are ordinarily regarded as continuous, but for purposes of computer analysis we must discretize them. This discretizing is also called digitizing or sampling. Discretizing continuous functions may at first be regarded as an evil that is necessary only because our data are not always known analytic functions. However, after gaining some experience with sampled functions, one realizes that many mathematical concepts are easier with sampled time than with continuous time. For example, in this chapter the concept of the Z transform is introduced and is shown to be equivalent to the Fourier transform. The Z transform is readily understood on a basis of elementary algebra, whereas the Fourier transform requires substantial experience in calculus.

1-1 SAMPLED DATA AND Z TRANSFORMS

Consider the time function graphed in Fig. 1-1.

To analyze such an observed time function in a computer it is necessary to approximate it in some way by a list of numbers. The usual way to do this is to evaluate

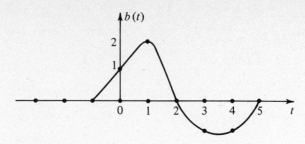

FIGURE 1-1
A continuous time function sampled at uniform time intervals.

or observe $b(t)$ at a uniform spacing of points in time. For this example, such a digital approximation to the continuous function could be denoted by the vector

$$b_t = (\ldots 0, 0, 1, 2, 0, -1, -1, 0, 0, \ldots)$$

Of course if time points were taken more closely together we would have a more accurate approximation. Besides a vector, a function can be represented as a polynomial where the *coefficients* of the polynomial represent the values of $b(t)$ at successive time points. In this example we have

$$B(Z) = 1 + 2Z + 0Z^2 - Z^3 - Z^4 \qquad (1\text{-}1\text{-}1)$$

This polynomial is called a Z transform. What is the meaning of Z in this polynomial? The meaning is not that Z should take on some numerical value; the meaning of Z is that it is the unit delay operator. For example the coefficients of $ZB(Z) = Z + 2Z^2 - Z^4 - Z^5$ are plotted in Fig. 1-2. It is the same waveform as in Fig. 1-1, but it has been delayed.
We see that the time function b_t is delayed n time units when $B(Z)$ is multiplied by Z^n. The delay operator Z is very important in analyzing waves simply because waves take a certain amount of time to get from place to place.

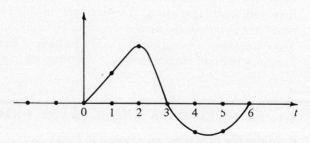

FIGURE 1-2
Coefficients of $Z B(Z)$ are a shifted version of the coefficients of $B(Z)$.

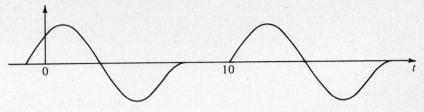

FIGURE 1-3
Response to two explosions.

Another value of the delay operator is that it may be used to build up more complicated time functions from simpler ones. Suppose $b(t)$ represents the acoustic pressure function or the seismogram observed after a distant explosion. Then $b(t)$ is called the *impulse response*. If another explosion occurs at $t = 10$ time units after the first, we expect the pressure function $y(t)$ depicted in Fig. 1-3.

In terms of Z transforms this would be expressed as $Y(Z) = B(Z) + Z^{10}B(Z)$. If the first explosion were followed by an implosion of half strength, we would have $B(Z) - \frac{1}{2}Z^{10}B(Z)$. If pulses overlap one another in time [as would be the case if $B(Z)$ was of degree greater than 10], the waveforms would just add together in the region of overlap. The supposition that they just add together without any interaction is called the *linearity assumption*. This linearity assumption is very often true in practical cases. In seismology we find that—although the earth is a very heterogeneous conglomeration of rocks of different shapes and types—when seismic waves (of usual amplitude) travel through the earth, they do not interfere with one another. They satisfy linear superposition. The plague of nonlinearity arises from large amplitude disturbances. Nonlinearity does not arise from geometrical complications.

Now suppose there was an explosion at $t = 0$, a half-strength implosion at $t = 1$, and another, quarter-strength explosion at $t = 3$. This sequence of events determines a "source" time series, $x_t = (1, -\frac{1}{2}, 0, \frac{1}{4})$. The Z transform of the source is $X(Z) = 1 - \frac{1}{2}Z + \frac{1}{4}Z^3$. The observed y_t for this sequence of explosions and implosions through the seismometer has a Z transform $Y(Z)$ given by

$$Y(Z) = B(Z) - \frac{Z}{2}B(Z) + \frac{Z^3}{4}B(Z)$$

$$= \left(1 - \frac{Z}{2} + \frac{Z^3}{4}\right)B(Z) \qquad (1\text{-}1\text{-}2)$$

$$= X(Z)B(Z)$$

The last equation illustrates the underlying basis of linear-system theory that the output $Y(Z)$ can be expressed as the input $X(Z)$ times the impulse response $B(Z)$.

There are many examples of linear systems. A wide class of electronic circuits is comprised of linear systems. Complicated linear systems are formed by taking the output of one system and plugging it into the input of another. Suppose

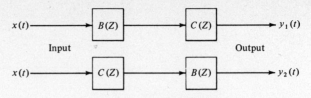

Input   Output

FIGURE 1-4
Two equivalent filtering systems.

we have two linear systems characterized by $B(Z)$ and $C(Z)$, respectively. Then the question arises whether the two combined systems of Fig. 1-4 are equivalent. The use of Z transforms makes it obvious that these two systems are equivalent since products of polynomials commute, i.e.,

$$Y_1(Z) = [X(Z)B(Z)]C(Z) = XBC \qquad (1\text{-}1\text{-}3)$$

$$Y_2(Z) = [X(Z)C(Z)]B(Z) = XCB = XBC \qquad (1\text{-}1\text{-}4)$$

Consider a system with an impulse response $B(Z) = 2 - Z - Z^2$. This polynomial can be factored into $2 - Z - Z^2 = (2 + Z)(1 - Z)$, and so we have the three equivalent systems in Fig. 1-5. Since any polynomial can be factored, any impulse response can be simulated by a cascade of two-term filters (impulse responses whose Z transforms are linear in Z).

What do we actually do in a computer when we multiply two Z transforms together? The filter $2 + Z$ would be represented in a computer by the storage in memory of the coefficients $(2, 1)$. Likewise, for $1 - Z$ the numbers $(1, -1)$ are stored. The polynomial multiplication program should take these inputs and produce the sequence $(2, -1, -1)$. Let us see how the computation proceeds in a general case, say

$$X(Z)B(Z) = Y(Z) \qquad (1\text{-}1\text{-}5)$$

$$(x_0 + x_1 Z + x_2 Z^2 + \cdots)(b_0 + b_1 Z + b_2 Z^2) = (y_0 + y_1 Z + y_2 Z^2 + \cdots) \qquad (1\text{-}1\text{-}6)$$

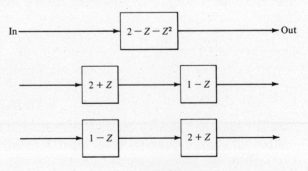

FIGURE 1-5
Three equivalent filtering systems.

```
DIMENSION X(LX),B(LB),Y(LY)
LY = LX+LB-1
DO 10 I=1,LY
10    Y(I) = 0.
DO 20 I=1,LX
DO 20 J=1,LB
20    Y(I+J-1) = Y(I+J-1) + X(I)*B(J)
```

FIGURE 1-6
A computer program to do convolution.

Identifying coefficients of successive powers of Z, we get

$$y_0 = x_0 b_0$$
$$y_1 = x_1 b_0 + x_0 b_1$$
$$y_2 = x_2 b_0 + x_1 b_1 + x_0 b_2 \qquad (1\text{-}1\text{-}7)$$
$$y_3 = x_3 b_0 + x_2 b_1 + x_1 b_2$$
$$y_4 = x_4 b_0 + x_3 b_1 + x_2 b_2$$

$$y_k = \sum_{i=0}^{2} x_{k-i} b_i \qquad (1\text{-}1\text{-}8)$$

Equation (1-1-8) is called a *convolution equation*. Thus, we may say that the product of two polynomials is another polynomial whose coefficients are found by convolution. A simple Fortran computer program which does convolution, including end effects on both ends, is shown in Fig. 1-6. The reader should notice that $X(Z)$ and $Y(Z)$ need not strictly be polynomials; they may contain both positive and negative powers of Z; that is,

$$X(Z) = \cdots \frac{x_{-2}}{Z^2} + \frac{x_{-1}}{Z} + x_0 + x_1 Z + \cdots$$

$$(1\text{-}1\text{-}9)$$

$$Y(Z) = \cdots \frac{y_{-2}}{Z^2} + \frac{y_{-1}}{Z} + y_0 + y_1 Z + \cdots$$

The effect of using negative powers of Z in $X(Z)$ and $Y(Z)$ is merely to indicate that data are defined before $t = 0$. The effect of using negative powers of Z in the filter is quite different. Inspection of (1-1-8) shows that the output y_k which occurs at time k is a linear combination of current and previous inputs; that is, $(x_i, i \leq k)$. If the filter $B(Z)$ had included a term like b_{-1}/Z, then the output y_k at time k would be a linear combination of current and previous inputs and x_{k+1}, an input which really has not arrived at time k. Such a filter is called a *nonrealizable* filter because it could not operate in the real world where nothing can respond now to an excitation which has not yet occurred. However, nonrealizable filters are occasionally useful in computer simulations where all of the data are prerecorded.

EXERCISES

1 Let $B(Z) = 1 + Z + Z^2 + Z^3 + Z^4$. Graph the coefficients of $B(Z)$ as a function of the powers of Z. Graph the coefficients of $[B(Z)]^2$.

2 If $x_t = \cos \omega_0 t$, where t takes on integral values $b_t = (b_0, b_1)$ and $Y(Z) = X(Z)B(Z)$, what are A and B in $y_t = A \cos \omega_0 t + B \sin \omega_0 t$?

3 Deduce that, if $x_t = \cos \omega_0 t$ and $b_t = (b_0, b_1, \ldots, b_n)$, then y_t always takes the form $A \cos \omega_0 t + B \sin \omega_0 t$.

1-2 Z-TRANSFORM TO FOURIER TRANSFORM

We have defined the Z transform as

$$B(Z) = \sum_t b_t Z^t \qquad (1\text{-}2\text{-}1)$$

If we make the substitution $Z = e^{i\omega}$ we have a "Fourier sum"

$$B(Z) = B(e^{i\omega}) = \sum_t b_t e^{i\omega t} \qquad (1\text{-}2\text{-}2)$$

This is like a Fourier integral, and we could obviously do a limiting operation to make it into an integral. Another point of view is that the Fourier integral

$$B(\omega) = \int_{-\infty}^{+\infty} b(t) e^{i\omega t} dt \qquad (1\text{-}2\text{-}3)$$

reduces to the sum (1-2-2) when $b(t)$ is not a continuous function of time but is defined as

$$b(t) = \sum_k b_k \delta(t - k) \qquad (1\text{-}2\text{-}4)$$

where δ is the Dirac delta function.

In the last section we saw that to multiply two polynomials the coefficients must be convolved. The same process in Fourier transform language is that a product in the frequency domain corresponds to a convolution in the time domain.

Although one thinks of a Fourier transform as an integral which may be difficult or impossible to do, the Z transform is always easy, in fact trivial. To do a Z transform one merely attaches powers of Z to successive data points. When one has $B(Z)$ one can refer to it either as a time function or a frequency function, depending on whether one graphs the polynomial coefficients or if one evaluates and graphs $B(Z = e^{i\omega})$ for various frequencies ω. The reader should observe that as ω goes from zero to 2π, $Z = e^{i\omega} = \cos \omega + i \sin \omega$ migrates once around the unit circle in the counterclockwise direction.

If taking a Z transform amounts to attaching powers of Z to successive points of a time function, then the inverse Z transform must be merely identifying coefficients of various powers of Z with different points in time. How can this simple "identification of coefficients" be the same as the apparently more complicated operation of inverse Fourier integrals? The inverse Fourier integral is

$$b(t) = \frac{1}{2\pi} \int_{-\infty}^{+\infty} B(\omega)e^{-i\omega t} \, d\omega \qquad (1\text{-}2\text{-}5)$$

First notice that the integration of Z^n about the unit circle or $e^{in\omega}$ over $-\pi \leq \omega < +\pi$ gives zero unless $n = 0$ because cosine and sine are oscillatory; that is,

$$
\begin{aligned}
\frac{1}{2\pi} \int_{-\pi}^{\pi} e^{in\omega} \, d\omega &= \frac{1}{2\pi} \int_{-\pi}^{\pi} (\cos n\omega + i \sin n\omega) \, d\omega \\
&= \begin{cases} 1 & \text{if } n = 0 \\ 0 & \text{if } n = \text{non-zero integer} \end{cases}
\end{aligned}
\qquad (1\text{-}2\text{-}6)
$$

In terms of our discretized time functions, the inverse Fourier integral (1-2-5) is

$$b_t = \frac{1}{2\pi} \int_{-\pi}^{\pi} (\cdots + b_{-1}e^{-i\omega} + b_0 + b_1 e^{+i\omega} + \cdots)e^{-i\omega t} \, d\omega \qquad (1\text{-}2\text{-}7)$$

Of all the terms in the integrand (1-2-7) we see by (1-2-6) that only the term with b_t will contribute to the integral; all the rest oscillate and cancel. In other words, it is only the coefficient of Z to the zero power which contributes to the integral, reducing (1-2-7) to

$$b_t = \frac{1}{2\pi} b_t \int_{-\pi}^{+\pi} d\omega = b_t \qquad (1\text{-}2\text{-}8)$$

This shows how inverse Fourier transformation is just like identifying coefficients of powers of Z.

In this book and many others, it is common to assume that the time span between data samples $\Delta t = 1$ is unity. To adapt given equations to other values of Δt, one only need replace ω by $\omega \, \Delta t$; that is,

$$\omega_{\text{book}} = \omega_{\text{book}} \Delta t_{\text{book}} = \omega_{\text{true}} \Delta t_{\text{true}} \qquad (1\text{-}2\text{-}9)$$

With Z transforms we have the spectrum given on a range of 2π for ω_{book}. In the limit Δt_{true} goes to zero, ω_{true} has the same infinite limits as the Fourier integral.

When a continuous function is approximated by a sampled function, it is necessary to take the sample spacing Δt_{true} small enough. The basic result of elementary texts is that, if there is no appreciable energy in a Fourier transform for frequencies higher than some frequency ω_{max}, then there is no appreciable loss of information if the sample spacing is $\Delta t = \pi/\omega_{\text{max}}$. In other words, a cosine wave must be sampled at least two points per wavelength. Figure 1-7a shows how insufficient sampling of a sine wave often causes it to appear as a sine wave of lower frequency.

Next we wish to examine odd/even symmetries to see how they are affected in Fourier transformation. The even part e_t of a time function b_t is defined as

$$e_t = \frac{b_t + b_{-t}}{2} \qquad (1\text{-}2\text{-}10)$$

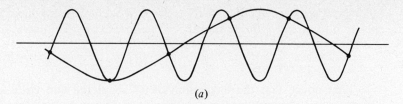

(a)

FIGURE 1-7a
If a high-frequency sinusoid is sampled insufficiently often, it becomes indistinguishable from a lower-frequency sinusoid. For this reason $\omega_{max} = \pi/\Delta t$ is said to be the folding frequency, as higher frequencies are folded down to look like lower frequencies. In practice, quasi-sinusoidal waves are always sampled more frequently than twice per wavelength. Good theoretical reasons for sampling eight or more points per wavelength are developed on pp. 44 to 47.

The odd part is

$$o_t = \frac{b_t - b_{-t}}{2} \quad (1\text{-}2\text{-}11)$$

A function is the sum of its even and odd parts. By adding (1-2-10) and (1-2-11), we get

$$b_t = e_t + o_t \quad (1\text{-}2\text{-}12)$$

Consider a simple, real, even time function such as $(b_{-1}, b_0, b_1) = (1, 0, 1)$. Its transform $Z + 1/Z = 2\cos\omega$ is an even function of ω since $\cos\omega = \cos(-\omega)$. Consider the real, odd time function $(b_{-1}, b_0, b_1) = (-1, 0, 1)$. Its transform $Z - 1/Z = 2(\sin\omega)/i$ is imaginary and odd, since $\sin\omega = -\sin(-\omega)$. Likewise, the transform of the imaginary even function $(i, 0, i)$ is the imaginary even function $i\cos\omega$ and the transform of the imaginary odd function $(-i, 0, i)$ is real and odd. Let r and i refer to real and imaginary, e and o refer to even and odd, and lower-case and upper-case refer to time and frequency functions. A summary of the symmetries of Fourier transformation is shown in Fig. 1-7b.

More elaborate time functions can be made up by adding together the two point functions we have considered. Since sums of even functions are even, and so on, the table of Fig. 1-7b applies to all time functions. Note that an arbitrary time function takes the form $b_t = (re + ro) + i(ie + io)_t$. On transformation of b_t, each of the four individual parts transforms according to the table.

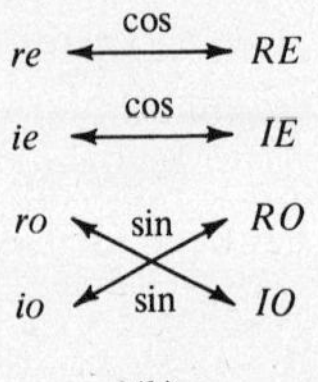

$$re \xleftarrow{\quad \cos \quad} RE$$

$$ie \xleftarrow{\quad \cos \quad} IE$$

$$ro \underset{\sin}{\overset{\sin}{\rlap{\diagup}\diagdown}} RO$$

$$io \qquad\qquad IO$$

FIGURE 1-7b
Mnemonic table illustrating how even/odd and real/imaginary properties are affected by Fourier transformation.

(b)

EXERCISES

1 Normally a function is specified entirely in the time domain or entirely in the frequency domain. When one is known, the other is determined by transformation. Now let us give half the information in the time domain by specifying that $b_t = 0$ for $t < 0$, and half in the frequency domain by giving the real part $RE + RO$ in the frequency domain. How can you determine the rest of the function?

1-3 THE FAST FOURIER TRANSFORM

When we write the expression

$$B(Z) = b_0 + b_1 Z + \cdots \qquad (1\text{-}3\text{-}1)$$

we have both a time function and its Fourier transform. If we plot the coefficients $(b_0, b_1, \ldots)$, we plot the time function. If we evaluate and plot (1-3-1) at numerous real ω, then we have plotted the transform. (Note that for real ω, Z is of unit magnitude; i.e., on the unit circle.) Since ω is a continuous variable and everything in a computer is finite, how do we select a finite number of values ω_k for plotting? The usual choice is to take evenly spaced frequencies. The lowest frequency can be zero. [Note $Z(\omega = 0) = e^{io} = 1$.] A frequency as high as $\omega = 2\pi$ [note $Z(\omega = 2\pi) = e^{i2\pi} = 1$ also] need not be considered, since (1-3-1) gives the same value for it as for zero frequency. Choosing uniformly spaced frequencies between these limits we have

$$\omega_k = \frac{(0, 1, 2, \ldots, M-1)2\pi}{M} \qquad (1\text{-}3\text{-}2)$$

where M is some integer. Now let us abbreviate $B(Z(\omega_k))$ as B_k.

For the special case of an N-point time function where $N = 4$, (1-3-1) may be expressed by the matrix multiplication

$$\begin{bmatrix} B_0 \\ B_1 \\ B_2 \\ B_3 \end{bmatrix} = \begin{bmatrix} 1 & 1 & 1 & 1 \\ 1 & W & W^2 & W^3 \\ 1 & W^2 & W^4 & W^6 \\ 1 & W^3 & W^6 & W^9 \end{bmatrix} \begin{bmatrix} b_0 \\ b_1 \\ b_2 \\ b_3 \end{bmatrix} \qquad (1\text{-}3\text{-}3)$$

where

$$W = e^{2\pi i/N} \qquad (1\text{-}3\text{-}4)$$

It is not essential to choose $N = M$ as we have done in (1-3-3), but it is a convenience. There is no loss of generality because one may always append zeros to a time function before inserting it into (1-3-3). A convenience of the choice $N = M$ is that the matrix in (1-3-3) will then be square and there will be an exact inverse. In fact, the inverse to (1-3-3) may be easily shown to be

$$\begin{bmatrix} b_0 \\ b_1 \\ b_2 \\ b_3 \end{bmatrix} = 1/N \begin{bmatrix} 1 & 1 & 1 & 1 \\ 1 & 1/W & 1/W^2 & 1/W^3 \\ 1 & 1/W^2 & 1/W^4 & 1/W^6 \\ 1 & 1/W^3 & 1/W^6 & 1/W^9 \end{bmatrix} \begin{bmatrix} B_0 \\ B_1 \\ B_2 \\ B_3 \end{bmatrix} \qquad (1\text{-}3\text{-}5)$$

Since $1/W$ is the complex conjugate of W, the matrices of (1-3-3) and (1-3-5) are just complex conjugates of one another. In fact, one observes no fundamental mathematical difference between time functions and frequency functions. This "duality" would be even more complete if we had used a scale factor of $N^{-1/2}$ in each of (1-3-3) and (1-3-5) rather than 1 in (1-3-3) and N^{-1} in (1-3-5). Note also that time functions and frequency functions could be interchanged in the mnemonic table describing symmetries. In fact, our earlier observation that the product of two frequency functions amounts to a convolution of the corresponding two time functions has a dual statement that the product of two time functions corresponds to the convolution of the corresponding two frequency functions. We will not "prove" this duality as it is standard fare in both mathematics and systems theory books. However we will occasionally call upon the reader to realize that in any theorem the meanings of "time" and "frequency" may be interchanged.

In making a plot of the transform B_k for ($k = 0, 1, \ldots, M - 1$) the frequency axis ranges as $0 \le \omega_k < 2\pi$. It is often more natural to display the interval $-\pi \le \omega < \pi$. Since the transform is periodic with period 2π, values of B_k on the interval $\pi \le \omega < 2\pi$ may simply be moved to the interval $-\pi \le \omega < 0$ for display.

Thus, for $N = 8$ one might plot successively

$$B_4 \quad B_5 \quad B_6 \quad B_7 \quad B_0 \quad B_1 \quad B_2 \quad B_3$$

corresponding to values of ω equal to

$$-\pi, \, -\frac{3\pi}{4}, \, -\frac{\pi}{2}, \, -\frac{\pi}{4}, \, 0, \frac{\pi}{4}, \frac{\pi}{2}, \frac{3\pi}{4}$$

One advantage of this display interval is that for continuous time series which are sampled sufficiently densely in time the transform values B_k get small on both ends. If the time series is real, the real part of B_k has even symmetry about B_0; the imaginary part has odd symmetry about B_0. Then, one need not bother to display half the values. Choice of an odd value of N would enable us to put $\omega = 0$ exactly in the middle of the interval, but the reader will soon see why we stick to an even number of data points.

The matrix times vector operation in (1-3-3) requires N^2 multiplications and additions. The rest of this section describes a trick method, called the fast Fourier transform, of accomplishing the matrix multiplication in $N \log_2 N$ multiplications and additions. Since, for example, $\log_2 1024$ is 10, this is a tremendous saving in effort.

A basic building block in the fast Fourier transform is called doubling. Given a series ($x_0, x_1, \ldots, x_{N-1}$) and its sampled Fourier transform ($X_0, X_1, \ldots, X_{N-1}$) and another series ($y_0, y_1, \ldots, y_{N-1}$) and its transform ($Y_0, Y_1, \ldots, Y_{N-1}$), one finds the transform of the interlaced double-length series

$$z_t = (x_0, y_0, x_1, y_1, \ldots, x_{N-1}, y_{N-1})$$

The process of doubling is used many times during the process of computing a fast Fourier transform. As the word *doubling* might suggest, it will be convenient to suppose that N is an integer formed by raising 2 to some integer power. Suppose

$N = 8 = 2^3$. We begin by dividing our eight-point series $x_0, x_1, \ldots, x_7$ into eight different series of one point each. The Fourier transform of each of the one-point series is just the point. Next, we use doubling four times to get the transforms of the four different two point series (x_0, x_4), (x_1, x_5), (x_2, x_6), and (x_3, x_7). We use doubling twice more to get the transforms of the two different four point series (x_0, x_2, x_4, x_6) and (x_1, x_3, x_5, x_7). Finally, we use doubling once more to get the transform of the original eight-point series $(x_0, x_1, x_2, \ldots, x_7)$.

It remains to look into the details of the doubling process.

Let

$$V = e^{i2\pi/2N} = W^{1/2} \qquad (1\text{-}3\text{-}6)$$

$$V^N = e^{i\pi} = -1 \qquad (1\text{-}3\text{-}7)$$

The transforms of two N-point series are by definition

$$X_k = \sum_{j=0}^{N-1} x_j V^{2jk} \qquad (k = 0, 1, \ldots, N-1)$$

$$Y_k = \sum_{j=0}^{N-1} y_j V^{2jk} \qquad (k = 0, 1, \ldots, N-1)$$

The transform of the interlaced series $z_j = (x_0, y_0, x_1, y_1, \ldots, x_{N-1}, y_{N-1})$ is by definition

$$Z_k = \sum_{l=0}^{2N-1} z_l V^{lk} \qquad (k = 0, 1, \ldots, 2N-1)$$

To make Z_k from X_k and Y_k we require two separate formulas: one for $k = 0, 1, \ldots, N-1$, and the other for $k = N, N+1, \ldots, 2N-1$.

First

$$Z_k = \sum_{l=0}^{2N-1} z_l V^{lk} \qquad (k = 0, 1, \ldots, N-1)$$

We split the sum into two parts, noting that x_j multiplies even powers of V and y_j multiplies odd powers.

$$Z_k = \sum_{j=0}^{N-1} x_j V^{2jk} + V^k \sum_{j=0}^{N-1} y_j V^{2jk}$$

$$= X_k + V^k Y_k \qquad (1\text{-}3\text{-}8)$$

We obtain the last half of the Z_k by

$$Z_k = \sum_{l=0}^{2N-1} z_l V^{lk} \qquad (k = N, N+1, \ldots, 2N-1)$$

$$= \sum_{l=0}^{2N-1} z_l V^{l(m+N)} \qquad (k - N = m = 0, 1, \ldots, N-1)$$

$$= \sum_{l=0}^{2N-1} z_l V^{lm}(V^N)^l$$

```
                  SUBROUTINE FORK(LX,CX,SIGNI)
C FAST FOURIER                                          2/15/69
C                      LX
C  CX(K) = SQRT(1/LX) SUM (CX(J)*EXP(2*PI*SIGNI*I*(J-1)*(K-1)/LX))
C                      J=1                 FOR K=1,2,...,(LX=2**INTEGER)
                  COMPLEX CX(LX),CARG,CEXP,CW,CTEMP
                  J=1
                  SC=SQRT(1./LX)
                  DO 30 I=1,LX
                  IF(I.GT.J) GO TO 10
                  CTEMP=CX(J)*SC
                  CX(J)=CX(I)*SC
                  CX(I)=CTEMP
   10             M=LX/2
   20             IF(J.LE.M) GO TO 30
                  J=J-M
                  M=M/2
                  IF(M.GE.1) GO TO 20
   30             J=J+M
                  L=1
   40             ISTEP=2*L
                  DO 50 M=1,L
                  CARG=(0.,1.)*(3.14159265*SIGNI*(M-1))/L
                  CW=CEXP(CARG)
                  DO 50 I=M,LX,ISTEP
                  CTEMP=CW*CX(I+L)
                  CX(I+L)=CX(I)-CTEMP
   50             CX(I)=CX(I)+CTEMP
                  L=ISTEP
                  IF(L.LT.LX) GO TO 40
                  RETURN
                  END
```

FIGURE 1-8

A program to do fast Fourier transform. Modified from Brenner. Calling this program twice returns the original data. SIGNI should be $+1.$ on one call and $-1.$ on the other. LX must be a power of 2.

$$= \sum_{l=0}^{2N-1} z_l V^{lm}(-1)^l$$

$$= \sum_{j=0}^{N-1} x_j V^{2jm} - V^m \sum_{j=0}^{N-1} y_j V^{2jm}$$

$$= X_m - V^m Y_m$$

$$Z_k = X_{k-N} - V^{k-N} Y_{k-N} \qquad (k = N, N+1, \ldots, 2N-1) \qquad (1\text{-}3\text{-}9)$$

The first machine computation with this algorithm known to the author was done by Vern Herbert, who used it extensively in the interpretation of reflection seismic data. He programmed it on an IBM 1401 computer at Chevron Standard Ltd., Calgary, Canada in 1962. Herbert never published the method. It was rediscovered and widely publicized by Cooley and Tukey in 1965. Thus it has come to be known as the Cooley and Tukey algorithm. (A good reference to literature on the subject is Ref. [9].)

EXERCISES

1 Verify that for an arbitrary $N \times N$ case the matrix of (1-3-5) is indeed the inverse of the matrix of (1-3-3).

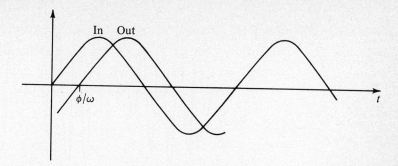

FIGURE 1-9.
A sinusoid sin ωt goes into a filter and a delayed sinusoid sin $(\omega t - \phi)$ comes out.

1-4 PHASE DELAY AND GROUP DELAY

Some filters make drastic changes to signals propagating through. Other filters do their best to make little or no change. In the latter category are transducers and recorders. In such cases, the principal form of signal change may be merely delay. One way to characterize the delay of a filter is to put in a sinusoid and compare its phase to that of the output. See Fig. 1-9.

If the input is sin ωt and the output is sin $(\omega t - \phi)$ then the so-called phase delay t_p is given by solving

$$\sin (\omega t - \phi) = \sin \omega (t - t_p)$$

$$\omega t - \phi = \omega t - \omega t_p$$

$$t_p = \frac{\phi}{\omega}$$

(1-4-1)

A more interesting kind of delay is called group delay. It is analogous to group velocity in wave propagation theory. Indeed, in the modeling of wave propagation on a computer the propagation of a wave from say point A to point B may be simulated with a filter.

When the waveshape observed at A differs from that at point B but the energy envelope at A resembles with delay that at B, then we have a situation where the idea of group velocity, meaning the energy envelope velocity, may be very useful. The sum of two cosine waves of slightly differing frequencies will beat together. Refer to Fig. 1-10.

When such a waveform goes through a filter, each frequency may suffer a different delay and the result will be that the envelope or beat will have a delay which differs from the phase delay of either frequency. The envelope delay, or group delay, may not even resemble the average of the phase delays of the two frequencies. We may understand this as follows: The input waveform x_t is

$$x_t = \cos \omega_1 t + \cos \omega_2 t$$

(1-4-2)

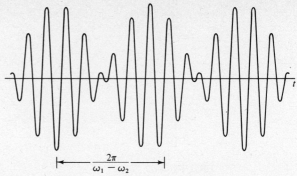

FIGURE 1-10
A graph of $\cos \omega_1 t + \cos \omega_2 t$ looks like an amplitude-modulated cosine of the average frequency.

By using a trigonometric identity

$$x_t = 2 \cos \left(\frac{\omega_1 + \omega_2}{2} t \right) \cos \left(\frac{\omega_1 - \omega_2}{2} t \right) \qquad (1\text{-}4\text{-}3)$$

we see that the sum of two cosines looks like a cosine of the average frequency multiplied by a cosine of half the difference frequency. Since the frequencies are taken close together, the difference frequency factor represents a slowly variable amplitude on the average frequency. Now let us take the output of the filter y_t to be

$$y_t = \cos (\omega_1 t - \phi_1) + \cos (\omega_2 t - \phi_2) \qquad (1\text{-}4\text{-}4)$$

In taking the output of the filter to be of the form of (1-4-4), we have assumed that neither frequency was attenuated. To allow differential attenuation of the two frequency components would greatly complicate the discussion. Utilizing the same trigonometric identity on (1-4-4), we get

$$y_t = 2 \cos \left(\frac{\omega_1 + \omega_2}{2} t - \frac{\phi_1 + \phi_2}{2} \right) \cos \left(\frac{\omega_1 - \omega_2}{2} t - \frac{\phi_1 - \phi_2}{2} \right) \qquad (1\text{-}4\text{-}5)$$

Rewriting the beat factor in terms of a time delay t_g, we have

$$\cos \left[\frac{\omega_1 - \omega_2}{2} (t - t_g) \right] = \cos \left(\frac{\omega_1 - \omega_2}{2} t - \frac{\phi_1 - \phi_2}{2} \right)$$

or

$$(\omega_1 - \omega_2) t_g = \phi_1 - \phi_2$$

or the group delay is given by

$$t_g = \frac{\phi_1 - \phi_2}{\omega_1 - \omega_2} = \frac{\Delta \phi}{\Delta \omega} \qquad (1\text{-}4\text{-}6)$$

In practice one never has two pure cosines but a band of frequencies. The group delay is then a frequency-dependent function given by $t_g = d\phi/d\omega$. The phase angle ϕ may be computed as the arctangent of the ratio of imaginary to real parts of the Fourier transform, namely $\phi(\omega) = \arctan [\operatorname{Im} B(\omega)/\operatorname{Re} B(\omega)]$. It is sometimes convenient to recall the definition of complex logarithm. Say,

$$B = re^{i\phi}$$

$$\ln B = \ln |r| + \ln e^{i\phi}$$

$$= \ln |r| + i\phi$$

So

$$\phi = \operatorname{Im} \ln B$$

$$t_g = \frac{d\phi}{d\omega} = \operatorname{Im} \frac{d}{d\omega} \ln B(\omega) \qquad (1\text{-}4\text{-}7)$$

$$= \operatorname{Im} \frac{1}{B} \frac{dB}{d\omega}$$

A convenient approximation when B is sampled in a computer is

$$t_g \approx \frac{2}{\Delta\omega} \operatorname{Im} \frac{B_{k+1} - B_k}{B_{k+1} + B_k} \qquad (1\text{-}4\text{-}8)$$

An important aspect of wave propagation theory is the distinction of phase velocity from group velocity. These are similar to phase delay and group delay. For example, if waves propagate along a two-dimensional surface, the phase function may be given by

$$\phi(x, y) = k_x(x - x_0) + k_y(y - y_0) \qquad (1\text{-}4\text{-}9)$$

Here (x_0, y_0) is the location of the filter input and (x, y) is where the phase is observed (like the filter output). The symbols k_x and k_y denote the "spatial frequencies," that is, k_x is 2π divided by the wavelength measured along the x axis. Methods of theoretical physics provide a relationship between ω and k_x and k_y. Often it can be explicitly given in the form

$$\omega = \omega(k_x, k_y) \qquad (1\text{-}4\text{-}10)$$

Since velocity is distance divided by time we can define the phase velocity along the x direction as

$$(V \text{ phase})_x = \frac{x - x_0}{\text{phase delay}}$$

$$= \frac{x - x_0}{\phi/\omega}$$

$$= \frac{\omega}{k_x}$$

For the x component of group velocity

$$(V \text{ group})_x = \frac{x - x_0}{\text{group delay}}$$

$$= \frac{x - x_0}{d\phi/d\omega}$$

$$= (x - x_0) \frac{d\omega}{d\phi} \qquad (1\text{-}4\text{-}11)$$

Say $y = y_0$, then (1-4-9) reduces to

$$\phi = k_x(x - x_0)$$

which gives

$$\frac{\partial k_x}{\partial \phi} = \frac{1}{x - x_0}$$

and together with (1-4-11) gives

$$(V \text{ group})_x = (x - x_0) \frac{\partial \omega}{\partial k_x} \frac{\partial k_x}{\partial \phi} = \frac{\partial \omega}{\partial k_x} \qquad (1\text{-}4\text{-}12)$$

Thus the vector group velocity is $(\partial \omega/\partial k_x, \partial \omega/\partial k_y)$. It sometimes happens that physical theory is so complicated that an explicit relationship like (1-4-10) cannot be found and one gets instead a complicated implicit relation, say $0 = F(\omega, k_x, k_y)$. In such a case it is useful to recall the relationship from the theory of partial derivatives:

$$\frac{\partial \omega}{\partial k_x} = -\frac{\partial F/\partial k_x}{\partial F/\partial \omega}$$

In observational geophysics the velocity one deals with is nearly always the group velocity. It is the velocity with which bundles of energy move. In the example shown in Fig. 1-11 there is an excessive amount of " noise " (not unusual in observational geophysics); however, it can be seen that the disturbance first displays the long-period oscillations and then the shorter-period oscillations. The group velocity is found by dividing the distance by the time of arrival. One could observe phase velocities by having two observation stations near each other and measuring the time delay of some particular zero crossing. The reason for having the stations near one another is that the waveforms are steadily changing, and if the stations are too far apart, it may not be possible to tell which zero crossings are to be compared.

1-5 CORRELATION AND SPECTRA

The spectrum of a time function is the magnitude squared of the Fourier transform of the function. In the case of a real function, the Fourier transform has an even

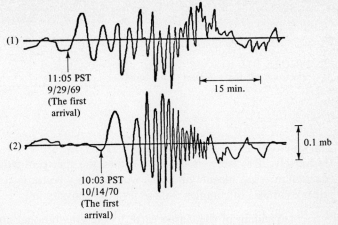

(1)

11:05 PST
9/29/69
(The first
arrival)

15 min.

(2)

10:03 PST
10/14/70
(The first
arrival)

0.1 mb

FIGURE 1-11
An example of a wave packet in which different frequencies may be seen propagating at different speeds. This example is of two air-pressure waves thought to result from nuclear explosions in Asia; they were recorded in California on one of the author's microbarographs.

real part RE and an imaginary odd part IO. Taking the squared magnitude, one has $(RE + iIO)(RE - iIO) = (RE)^2 + (IO)^2$. The square of an even function is obviously even and the square of an odd function is also even. Thus, the spectrum of a real time function is even so that its values at plus frequencies are the same as its values at minus frequencies. In other words, there is no special meaning to be attached to negative frequencies.

Although most time functions which arise in applications are real time functions, a discussion of correlation and spectra is not mathematically complete without considering complex-valued time functions. Furthermore, complex-valued time functions can be extremely useful in many physical problems in which rotation occurs. For example consider two vector-component wind-speed indicators: one pointing north, recording n_t, and the other pointing west, recording w_t. Now if one makes up a complex-valued time series $v_t = n_t + iw_t$, the magnitude and phase angle of the complex numbers have obvious physical interpretation. The $(RE+iIO)$ part of the transform relates to n_t and the $(RO + iIE)$ part relates to w_t. The spectrum, however, is $(RE + RO)^2 + (IE + IO)^2$, which is neither even nor odd, and the fact that $V(+\omega) \neq V(-\omega)$ must have some interpretation. Indeed it does, and the meaning is that $+\omega$ corresponds to rotation in one sense (counterclockwise) and $(-\omega)$ to rotation in the other direction. To see this, suppose $n_t = \cos(\omega_0 t + \phi)$ and $w_t = -\sin(\omega_0 t + \phi)$. Then $v_t = e^{-i(\omega_0 t + \phi)}$. The transform is

$$V(\omega) = \int_{-\infty}^{+\infty} e^{-i(\omega_0 t + \phi)} e^{i\omega t} \, dt \qquad (1\text{-}5\text{-}1)$$

$$= \delta(\omega - \omega_0) e^{-i\phi} \qquad (1\text{-}5\text{-}2)$$

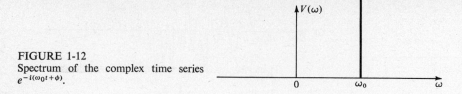

FIGURE 1-12
Spectrum of the complex time series
$e^{-i(\omega_0 t + \phi)}$.

The spectrum $\delta^2(\omega - \omega_0)$ is shown in Fig. 1-12.

Conversely, if w_t were $\sin(\omega_0 t + \phi)$, then the spectrum would have been a delta function at $-\omega_0$, meaning that the wind velocity vector is rotating the other way. Other examples of complex time series in geophysics are

1 Yielding of the elastic earth to the gravitational attraction of the moon causes local ground tilt. The north-south tilt could determine an x_t time series, and the east-west tilt could determine a y_t time series. Then $x_t + iy_t$ would tend to have one rotational sense in the northern hemisphere and the opposite sense in the southern hemisphere.

2 Vertical and horizontal seismograph motions could make up a complex time series.

3 Nutation of the earth's figure axis about the angular momentum axis (Chandler Wobble).

4 Rotational polarizations of an electromagnetic wave.

Let us look at the spectrum in terms of Z transforms. Let the spectrum be $R(\omega)$, where

$$R(\omega) = |B(\omega)|^2 = \overline{B(\omega)}B(\omega) \qquad (1\text{-}5\text{-}3)$$

Let us express this in terms of a three-point Z transform:

$$R(\omega) = (\bar{b}_0 + \bar{b}_1 e^{-i\omega} + \bar{b}_2 e^{-i2\omega})(b_0 + b_1 e^{i\omega} + b_2 e^{i2\omega}) \qquad (1\text{-}5\text{-}4)$$

$$B(\omega) = \left(\bar{b}_0 + \frac{\bar{b}_1}{Z} + \frac{\bar{b}_2}{Z^2}\right)(b_0 + b_1 Z + b_2 Z^2) \qquad (1\text{-}5\text{-}5)$$

$$B(\omega) = \bar{B}\left(\frac{1}{Z}\right) B(Z) \qquad (1\text{-}5\text{-}6)$$

It is of interest to multiply out the polynomials $\bar{B}(1/Z)$ with $B(Z)$ in order to examine the coefficients of $R(Z)$.

$$R(Z) = \frac{\bar{b}_2 b_0}{Z^2} + \frac{(\bar{b}_1 b_0 + \bar{b}_2 b_1)}{Z} + (\bar{b}_0 b_0 + \bar{b}_1 b_1 + \bar{b}_2 b_2)$$
$$+ (\bar{b}_0 b_1 + \bar{b}_1 b_2)Z + \bar{b}_0 b_2 Z^2 \qquad (1\text{-}5\text{-}7)$$

$$R(Z) = \frac{r_{-2}}{Z^2} + \frac{r_{-1}}{Z} + r_0 + r_1 Z + r_2 Z^2 \qquad (1\text{-}5\text{-}8)$$

The coefficient r_k of Z^k is given by

$$r_k = \sum_i \bar{b}_i b_{i+k} \qquad (1\text{-}5\text{-}9)$$

Equation (1-5-9) is known as the *autocorrelation* formula. The autocorrelation value r_k at lag 10 is r_{10}. It is a measure of the similarity of b_i with itself shifted 10 units in time. In the most frequently occurring case, b_i is real; then by inspection of (1-5-7) or (1-5-9) one sees that the autocorrelation coefficients are real and $r_k = r_{-k}$. With the specialization to real time series, then, we have

$$R(Z) = r_0 + r_1 \left(Z + \frac{1}{Z} \right) + r_2 \left(Z^2 + \frac{1}{Z^2} \right) \qquad (1\text{-}5\text{-}10)$$

$$R(Z) = r_0 + r_1(e^{i\omega} + e^{-i\omega}) + r_2(e^{i2\omega} + e^{-i2\omega}) \qquad (1\text{-}5\text{-}11)$$

$$R(Z) = r_0 + 2r_1 \cos \omega + 2r_2 \cos 2\omega \qquad (1\text{-}5\text{-}12)$$

$$R(Z) = \sum_k r_k \cos k\, \omega \qquad (1\text{-}5\text{-}13)$$

$$R(Z) = \text{cosine transform of } r_k \qquad (1\text{-}5\text{-}14)$$

We have just shown what is a fairly difficult theorem in continuous time textbooks, namely that the cosine transform of the autocorrelation equals the magnitude squared of the Fourier transform. There are two computationally distinct methods to compute a spectrum: (1) Compute the r_k coefficients from (1-5-9) once, then form the cosine sum (1-5-13); or (2) evaluate $B(Z)$ for some value of Z on the unit circle, and multiply the resulting number by its complex conjugate. Repeat for many values of Z on the unit circle. The second method is the cheapest because the fast Fourier transform may be used.

The concept of autocorrelation and spectrum is easily generalized to cross-correlation and cross spectrum. Consider two Z transforms $A(Z)$ and $B(Z)$. Then the cross spectrum $C(Z)$ is defined by

$$C(Z) = \bar{A}\left(\frac{1}{Z}\right) B(Z) \qquad (1\text{-}5\text{-}15)$$

If some particular coefficient c_k in $C(Z)$ is greater than any of the others, then it may be said that the waveform a_t most resembles the waveform b_t if one is delayed k time units with respect to the other.

EXERCISES

1 Suppose a wavelet is made up of complex numbers. Is the autocorrelation relation $r_k = r_{-k}$ true? Is r_k real or complex? Is $R(\omega)$ real or complex?

2 Let x_t be some real time function. Let $y_t = x_{t+3}$ be another real time function. Sketch the phase as a function of frequency of the cross spectrum $X(1/Z)Y(Z)$ as computed by a computer which put all arctangents in the principal quadrants $-\pi/2 < \arctan < \pi/2$. Label axis scales.

3 If concepts of time and frequency are interchanged, what does the meaning of spectrum become?

1-6 HILBERT TRANSFORM

A filter which converts sines into cosines is called a 90° phase shift filter or a quadrature filter. More specifically if the input is $\cos(\omega t + \phi_1)$, then the output should be $\cos(\omega t + \phi_1 + \pi/2)$. Such a filter can be useful in constructing the envelope of a time function. Let $X(Z)$ denote the Z transform of a real data series, $Q(Z)$ denote a quadrature filter, and let $Y(Z) = Q(Z)X(Z)$ be the output of the quadrature filter. Then the envelope time function may be defined by $e_t = (x_t^2 + y_t^2)^{1/2}$. Alternatively, one could construct a complex time function $u_t = x_t + iy_t$. In terms of Z transforms we have

$$U(Z) = [1 + iQ(Z)]X(Z)$$

Now $u_t \bar{u}_t$ represents the squared envelope function. Likewise the phase ϕ_t as a function of time may be defined as $\phi_t = \arctan(y_t/x_t)$. The instantaneous frequency is $d\phi/dt$. This may be approximated in the following way.

$$\phi_t = \text{Im} \ln u_t$$

$$\frac{d\phi}{dt} = \text{Im} \frac{1}{u} \frac{du}{dt}$$

$$\approx \text{Im} \frac{2}{\Delta t} \frac{u_t - u_{t-1}}{u_t + u_{t-1}}$$

Now that we have some idea what a 90° phase shift filter can be used for, let us find out the numerical values of q_t. The time derivative operation has the desired 90° phase-shifting property we seek. The trouble with a differentiator is that higher frequencies are amplified with respect to lower frequencies. Specifically

$$f(t) = \int F(\omega)e^{-i\omega t}\,d\omega$$

$$\frac{df}{dt} = \int -i\omega F(\omega)e^{-i\omega t}\,d\omega$$

Thus we see that time differentiation corresponds to the weight factor $-i\omega$ in the frequency domain. The weight $-i\omega$ has the proper phase but the wrong amplitude. The desired weight factor is $Q(\omega) = -i\omega/|\omega|$. It is the step function shown in Fig. 1-13.

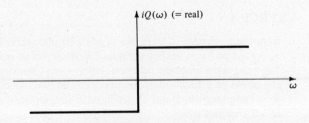

FIGURE 1-13
Frequency response of 90° phase-shifting filter.

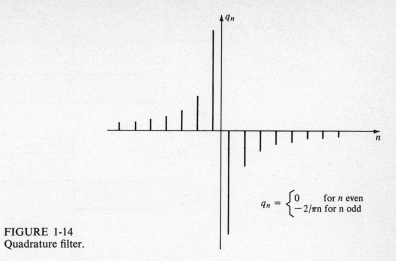

$$q_n = \begin{cases} 0 & \text{for } n \text{ even} \\ -2/\pi n & \text{for n odd} \end{cases}$$

FIGURE 1-14
Quadrature filter.

Let us transform $Q(\omega)$ into the time domain

$$q_n = \frac{1}{2\pi} \int_{-\pi}^{\pi} Q(\omega)e^{-i\omega n}\, d\omega$$

$$= \frac{i}{2\pi} \int_{-\pi}^{0} e^{-i\omega n}\, d\omega - \frac{i}{2\pi} \int_{0}^{\pi} e^{-i\omega n}\, d\omega$$

$$= \frac{i}{2\pi} \left(\frac{e^{-i\omega n}}{-in} \bigg|_{-\pi}^{0} - \frac{e^{-i\omega n}}{-in} \bigg|_{0}^{\pi} \right)$$

$$= \frac{1}{2\pi n} (-1 + e^{+in\pi} + e^{-in\pi} - 1)$$

$$= \begin{cases} 0 & \text{for } n \text{ even} \\ -2/\pi n & \text{for } n \text{ odd} \end{cases}$$

The result is shown in Fig. 1-14.

Since the filter does not vanish for negative n, this is obviously a nonrealizable filter (one which requires future inputs to create its present output). If the discussion were in continuous time rather than sampled time, the filter would be of the form $1/t$, a function which has a singularity at $t = 0$ and whose integral over $+t$ is divergent. Convolution with the filter coefficients q_n is therefore very awkward because the infinite sequence drops off very slowly. Convolution with the filter q is called *Hilbert transformation*.

Let us return to the filter $1 + iQ(Z)$ mentioned earlier. As shown in Fig. 1-15, this filter is simply a step function in the frequency domain. A cheap way to achieve the 90° phase shift operation is to do it in the frequency domain. One begins with $x_t + i \cdot 0$ and transforms it to the frequency domain. Then multiply by the step of Fig. 1-15. Finally, inverse transformation gives $x_t + iy_t$. The progress of even, odd, real, and imaginary parts is detailed in Fig. 1-16.

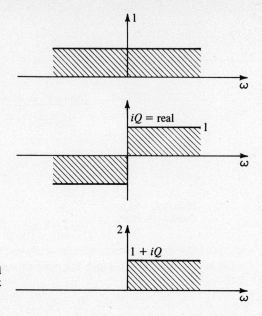

FIGURE 1-15
The filter $1 + iQ(Z)$ is real and one-sided in the frequency domain but complex and two-sided in the time domain.

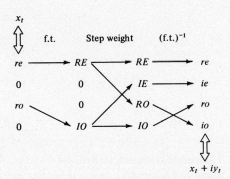

FIGURE 1-16
Hilbert transform or quadrature filtering by step weight in the frequency domain.

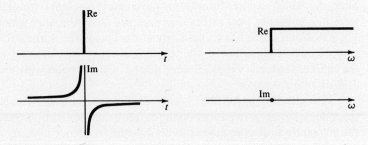

FIGURE 1-17
Impulse plus i times a 90° phase-shift filter becomes a real step in the frequency domain.

The function $1 + iQ$ plays a special role in theoretical time series analysis which, in later chapters, will be shown to be related to the principle of causality. For future reference we summarize the properties of this function in Fig. 1-17.

EXERCISES

1 By means of partial fractions convolve the waveform

$$(2/\pi)(\ldots, \ -\tfrac{1}{5}, 0, -\tfrac{1}{3}, 0, -1, 0, 1, 0, \tfrac{1}{3}, 0, \tfrac{1}{5}, \ldots)$$

with itself. What is the interpretation of the fact that the result is $(\ldots, 0, 0, -1, 0$ $0, \ldots)$? (HINT: $\pi^2/8 = 1 + \tfrac{1}{9} + \tfrac{1}{25} + \tfrac{1}{49} + \cdots$).

2 In terms of the fast Fourier transform matrix the quadrature filter $Q(\omega)$ may be represented by the column vector

$$-i(0, 1, 1, 1, \ldots, 0, -1, -1, -1, \ldots, -1)^T$$

Multiply this into the inverse transform matrix to show that the transform is proportional to $(\cos \pi k/N)/(\sin \pi k/N)$. What is the scale factor? Sketch it for $k \ll N$ indicating the limit $N \to \infty$. [HINT: $1 + x + x^2 + \cdots x^N = (1 - x^{N+1})/(1 - x)$.]

2

ONE-SIDED FUNCTIONS

All physical systems share the property that they do not respond before they are excited. Thus the impulse response of any physical system is a one-sided time function (it vanishes before $t = 0$). In system theory such a filter function is called realizable. In wave propagation this property is associated with causality in that no wave may begin to arrive before it is transmitted. The lag-time point $t = 0$ plays a peculiar and an important role. For this reason, many subtle matters will be much more clearly understood with sampled time than with continuous time. When a filter responds at and after lag time $t = 0$, we will say the filter is realizable or causal. The word *causal* is appropriate in physics where stress may cause (practically) instantaneous strain and vice versa, but one should revert to the more precise words *realizable* or *one-sided* when using filter theory to describe economic or social systems where simultaneity is quite different from cause and effect.

2-1 INVERSE FILTERS

To understand causal filters better, we now take up the task of undoing what a causal filter has done. Consider that the output y_t of a filter b_t is known but the input x_t is unknown. See Fig. 2-1.

FIGURE 2-1
Sometimes the input to a filter is unknown.

$X(Z) = \text{unknown}$ → $B(Z)$ → $Y(Z) = \text{known}$

This is the problem that one always has with a transducer/recorder system. For example, the output of a seismometer is a wiggly line on a piece of paper from which the seismologist may wish to determine the displacement, velocity, or acceleration of the ground. To undo the filtering operation of the filter $B(Z)$, we will try to find another filter $A(Z)$ as indicated in Fig. 2-2.

To solve for the coefficients of the filter $A(Z)$, we merely identify coefficients of powers of Z in $B(Z)A(Z) = 1$. For $B(Z)$, a three-term filter, this is

$$(a_0 + a_1 Z + a_2 Z^2 + a_3 Z^3 + \cdots)(b_0 + b_1 Z + b_2 Z^2) = 1 \qquad (2\text{-}1\text{-}1)$$

The coefficients of Z^0, Z^1, Z^2, ... in (2-1-1) are

$$a_0 b_0 \qquad\qquad\qquad = 1 \qquad (2\text{-}1\text{-}2)$$

$$a_1 b_0 + a_0 b_1 \qquad\qquad = 0 \qquad (2\text{-}1\text{-}3)$$

$$a_2 b_0 + a_1 b_1 + a_0 b_2 \qquad = 0 \qquad (2\text{-}1\text{-}4)$$

$$a_3 b_0 + a_2 b_1 + a_1 b_2 \qquad = 0 \qquad (2\text{-}1\text{-}5)$$

$$a_4 b_0 + a_3 b_1 + a_2 b_2 \qquad = 0 \qquad (2\text{-}1\text{-}6)$$

$$\cdots\cdots\cdots\cdots\cdots\cdots\cdots$$

$$a_k b_0 + a_{k-1} b_1 + a_{k-2} b_2 = 0 \qquad (2\text{-}1\text{-}7)$$

From (2-1-2) one may get a_0 from b_0. From (2-1-3) one may get a_1 from a_0 and the b_k. From (2-1-4) one may get a_2 from a_1, a_0, and the b_k. Likewise, in the general case a_k may be found from a_{k-1}, a_{k-2}, and the b_k. Specifically, from (2-1-7) the a_k may be determined recursively by

$$a_k = \frac{-\displaystyle\sum_{i=1}^{2} a_{k-i} b_i}{b_0} \qquad (2\text{-}1\text{-}8)$$

Consider the example where $B(Z) = 1 - Z/2$; then, by equations like (2-1-2) to (2-1-7), by the binomial theorem, by polynomial division, or by Taylor's power series formula we obtain

$$A(Z) = \frac{1}{1 - Z/2} = 1 + \frac{Z}{2} + \frac{Z^2}{4} + \frac{Z^3}{8} + \cdots \qquad (2\text{-}1\text{-}9)$$

FIGURE 2-2
The filter $A(Z)$ is inverse to the filter $B(Z)$.

$X(Z)$ → $B(Z)$ → $Y(Z)$ → $A(Z)$ → $X(Z)$

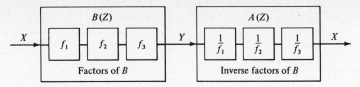

FIGURE 2-3
Factoring the polynomial $B(Z)$ breaks the filter into many two-term filters. Each one should have a bounded inverse.

We see that there are an infinite number of filter coefficients but that they drop off rapidly in size so that approximation in a computer presents no problem. The situation is not so rosy with the filter $B(Z) = 1 - 2Z$. Here we obtain

$$A(Z) = \frac{1}{1 - 2Z} = 1 + 2Z + 4Z^2 + 8Z^3 + 16Z^4 + 32Z^5 + \cdots \qquad (2\text{-}1\text{-}10)$$

The coefficients of the series increase without bound. The outputs of the filter $A(Z)$ depend infinitely strongly on inputs of the infinitely distant past. [Recall that the present output of $A(Z)$ is a_0 times the present input x_t plus a_1 times the previous input x_{t-1}, etc., so a_n represents memory of n time units earlier.] The implication of this is that some filters $B(Z)$ will not have useful finite approximate inverses $A(Z)$ determined from (2-1-2) to (2-1-8). We now seek ways to identify the good filters from the bad ones. With a two-pulse filter, the criterion is merely that the first pulse in $B(Z)$ be larger than the second. A more mathematical description of the state of affairs results from solving for the roots of $B(Z)$, that is, find values of Z_0 for which $B(Z_0) = 0$. For the example $1 - Z/2$ we find $Z_0 = 2$. For the example $1 - 2Z$, we find $Z_0 = \frac{1}{2}$. The general case for wavelets with complex coefficients is that, if the solution value Z_0 of $B(Z_0) = 0$ lies inside the unit circle in the complex plane, then $1/B(Z)$ will have coefficients which blow up; and if the root lies outside the unit circle, then the inverse $1/B(Z)$ will be bounded.

Recalling earlier discussion that a polynomial $B(Z)$ of degree N may be factored into N subsystems and that the ordering of subsystems is unimportant (see Fig. 2-3), we suspect that if any of the N roots of $B(Z)$ lies inside the unit circle we may have difficulty with $A(Z)$. Actual proof of this suspicion relies on a theorem from complex-variable theory about absolutely convergent series. The theorem is that the product of absolutely convergent series is convergent, and conversely the product of any convergent series with a divergent series is divergent. Another proof may be based upon the fact that a power series for $1/B(Z)$ converges in a circle about the origin with a radius from the origin out to the first pole [the zero of $B(Z)$ of smallest magnitude]. Convergence of $A(Z)$ on the unit circle means, in terms of filters, that the coefficients of $A(Z)$ are decreasing. Thus, if all the zeros of $B(Z)$ are outside the unit circle, we will get a convergent filter from (2-1-8).

Can anything at all be done if there is one root or more inside the circle? An answer is suggested by the example

$$\frac{1}{1 - 2Z} = -\frac{1}{2Z} \frac{1}{1 - 1/2Z} = -\frac{1}{2Z} \left[1 + \frac{1}{2Z} + \frac{1}{(2Z)^2} + \cdots \right] \qquad (2\text{-}1\text{-}11)$$

Equation (2-1-11) is a series expansion in $1/Z$, that is, a Taylor series about infinity. It converges from $Z = \infty$ all the way in to a circle of radius $1/2$. This means that the inverse converges on the unit circle where it must, if the coefficients are to be bounded. In terms of filters it means that the inverse filter must be one of those filters which responds to future inputs and hence is not physically realizable but may be used in computer simulation.

In the general case, then, one must factor $B(Z)$ into two parts: $B(Z) = B_{out}(Z)B_{in}(Z)$ where B_{out} contains roots outside the unit circle and B_{in} contains the roots inside. Then the inverse of B_{out} is expressed as a Taylor series about the origin and the inverse of B_{in} is expressed as a Taylor series about infinity. The final expression for $1/B(Z)$ is called a Laurent expansion for $1/B(Z)$, and it converges on a ring surrounding the unit circle. Cases with zeros exactly on the unit circle present special problems. Sometimes you can argue yourself out of the difficulty but at other times roots on or even near the circle may mean that a certain computing scheme won't work out well in practice.

Finally, let us consider a mechanical interpretation. The stress (pressure) in a material may be represented by x_t, and the strain (volume change) may be represented by y_t. The following two statements are equivalent; that is, in some situations they are both true, and in other situations they are both false:

STATEMENT A   The stress in a material may be expressed as a linear combination of present and past strains. Likewise, the strain may be deduced from present and past stresses.

STATEMENT B   The filter which relates stress to strain and vice versa has all poles and zeros outside the unit circle.

EXERCISES

1 Find the filter which is inverse to $(2 - 5Z + 2Z^2)$. You may just drop higher-order powers of Z, but an exact expression for the coefficients of any power of Z is preferred. (Partial fractions is a useful, though not necessary, technique.) Sketch the impulse response.

2 Show that multiplication by $(1 - Z)$ in discretized time is analogous to time differentiation in continuous time. Show that dividing by $(1 - Z)$ is analogous to integration. What are the limits on the integral?

3 Describe a general method for determining $A(Z)$ and $B(Z)$ from a Taylor series of $B(Z)/A(Z) = C_0 + C_1 Z + C_2 Z^2 + \cdots + C_\infty Z^\infty$ where $B(Z)$ and $A(Z)$ are polynomials of unknown degree n and m, respectively. Work out the case $C(Z) = \frac{1}{2} - \frac{3}{4}Z - \frac{3}{8}Z^2 - \frac{3}{16}Z^3 - \frac{3}{32}Z^4 - \cdots$. Don't try this problem unless you are quite familiar with determinants. [HINT: Identify coefficients of $B(Z) = A(Z)C(Z)$.]

2-2 MINIMUM PHASE

In Sec. 2-1 we learned that knowledge of convergence of the Taylor series of $1/B(Z)$ on $|Z| = 1$ is equivalent to knowledge that $B(Z)$ has no roots inside the

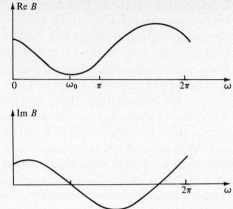

FIGURE 2-4
Real and imaginary parts of the Z transform $1 - Z/(1.25\, e^{i2\pi/3})$.

unit circle. Now we will see that these conditions are also equivalent to a certain behavior of the phase of $B(Z)$ on the unit circle.

Let us consider the phase shift of the two-term filter

$$B = 1 - \frac{Z}{Z_0} \qquad\qquad (Z_0 = \rho e^{i\omega_0})$$

$$= 1 - \rho^{-1} e^{i(\omega - \omega_0)}$$

$$= 1 - \rho^{-1} \cos(\omega - \omega_0) - i\rho^{-1} \sin(\omega - \omega_0)$$

By definition, phase is the arctangent of the ratio of the imaginary part to the real part.

A graph of phase as a function of frequency looks radically different for $\rho < 1$ than for $\rho > 1$. See Fig. 2-4 for the case $\rho > 1$.

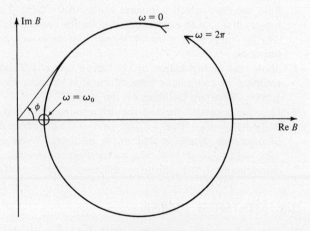

FIGURE 2-5
Phase of the two-term filter of Fig. 2-4.

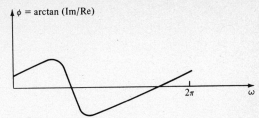

FIGURE 2-6
The phase of a two-term minimum-phase filter.

The phase is the arctangent of Im B/Re B. The easiest way to keep track of the phase is in the complex B plane. This is shown in Fig. 2-5.

Thus phase as a function of frequency is shown in Fig. 2-6. Notice that the phase ϕ at $\omega = 0$ is the same as the phase at $\omega = 2\pi$. This follows because the real and imaginary parts are periodic with 2π. The situation will be different when there is a zero inside the unit circle; that is, $\rho < 1$. The real and imaginary parts are shown in Fig. 2-7 and the complex plane in Fig. 2-8.

The phase ϕ increases by 2π as ω goes from zero to 2π because the circular path surrounds the origin. The phase curve is shown in Fig. 2-9. The case $\rho > 1$ where $\phi(\omega) = \phi(\omega + 2\pi)$ has come to be called *minimum phase* or *minimum delay*.

Now we are ready to consider a complicated filter like

$$B(Z) = \frac{(Z - c_1)(Z - c_2)\cdots}{(Z - a_1)(Z - a_2)\cdots} \qquad (2\text{-}2\text{-}1)$$

By the rules of complex-number multiplication the phase of $B(Z)$ is the sum of the phases in the numerator minus the sum of the phases in the denominator. Since we are discussing realizable filters the denominator factors must all be minimum phase, and so the denominator phase curve is a sum of curves like Fig. 2-6. The numerator factors may or may not be minimum phase. Thus the numerator phase curve is a sum of curves like either Fig. 2-6 or Fig. 2-9. If any factors at all are like Fig. 2-9, then the total phase will resemble Fig. 2-9 in that the phase at $\omega = 2\pi$ will be greater than the phase at $\omega = 0$. Then the filter will be nonminimum phase.

2-3 FILTERS IN PARALLEL

We have seen that in a cascade of filters the filter polynomials are multiplied together. One might conceive of adding two polynomials $A(Z)$ and $G(Z)$ when they correspond to filters which operate in parallel. See Fig. 2-10.

When filters operate in parallel their Z transforms add together. We have seen that a cascade of filters is minimum phase if, and only if, each element of the product is minimum phase. Now we will see a sufficient (but not necessary) condition that the sum $A(Z) + G(Z)$ be minimum phase. First of all, let us assume that $A(Z)$ is minimum phase. Then we may write

$$A(Z) + G(Z) = A(Z)\left[1 + \frac{G(Z)}{A(Z)}\right]$$

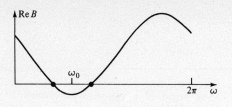

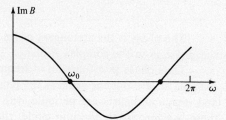

FIGURE 2-7
Real and imaginary parts of the two-term nonminimum-phase filter, $1 - 1.25\, Z\, e^{-i2\pi/3}$.

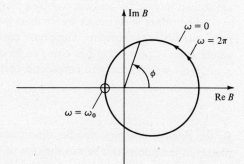

FIGURE 2-8
Phase in complex plane.

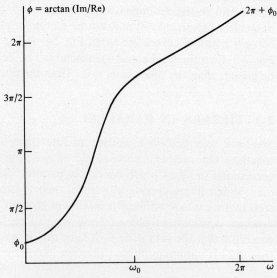

FIGURE 2-9
The phase of a two-term nonminimum-phase filter.

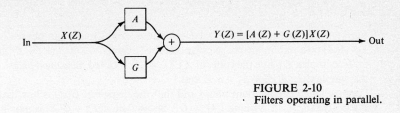

FIGURE 2-10
· Filters operating in parallel.

The question of whether $A(Z) + G(Z)$ is minimum phase is now reduced to determining whether $A(Z)$ and $1 + G(Z)/A(Z)$ are both minimum phase. We have assumed that $A(Z)$ is minimum phase. Before we ask whether $1 + G(Z)/A(Z)$ is minimum phase we need to be sure that it's causal. Since $1/A(Z)$ is expandable in positive powers of Z only, then $G(Z)/A(Z)$ is also causal. We will next see that a sufficient condition for $1 + G(Z)/A(Z)$ to be minimum phase is that the spectrum of A exceeds that of G at all frequencies. In other words, for any real ω, $|A| > |G|$. Thus, if we plot the curve of $G(Z)/A(Z)$ in the complex plane, for real $0 \le \omega \le 2\pi$ it lies everywhere inside the unit circle. Now if we add unity—getting $1 + G(Z)/A(Z)$, the curve will always have a positive real part. See Fig. 2-11. Since the curve cannot enclose the origin, the phase must be that of a minimum-phase function. In words, "You can add garbage to a minimum-phase wavelet if you do not add too much." This somewhat abstract theorem has an immediate physical consequence. Suppose a wave characterized by a minimum phase $A(Z)$ is emitted from a source and detected at a receiver some time later. At a still later time an echo bounces off a nearby object and is also detected at the receiver. The receiver sees the signal $Y(Z) = A(Z) + Z^n\alpha A(Z)$ where n measures the delay from the first arrival to the echo and α represents the amplitude attenuation of the echo. To see that $Y(Z)$ is minimum phase, we note that the magnitude of Z^n is unity and that the reflection coefficient α must be less than unity (to avoid perpetual motion) so that $Z^n\alpha A(Z)$ takes the role of $G(Z)$. Thus a minimum-phase wave along with its echo is minimum phase. We will later consider wave propagation situations with echoes of the echoes ad infinitum.

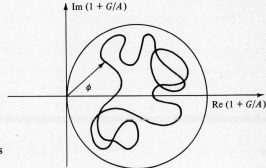

FIGURE 2-11
Phase of a positive real function lies
between $\pm \pi/2$.

EXERCISES

1 Find two nonminimum-phase wavelets whose sum is minimum phase.
2 Let $A(Z)$ be a minimum-phase polynomial of degree N. Let $A'(Z) = Z^N \bar{A}(1/Z)$. Locate in the complex Z plane the roots of $A'(Z)$. $A'(Z)$ is called *maximum phase*. [HINT: Work the simple case $A(Z) = a_0 + a_1 Z$ first.]
3 Suppose $A(Z)$ is maximum phase and that the degree of $G(Z)$ is less than or equal to the degree of $A(Z)$. Assume $|A| > |G|$. Show that $A(Z) + G(Z)$ is maximum phase.
4 Let $A(Z)$ be minimum phase. Where are the roots of $A(Z) + cZ^N \bar{A}(1/Z)$ in the three cases $|c| < 1$, $|c| > 1$, $|c| = 1$? (HINT: The roots of a polynomial are continuous functions of the polynomial coefficients.)

2-4 POSITIVE REAL FUNCTIONS

Two similar types of functions called *admittance functions* $Y(Z)$ and *impedance functions* $I(Z)$ occur in many physical problems. In electronics, they are ratios of current to voltage and of voltage to current; in acoustics, impedance is the ratio of pressure to velocity. When the appropriate electrical network or acoustical region contains no sources of energy, then these ratios have the positive real property. To see this in a mechanical example, we may imagine applying a known force $F(Z)$ and observing the resulting velocity $V(Z)$. In filter theory, it is like considering that $F(Z)$ is input to a filter $Y(Z)$ giving output $V(Z)$. We have

$$V(Z) = Y(Z)F(Z) \qquad (2\text{-}4\text{-}1)$$

This filter $Y(Z)$ is obviously causal. Since we believe we can do it the other way around, that is, prescribe the velocity and observe the force, there must exist a convergent causal $I(Z)$ such that

$$F(Z) = I(Z)V(Z) \qquad (2\text{-}4\text{-}2)$$

Since Y and I are inverses of one another and since they are both presumed bounded and causal, then they both must be minimum phase.

First, before we consider any physics, note that if the complex number $a + ib$ has a positive real part a, then the real part of $(a + ib)^{-1}$ namely $a/(a^2 + b^2)$ is also positive. Taking $a + ib$ to represent a value of $Y(Z)$ or $I(Z)$ on the unit circle, we see the obvious fact that if either Y or I has the positive real property, then the other does, too.

Power dissipated is the product of force times velocity, that is

$$\text{Power} = \cdots + f_0 v_0 + f_1 v_1 + f_2 v_2 + \cdots \qquad (2\text{-}4\text{-}3)$$

This may be expressed in terms of Z transforms as

$$\text{Power} = \frac{1}{2} \text{ coeff of } Z^0 \text{ of } V\left(\frac{1}{Z}\right) F(Z) + F\left(\frac{1}{Z}\right) V(Z)$$

$$= \frac{1}{2} \frac{1}{2\pi} \int_{-\pi}^{+\pi} \left[V\left(\frac{1}{Z}\right) F(Z) + F\left(\frac{1}{Z}\right) V(Z) \right] d\omega \qquad (2\text{-}4\text{-}4)$$

Using (2-4-1) to eliminate $V(Z)$ we get

$$\text{Power} = \frac{1}{2} \frac{1}{2\pi} \int_{-\pi}^{+\pi} F\left(\frac{1}{Z}\right) \left[Y\left(\frac{1}{Z}\right) + Y(Z) \right] F(Z) \, d\omega$$

We note that $Y(Z) + Y(1/Z)$ looks superficially like a spectrum because the coefficient of Z^k equals that of Z^{-k}, which shows the symmetry of an autocorrelation function. Defining

$$R(Z) = Y(Z) + Y\left(\frac{1}{Z}\right) \qquad (2\text{-}4\text{-}5)$$

(2-4-4) becomes

$$\text{Power} = \frac{1}{2} \frac{1}{2\pi} \int_{-\pi}^{+\pi} R(Z) F\left(\frac{1}{Z}\right) F(Z) \, d\omega \qquad (2\text{-}4\text{-}6)$$

The integrand is the product of the arbitrary positive input force spectrum and $R(Z)$. If the power dissipation is expected to be positive at all frequencies (for all $\bar{F}F$), then obviously $R(Z)$ must be positive at all frequencies; thus R is indeed a spectrum. Since we have now discovered that $Y(Z) + Y(1/Z)$ must be positive for all frequencies, we have discovered that $Y(Z)$ is not an arbitrary minimum-phase filter. The real part of both $Y(Z)$ and $Y(1/Z)$ is

$$\text{Re}[Y(Z)] = \text{Re}\left[Y\left(\frac{1}{Z}\right) \right] = y_0 + y_1 \cos \omega + y_2 \cos 2\omega + \cdots$$

Since the real part of the sum must be positive, then obviously the real part of each of the equal parts must be positive.

Now if the material or mechanism being studied is passive (contains no energy sources) then we must have positive dissipation over any time gate from minus infinity up to any time t. Let us find an expression for dissipation in such a time gate. For simplicity take both the force and velocity vanishing before $t = 0$. Let the end of the time gate include the point $t = 2$ but not $t = 3$. Define

$$f'_t = \begin{cases} f_t & t \leqslant 2 \\ 0 & t > 2 \end{cases} \qquad (2\text{-}4\text{-}7)$$

To find the work done over all time we may integrate (2-4-6) over all frequencies. To find the work done in the selected gate we may replace F by F' and integrate over all frequencies, namely

$$W_2 = \frac{1}{2} \frac{1}{2\pi} \int_{-\pi}^{+\pi} \frac{1}{2} F'\left(\frac{1}{Z}\right) R(Z) F'(Z) \, d\omega \qquad (2\text{-}4\text{-}8)$$

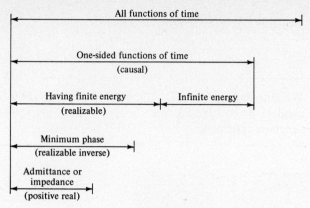

FIGURE 2-12
Important classes of time functions.

As we have seen, this integral merely selects the coefficient of Z^0 of the integrand. Let us work this out. First, collect coefficients of powers of Z in $R(Z)F'(Z)$. We have

$$Z^0: \quad r_0 f'_0 + r_1 f'_1 + r_{-2} f'_2$$
$$Z^1: \quad r_1 f'_1 + r_0 f'_1 + r_{-1} f'_2$$
$$Z^2: \quad r_2 f'_0 + r_1 f'_1 + r_2 f'_2$$

To obtain the coefficient of Z^0 in $F'(1/Z)[R(Z) F'(Z)]$ we must multiply the top row above by f'_0, the second row by f'_1 and the third row by f'_2. The result can be arranged in a very orderly fashion by

$$W_2 = \tfrac{1}{2}[f_0 \quad f_1 \quad f_2] \begin{bmatrix} r_0 & r_{-1} & r_{-2} \\ r_1 & r_0 & r_{-1} \\ r_2 & r_1 & r_0 \end{bmatrix} \begin{bmatrix} f_0 \\ f_1 \\ f_2 \end{bmatrix}$$

$$= \tfrac{1}{2}[f_0 \quad f_1 \quad f_2] \begin{bmatrix} 2y_0 & y_1 & y_2 \\ y_1 & 2y_0 & y_1 \\ y_2 & y_1 & 2y_0 \end{bmatrix} \begin{bmatrix} f_0 \\ f_1 \\ f_2 \end{bmatrix} \qquad (2\text{-}4\text{-}9)$$

Not only must the 3×3 quadratic form (2-4-9) be positive (i.e., $W_2 \geqslant 0$ for arbitrary f_t) but all $t \times t$ similar quadratic forms W_t must be positive.

In conclusion, the positive real property in the frequency domain means that $Y(Z) + Y(1/Z)$ is positive for any real ω and the positive real property in the time domain means that all $t \times t$ matrices like that of (2-4-9) are positive definite. Figure 2-12 summarizes the function types which we have considered.

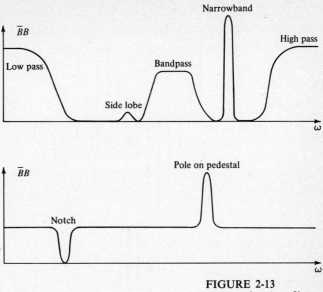

FIGURE 2-13
Spectra of various filters.

EXERCISES

1 In mechanics we have force and velocity of a free unit mass related by $dv/dt = f$ or $v = \int_{-\infty}^{t} f\, dt$. Compute the power dissipated as a function of frequency if integration is approximated by convolution with $(.5, 1., 1., 1., \ldots)$. [HINT: Expand $(1 + Z)/2(1 - Z)$ in positive powers of Z.]

2 Construct an example of a simple function which is minimum phase but not positive real.

2-5 NARROW-BAND FILTERS

Filters are often used to modify the spectrum of given data. With input $X(Z)$, filter $B(Z)$, and output $Y(Z)$ we have $Y(Z) = B(Z)X(Z)$ and the Fourier conjugate $\bar{Y}(1/Z) = \bar{B}(1/Z)\bar{X}(1/Z)$. Multiplying these two relations together we get

$$\bar{Y}Y = (\bar{B}B)(\bar{X}X)$$

which says that the spectrum of the input times the spectrum of the filter equals the spectrum of the output. Filters are often characterized by the shape of their spectra. Some examples are shown in Fig. 2-13.

We will have frequent occasion to deal with sinusoidal time functions. A simple way to represent a sinusoid by Z transforms is

$$\frac{1}{1 - Ze^{i\omega_0}} = 1 + Ze^{i\omega_0} + Z^2 e^{i2\omega_0} + \cdots \qquad (2\text{-}5\text{-}1)$$

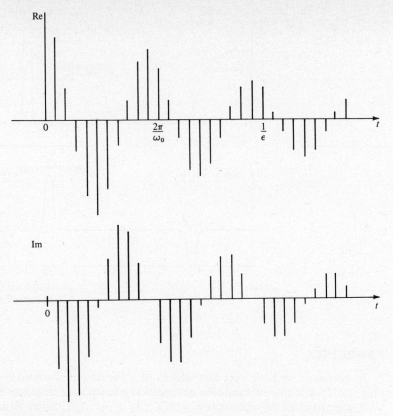

FIGURE 2-14
The time function associated with a simple pole just outside the unit circle at
$Z_0 = 1.1 \, e^{i\pi/5}$.

The time function associated with this Z transform is $e^{i\omega_0 t}$, but it is "turned on" at
$t = 0$. Actually, the left-hand side of (2-5-1) contains a pole exactly on the unit
circle, so that the series sits on the borderline between convergence and divergence.
This can cause paradoxical situations [you could expand (2-5-1) so that the sinusoid
turns off at $t = 0$] which we will avoid by pushing the pole from the unit circle to a
small distance ε outside the unit circle. Let $Z_0 = (1 + \varepsilon)e^{i\omega_0}$. Then define

$$B(Z) = \frac{1}{A(Z)} = \frac{1}{1 - Z/Z_0}$$

$$= 1 + \frac{Z}{Z_0} + \left(\frac{Z}{Z_0}\right)^2 + \cdots \qquad (2\text{-}5\text{-}2)$$

The time function corresponding to $B(Z)$ is zero before $t = 0$ and is $e^{-i\omega_0 t}/(1 + \varepsilon)^t$
after $t = 0$. It is a sinusoidal function which decreases gradually with time according to $(1 + \varepsilon)^{-t}$. The coefficients are shown in Fig. 2-14.

It is intuitively obvious, although we will prove it later, that convolution with the coefficients of (2-5-2), which are sketched in Fig. 2-14, is a narrow-banded filtering operation. If the pole is chosen very close to the unit circle, the filter bandpass becomes narrower and narrower and the coefficients of $B(Z)$ drop off more and more slowly. To actually perform the convolution it is necessary to truncate, that is, to drop powers of Z beyond a certain practical limit. It turns out that there is a very much cheaper method of narrow-band filtering than convolution with the coefficients of $B(Z)$. This method is polynomial division by $A(Z)$. We have for the output $Y(Z)$

$$Y(Z) = B(Z)X(Z) \qquad (2\text{-}5\text{-}3)$$

$$Y(Z) = \frac{X(Z)}{A(Z)} \qquad (2\text{-}5\text{-}4)$$

Multiply both sides of (2-5-4) by $A(Z)$

$$Y(Z)A(Z) = X(Z) \qquad (2\text{-}5\text{-}5)$$

For definiteness, let us suppose the x_t and y_t vanish before $t = 0$. Now identify coefficients of successive powers of Z. We get

$$
\begin{aligned}
y_0 a_0 &= x_0 \\
y_1 a_0 + y_0 a_1 &= x_1 \\
y_2 a_0 + y_1 a_1 + y_0 a_2 &= x_2 \\
y_3 a_0 + y_2 a_1 + y_1 a_2 + y_0 a_3 &= x_3
\end{aligned}
\qquad (2\text{-}5\text{-}6)
$$

etc.

A general equation is

$$y_k a_0 + \sum_{i=1}^{\infty} y_{k-i} a_i = x_k \qquad (2\text{-}5\text{-}7)$$

Solving for y_k we get

$$y_k = \frac{x_k - \sum_{i=1}^{\infty} y_{k-i} a_i}{a_0} \qquad (2\text{-}5\text{-}8)$$

Equation (2-5-8) may be used to solve for y_k once $y_{k-1}, y_{k-2}, \ldots$ are known. Thus the solution is recursive, and it will not diverge if the a_i are coefficients of a minimum-phase polynomial. In practice the infinite limit on the sum is truncated whenever you run out of coefficients of either $A(Z)$ or $Y(Z)$. For the example we have been considering, $B(Z) = 1/A(Z) = 1/(1 - Z/Z_0)$, there will be only one term in the sum. Filtering in this way is called *feedback filtering*, and for narrowband filtering it will be vastly more economical than filtering by convolution, since there are much fewer coefficients in $A(Z)$ than $B(Z) = 1/A(Z)$. Finally, let us examine the spectrum of $B(Z)$. We have

$$A(Z) = 1 - \frac{Z}{Z_0}$$

$$= 1 - \frac{e^{i\omega}}{(1 + \varepsilon)e^{i\omega_0}}$$

$$= 1 - \frac{e^{i(\omega - \omega_0)}}{(1 + \varepsilon)}$$

and

$$\bar{A}\left(\frac{1}{Z}\right) = 1 - \frac{e^{-i(\omega - \omega_0)}}{1 + \varepsilon}$$

so

$$\bar{A}\left(\frac{1}{Z}\right)A(Z) = \left(1 - \frac{e^{-i(\omega - \omega_0)}}{1 + \varepsilon}\right)\left(\frac{1 - e^{i(\omega - \omega_0)}}{1 + \varepsilon}\right)$$

$$= 1 + \frac{1}{(1 + \varepsilon)^2} - \frac{1}{1 + \varepsilon}(e^{-i(\omega - \omega_0)} + e^{i(\omega - \omega_0)})$$

$$= 1 + \frac{1}{(1 + \varepsilon)^2} - \frac{2\cos(\omega - \omega_0)}{1 + \varepsilon}$$

$$= 1 + \frac{1}{(1 + \varepsilon)^2} - \frac{2}{1 + \varepsilon} + \frac{2}{1 + \varepsilon}[1 - \cos(\omega - \omega_0)]$$

$$= \left(1 - \frac{1}{1 + \varepsilon}\right)^2 + \frac{4}{1 + \varepsilon}\sin^2\frac{\omega - \omega_0}{2}$$

$$\bar{B}\left(\frac{1}{Z}\right)B(Z) = \frac{(1 + \varepsilon)^2}{\varepsilon^2 + 4(1 + \varepsilon)\sin^2\left(\dfrac{\omega - \omega_0}{2}\right)} \tag{2-5-9}$$

To a good approximation this function may be thought of as $1/[\varepsilon^2 + (\omega - \omega_0)^2]$. A plot of (2-5-9) is shown in Fig. 2-15.

Now it should be apparent why this is called a narrowband filter. It amplifies a very narrow band of frequencies and attenuates all others. The frequency window of this filter is said to be $\Delta\omega \approx 2\varepsilon$ in width. The time window is $\Delta t = 1/\varepsilon$, the damping time constant of the damped sinusoid b_t.

One practical disadvantage of the filter under discussion is that although its input may be a real time series its output will be a complex time series. For many applications a filter with real coefficients may be preferred.

One approach is to follow the filter $[1, e^{i\omega_0}/(1 + \varepsilon)]$ by the time-domain, complex conjugate filter $[1, e^{-i\omega_0}/(1 + \varepsilon)]$. The composite time-domain operator is

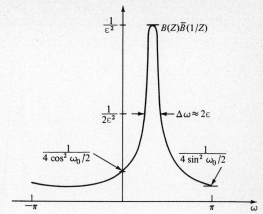

FIGURE 2-15
Spectrum associated with a single pole at $Z_0 = (1 + \varepsilon) e^{i\omega_0}$.

now $[1, (2 \cos \omega_0)/(1 + \varepsilon), 1/(1 + \varepsilon)^2]$ which is real. [Note that the complex conjugate in the frequency domain is $\bar{B}(1/Z)$ but in the time domain it is $\bar{B}(Z) = \bar{b}_0 + \bar{b}_1 Z + \cdots$]. The composite filter may be denoted by $B(Z)\bar{B}(Z)$. The spectrum of this filter is $[B(Z)\bar{B}(1/Z)][\bar{B}(Z)B(1/Z)]$. One may quickly verify that the spectrum of $\bar{B}(Z)$ is like that of $B(Z)$, but the peak is at $-\omega_0$ instead of $+\omega_0$. Thus, the composite spectrum is the product of Fig. 2-15 with itself reversed along the frequency axis. This is shown in Fig. 2-16.

EXERCISES

1 A simple feedback operation is $y_t = (1 - \varepsilon)y_{t-1} + x_t$. This operation is called leaky integration. Give a closed form expression for the output y_t if x_t is an impulse. What is the decay time τ of your solution (the time it takes for y_t to drop to $e^{-1}y_0$)? For small ε, say $= 0.1, .001,$ or 0.0001, what is τ?

2 How far from the unit circle are the poles of $1/(1 - .1 Z + .9 Z^2)$? What is the decay time of the filter and its resonant frequency?

3 Find a three-term real feedback filter to pass 59–61 Hz on data which are sampled at 500 points/sec. Where are the poles? What is the decay time of the filter?

2-6 ALL-PASS FILTERS

In this section we consider filters with constant unit spectra, that is, $B(Z)\bar{B}(1/Z) = 1$. In other words, in the frequency domain $B(Z)$ takes the form $e^{i\phi(\omega)}$ where ϕ is real and is called the *phase shift*. Clearly $B\bar{B} = 1$ for all real ϕ. It is an easy matter to construct a filter with any desired phase shift; one merely Fourier transforms $e^{i\phi(\omega)}$ into the time domain. If $\phi(\omega)$ is arbitrary, the resulting time function is likely to be two-sided. Since we are interested in physical processes which are

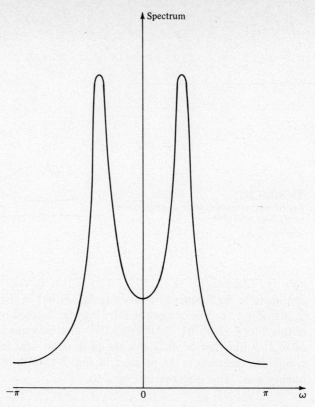

FIGURE 2-16
Spectrum of a two-pole filter where one pole is like Fig. 2-15 and the other is at the conjugate position.

causal, we may wonder what class of functions $\phi(\omega)$ corresponds to one-sided time functions. The easiest way to proceed is to begin with a simple case of a single-pole, single-zero all-pass filter. Then more elaborate all-pass filters can be made up by cascading these simple filters. Consider the filter

$$P(Z) = \frac{Z - 1/\bar{Z}_0}{1 - Z/Z_0} \qquad (2\text{-}6\text{-}1)$$

Note that this is a simple case of functions of the form $Z^N \bar{A}(1/Z)/A(Z)$, where $A(Z)$ is a polynomial of degree N or less. Now observe that the spectrum of the filter p_t is indeed a frequency-independent constant. The spectrum is

$$\bar{P}\left(\frac{1}{Z}\right) P(Z) = \frac{1/Z - 1/Z_0}{1 - 1(Z/\bar{Z}_0)} \frac{Z - 1/\bar{Z}_0}{Z - 1/Z_0} \qquad (2\text{-}6\text{-}2)$$

Multiply top and bottom on the left by Z. We now have

$$\bar{P}\left(\frac{1}{Z}\right) P(Z) = \frac{1 - Z/Z_0}{Z - 1/\bar{Z}_0} \frac{Z - 1/\bar{Z}_0}{1 - Z/Z_0} = 1 \qquad (2\text{-}6\text{-}3)$$

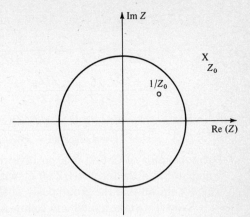

FIGURE 2-17
The pole of the all-pass filter lies outside
the unit circle and the zero is inside.
They lie on the same radius line.

It is easy to show that $\bar{P}(1/Z)P(Z) = 1$ for the general form $P(Z) = Z^N\bar{A}(1/Z)/A(Z)$. If Z_0 is chosen outside the unit circle, then the denominator of (2-6-1) can be expanded in positive powers of Z and the expansion is convergent on the unit circle. This means that causality is equivalent to Z_0 outside the unit circle. Setting the numerator of $P(Z)$ equal to zero, we discover that the zero $Z = 1/\bar{Z}_0$ is then inside the unit circle. The situation is depicted in Fig. 2-17. To see that the pole and zero are on the same radius line, express Z_0 in polar form $r_0 e^{i\phi_0}$.

From Sec. 2-2 (on minimum phase) we see that the numerator of P is not minimum phase and its phase is augmented by 2π as ω goes from 0 to 2π. Thus the average group delay $d\phi/d\omega$ is positive. Not only is the average positive but, in fact, the group delay turns out to be positive at every frequency. To see this, first note that

$$Z = e^{i\omega}$$

$$\frac{dZ}{d\omega} = ie^{i\omega} = iZ$$

$$\frac{d}{d\omega} = \frac{dZ}{d\omega}\frac{d}{dZ} = iZ\frac{d}{dZ} \qquad (2\text{-}6\text{-}4)$$

The phase of the all-pass filter (or any complex number) may be written as

$$\phi = \text{Im ln } P(Z) \qquad (2\text{-}6\text{-}5)$$

Since $|P| = 1$ the real part of the log vanishes; and so, for the all-pass filter (only) we may specialize (2-6-5) to

$$\phi = \frac{1}{i}\ln P(Z) = \frac{1}{i}\ln\frac{Z - 1/\bar{Z}_0}{1 - Z/Z_0}$$

$$= \frac{1}{i}\left[\ln\left(Z - \frac{1}{\bar{Z}_0}\right) - \ln\left(1 - \frac{Z}{Z_0}\right)\right] \qquad (2\text{-}6\text{-}6)$$

Using (2-6-4) the group delay is now found to be

$$\tau_g = \frac{d\phi}{d\omega} = iZ\frac{d\phi}{dZ} = Z\left(\frac{1}{Z - 1/\bar{Z}_0} + \frac{1/Z_0}{Z - Z/Z_0}\right)$$

$$= \frac{1}{1 - 1/\bar{Z}_0 Z} + \frac{Z/Z_0}{1 - Z/Z_0}$$

$$= \frac{1 - Z/Z_0 + (1 - 1/\bar{Z}_0 Z)(Z/Z_0)}{(1 - 1/\bar{Z}_0 Z)(1 - Z/Z_0)} = \frac{1 - 1/Z_0\bar{Z}_0}{(1 - 1/\bar{Z}_0 Z)(1 - Z/Z_0)} \qquad (2\text{-}6\text{-}7)$$

The numerator of (2-6-7) is a positive real number (since $|Z_0| > 1$), and the denominator is of the form $\bar{A}(1/Z)A(Z)$, which is a spectrum and also positive. Thus we have shown that the group delay of this causal all-pass filter is always positive.

Now if we take a filter and follow it with an all-pass filter, the phases add and the group delay of the composite filter must necessarily be greater than the group delay of the original filter. By the same reasoning the minimum-phase filter must have less group delay than any other filter with the same spectrum.

In summary, a single-pole, single-zero all-pass filter passes all frequency components with constant gain and a phase shift which may be adjusted by the placement of a pole. Taking Z_0 near the unit circle causes most of the phase shift to be concentrated near the frequency where the pole is located. Taking the pole further away causes the delay to be spread over more frequencies. Complicated phase shifts or group delays may be built up by cascading several single-pole filters.

EXERCISES

1 An example of an all-pass filter is the time function $p_t = (\frac{1}{2}, -\frac{3}{4}, -\frac{3}{8}, -\frac{3}{16} \cdots)$. Calculate a few lags of its autocorrelation by summing some infinite series.
2 Sketch the amplitude, phase, and group delay of the all-pass filter $(1 - \bar{Z}_0 Z)/(Z_0 - Z)$ where $Z_0 = (1 + \varepsilon)e^{i\omega_0}$ and ε is small. Indicate important parameters on the curve.
3 Show that the coefficients of an all-pass, phase-shifting filter made by cascading $(1 - \bar{Z}_0 Z)/(Z_0 - Z)$ with $(1 - Z_0 Z)/(\bar{Z}_0 - Z)$ are real.
4 A continuous time function is the impulse response of a continuous-time, all-pass filter. Describe the function in both time domain and frequency domain. Interchange the words *time* and *frequency* in your description of the function. What is a physical example of such a function? What happens to the statement: "The group delay of an all-pass filter is positive."?
5 A graph of the group delay $\tau_g(\omega)$ in equation (2-6-7) shows τ_g to be positive for all ω. What is the area under τ_g in the range $0 < \omega < 2\pi$. (HINT: This is a trick question you can solve in your head.)

2-7 NOTCH FILTER AND POLE ON PEDESTAL

In some applications it is desired to reject a very narrow frequency band leaving the rest of the spectrum little changed. The most common example is 60-Hz noise from power lines. Such a filter can easily be made with a slight variation on the

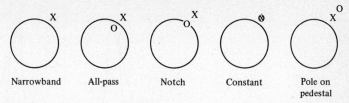

FIGURE 2-18
Pole and zero locations for some simple filters. Circles are unit circles in the Z plane. Poles are marked by X and zeros by 0.

all-pass filter. In the all-pass filter the pole and zero have an equal (logarithmic) relative distance from the unit circle. All we need to do is to put the zero closer to the circle. In fact, there is no reason why we should not put the zero right on the circle. Then the frequency at which the zero is located is exactly canceled from the spectrum of input data. If the undesired frequency need not be completely rejected, then the zero can be left just inside or outside the circle. As the zero is moved farther away from the circle, the notch becomes less deep until finally the zero is farther from the circle than the pole and the notch has become a hump. The resulting filter which will be called *pole on pedestal* is in many respects like the narrowband filter discussed earlier. Some of these filters are illustrated in Figs. 2-18 and 2-19. The difference between the pole-on-pedestal and the narrowband filters is in the asymptotic behavior away from ω_0. The former is flat, while the latter continues to decay with increasing $|\omega - \omega_0|$. This makes the pole on pedestal more convenient for creating complicated filter shapes by cascades of single-pole filters.

Narrowband filters and sharp cutoff filters should be used with caution. An ever-present penalty for such filters is that they do not decay rapidly in time. Although this may not present problems in some applications, it will do so in others. Obviously, if the data collection duration is shorter or comparable to the impulse response of the narrowband filter, then the transient effects of starting up the experiment will not have time to die out. Likewise, the notch should not be too narrow in a 60-Hz rejection filter. Even a bandpass filter (easier to implement with fast Fourier transform than with a few poles) has a certain decay rate in the time domain which may be too slow for some experiments. In radar and in reflection seismology the importance of a signal is not related to its strength. Late-arriving echoes may be very weak, but they contain information not found in earlier echoes. If too sharp a frequency characteristic is used, then filter resonance from early strong arrivals may not have decayed sufficiently by the time that the weak late echoes arrive.

EXERCISES

1 Consider a symmetric (nonrealizable) filter which passes all frequencies less than ω_0 with unit gain. Frequencies above ω_0 are completely attenuated. What is the rate of decay of amplitude with time for this filter?

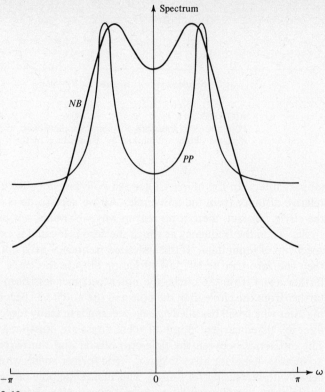

FIGURE 2-19
Amplitude vs. frequency for narrowband filter (NB) and pole-on-pedestal filter (PP). Each has one pole at $Z_0 = 1.2\,e^{i\pi/3}$. A second pole at $Z_0 = 1.2\,e^{-i\pi/3}$ enables the filters to be real in the time domain.

2 Waves spreading from a point source decay in energy as the area on a sphere. The amplitude decays as the square root of the energy. This implies a certain decay rate in time. The time-decay rate is the same if the waves reflect from planar interfaces. To what power of time t do the signal amplitudes decay? For waves backscattered to the source from point reflectors, energy decays as distance to the minus fourth power. What is the associated decay with time?

3 Discuss the use of the filter of Exercise 1 on the data of Exercise 2.

4 Design a single-pole, single-zero notch filter to reject 59 to 61 Hz on data which are sampled at 500 points per second.

2-8 THE BILINEAR TRANSFORM

Z transforms and Fourier transforms are related by the relations $Z = e^{i\omega}$ and $i\omega = \ln Z$. A problem with these relations is that simple ratios of polynomials in Z do not translate to ratios of polynomials in ω and vice versa. The approximation

$$-i\hat{\omega} = 2\,\frac{1 - Z}{1 + Z} \qquad (2\text{-}8\text{-}1)$$

is easily solved for Z as

$$Z = \frac{1 + i\hat{\omega}/2}{1 - i\hat{\omega}/2} \qquad (2\text{-}8\text{-}2)$$

These approximations are often useful. They are truncations of the exact power series expansions

$$-i\omega = -\ln e^{i\omega} = -\ln Z = 2\left[\frac{1 - Z}{1 + Z} + \frac{1}{3}\frac{(1 - Z)^3}{(1 + Z)^3} + \frac{1}{5}\cdots\right] \qquad (2\text{-}8\text{-}3)$$

and

$$Z = e^{i\omega} = \frac{e^{i\omega/2}}{e^{-i\omega/2}} = \frac{1 + i\omega/2 + (i\omega/2)^2/2! + \cdots}{1 - i\omega/2 + (i\omega/2)^2/2! + \cdots} \qquad (2\text{-}8\text{-}4)$$

For a Z transform $B(Z)$ to be minimum phase, any root Z_0 of $0 = B(Z_0)$ should be outside the unit circle. Since $Z_0 = \exp\{i[\text{Re}\,(\omega_0) + i\,\text{Im}\,(\omega_0)]\}$ and $|Z_0| = e^{-\text{Im}(\omega_0)}$, it means that for minimum phase $\text{Im}\,(\omega_0)$ should be negative. (In other words, ω_0 is in the lower half-plane.) Thus it may be said that $Z = e^{i\omega}$ maps the exterior of the unit circle to the lower half-plane. By inspection of Figs. 2-20 and 2-21, it is found that the bilinear approximation (2-8-1) or (2-8-2) also maps the exterior of the unit circle into the lower half-plane.

Thus, although the bilinear approximation is an approximation, it turns out to exactly preserve the minimum-phase property. This is very fortunate because if a stable differential equation is converted to a difference equation via (2-8-1), the resulting difference equation will be stable. (Many cases may be found where the approximation of a time derivative by multiplication with $1 - Z$ would convert a stable differential equation into an unstable difference equation.)

A handy way to remember (2-8-1) is that $-i\omega$ corresponds to time differentiation of a Fourier transform and $(1 - Z)$ is the first differencing operator. The $(1 + Z)$ in the denominator gets things "centered" at $Z^{1/2}$.

To see that the bilinear approximation is a low-frequency approximation, multiply top and bottom of (2-8-1) by $Z^{-1/2}$

$$-i\hat{\omega} = 2\frac{Z^{-1/2} - Z^{1/2}}{Z^{-1/2} + Z^{1/2}}$$

$$= -2i\frac{\sin \omega/2}{\cos \omega/2}$$

$$\hat{\omega} = 2\tan \omega/2 \qquad (2\text{-}8\text{-}5)$$

Equation (2-8-5) implicitly refers to a sampling rate of one sample per second. Taking an arbitrary sampling rate Δt, the approximation (2-8-5) becomes

$$\omega\,\Delta t \approx 2\tan \omega\,\Delta t/2 \qquad (2\text{-}8\text{-}6)$$

This approximation is plotted in Fig. 2-22. Clearly, the error can be made as small as one wishes merely by sampling often enough; that is, taking Δt small enough.

	Z	$\omega = 2\pi n - i \ln Z$	$\hat{\omega} = 2i\dfrac{1-Z}{1+Z}$
A	1	$2\pi n + 0$	0
B	i	$2\pi n + \pi/2$	2
C	-1	$2\pi n + \pi$	$\pm\infty$
D	$-i$	$2\pi n - \pi/2$	-2
E	$\frac{1}{2}$	$2\pi n + .693i$	$i\frac{2}{3}$
F	2	$2\pi n - .693i$	$-i\frac{2}{3}$

FIGURE 2-20
Some typical points in the Z-plane, the ω-plane, and the $\hat{\omega}$-plane.

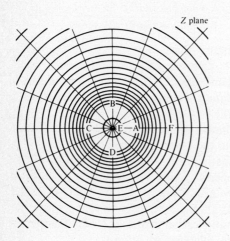

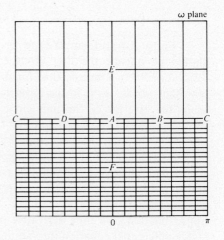

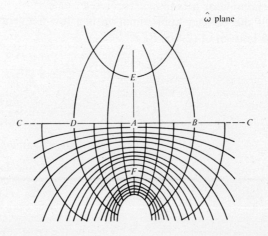

FIGURE 2-21
The points of Fig. 2-20 displayed in the Z plane, the ω plane, and the $\hat{\omega}$-plane.

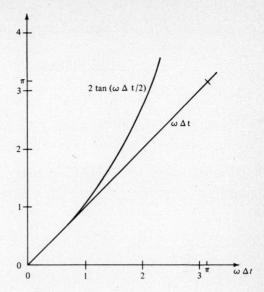

FIGURE 2-22
The accuracy of the bilinear transformation approximation.

From Fig. 2-22 we see that the error will be only a few percent if we choose Δt small enough so that $\omega_{max}\Delta t \leq 1$. Readers familiar with the folding theorem will recall that it gives the less severe restraint $\omega_{max}\Delta t < \pi$. Clearly, the folding theorem is too generous for applications involving the bilinear transform.

Now, by way of example, let us take up the case of a pole $1/-i\omega$ at zero frequency. This is integration. For reasons which will presently be clear, we will consider the slightly different pole

$$P = \frac{1}{-i\omega + \varepsilon} \qquad (2\text{-}8\text{-}7)$$

where ε is small. Inserting the bilinear transform, we get

$$P = \frac{1}{2[(1-Z)/(1+Z)] + \varepsilon} = \frac{0.5(1+Z)}{1 - Z + \varepsilon[(1+Z)/2]}$$

$$= \frac{0.5(1+Z)}{(1+\varepsilon/2) - Z(1-\varepsilon/2)} \qquad (2\text{-}8\text{-}8)$$

By inspection of (2-8-8) we see that the time-domain function is real, and as ε goes to zero it takes the form $(.5, 1, 1, 1, \ldots)$. (Taking ε positive forces the step to go out into positive time, whereas ε negative would cause the step to rise at negative time.) The properties of this function are summarized in Fig. 2-23. It is curious to note that if time domain and frequency domain are switched around, we have the quadrature filter described in Fig. 1–17.

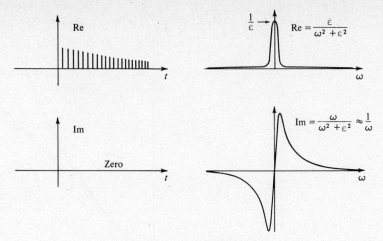

FIGURE 2-23
Properties of the integration operator.

EXERCISE

1 In the solution to diffusion problems, the factor $F(\omega) = 1/(-i\omega)^{1/2}$ often arises as a multiplier. To see the equivalent convolution operation, find a causal, sampled-time representation f_t of $F(\omega)$ by identification of powers of Z in

$$(f_0 + f_1 Z + f_2 Z^2 + \cdots)^2 = 1/(-i\omega) \simeq \tfrac{1}{2}(1 + Z)/(1 - Z)$$

Solve numerically for f_0 through f_7.

3

SPECTRAL FACTORIZATION

As we will see, there is an infinite number of time functions with any given spectrum. Spectral factorization is a method of finding the one time function which is also minimum phase. The minimum-phase function has many uses. It, and it alone, may be used for feedback filtering. It will arise frequently in wave propagation problems of later chapters. It arises in the theory of prediction and regulation for the given spectrum. We will further see that it has its energy squeezed up as close as possible to $t = 0$. It determines the minimum amount of dispersion in viscous wave propagation which is implied by causality. It finds application in two-dimensional potential theory where a vector field magnitude is observed and the components are to be inferred.

This chapter contains four computationally distinct methods of computing the minimum-phase wavelet from a given spectrum. Being distinct, they offer separate insights into the meaning of spectral factorization and minimum phase.

3-1 ROOT METHOD

The time function $(2, 1)$ has the same spectrum as the time function $(1, 2)$. The autocorrelation is $(2, 5, 2)$. We may utilize this observation to explore the multiplicity of all time functions with the same autocorrelation and spectrum. It would seem that the time reverse of any function would have the same autocorrelation as the function. Actually, certain applications will involve complex time series; therefore we should make the more precise statement that any wavelet and its complex-conjugate time-reverse share the same autocorrelation and spectrum. Let us verify this for simple two-point time functions. The spectrum of (b_0, b_1) is

$$\bar{B}\left(\frac{1}{Z}\right) B(Z) = \left(\bar{b}_0 + \frac{\bar{b}_1}{Z}\right)(b_0 + b_1 Z)$$

$$= \frac{\bar{b}_1 b_0}{Z} + (\bar{b}_0 b_0 + \bar{b}_1 b_1) + \bar{b}_0 b_1 Z \qquad (3\text{-}1\text{-}1)$$

The conjugate-reversed time function $(\bar{b}_1, \bar{b}_0)$ with Z transform $B_r(Z) = \bar{b}_1 + \bar{b}_0 Z$ has a spectrum

$$\bar{B}_r\left(\frac{1}{Z}\right) B_r(Z) = \left(b_1 + \frac{b_0}{Z}\right)(\bar{b}_1 + \bar{b}_0 Z)$$

$$= \frac{b_0 \bar{b}_1}{Z} + (b_0 \bar{b}_0 + b_1 \bar{b}_1) + b_1 \bar{b}_0 Z \qquad (3\text{-}1\text{-}2)$$

We see that the spectrum (3-1-1) is indeed identical to (3-1-2). Now we wish to extend the idea to time functions with three and more points. Full generality may be observed for three-point time functions, say $B(Z) = b_0 + b_1 Z + b_2 Z^2$. First, we call upon the fundamental theorem of algebra (which states that a polynomial of degree n has exactly n roots) to write $B(Z)$ in factored form

$$B(Z) = b_2(Z_1 - Z)(Z_2 - Z) \qquad (3\text{-}1\text{-}3)$$

Its spectrum is

$$R(Z) = \bar{B}\left(\frac{1}{Z}\right) B(Z) = \bar{b}_2 b_2 \left(\bar{Z}_1 - \frac{1}{Z}\right)(Z_1 - Z)\left(\bar{Z}_2 - \frac{1}{Z}\right)(Z_2 - Z) \qquad (3\text{-}1\text{-}4)$$

Now, what can we do to change the wavelet (3-1-3) which will leave its spectrum (3-1-4) unchanged? Clearly, b_2 may be multiplied by any complex number of unit magnitude. What is left of (3-1-4) can be broken up into a product of factors of form $(\bar{Z}_i - 1/Z)(Z_i - Z)$. But such a factor is just like (3-1-1). The time function of $(Z_i - Z)$ is $(Z_i, -1)$, and its complex-conjugate time-reverse is $(-1, \bar{Z}_i)$. Thus, any factor $(Z_i - Z)$ in (3-1-3) may be replaced by a factor $(-1 + \bar{Z}_i Z)$. In a generalization of (3-1-3) there could be N factors $[(Z_i - Z), i = 1, 2, \ldots, N)]$. Any combination of them could be reversed. Hence there are 2^N different wavelets which may be formed by reversals, and all of the wavelets have the same spectrum. Let us look off the unit circle in the complex plane. The factor $(Z_i - Z)$ means that Z_i is a root of both $B(Z)$ and $R(Z)$. If we replace $(Z_i - Z)$ by $(-1 + \bar{Z}_i Z)$ in $B(Z)$, we have removed a root at Z_i from $B(Z)$ and replaced it by another at $Z = 1/\bar{Z}_i$. The roots of $R(Z)$ have not changed a bit because there were originally roots at both Z_i and $1/\bar{Z}_i$ and the reversal has merely switched them around. Summarizing the situation in the complex plane, $B(Z)$ has roots Z_i which occur anywhere, $R(Z)$ must

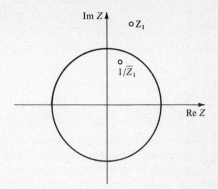

FIGURE 3-1
Roots of $\bar{B}(1/Z)\,B(Z)$.

have all the roots Z_i and, in addition, the roots $1/\bar{Z}_i$. Replacing some particular root Z_i by $1/\bar{Z}_i$ changes $B(Z)$ but not $R(Z)$. The operation of replacing a root at Z_i by one at $1/\bar{Z}_i$ may be written as

$$B'(Z) = \frac{Z - 1/\bar{Z}_i}{1 - Z/Z_i}\,B(Z) \qquad (3\text{-}1\text{-}5)$$

The multipyling factor is none other than the all-pass filter considered in an earlier chapter. With that in mind, it is obvious that $B'(Z)$ has the same spectrum as $B(Z)$. In fact, there is really no reason for Z_i to be a root of $B(Z)$. If Z_i is a root of $B(Z)$, then $B'(Z)$ will be a polynomial; otherwise it will be an infinite series.

Now let us discuss the calculation of $B(Z)$ from a given $R(Z)$. First, the roots of $R(Z)$ are by definition the solutions to $R(Z) = 0$. If we multiply $R(Z)$ by Z^N (where $R(Z)$ has been given up to degree N), then $Z^N R(Z)$ is a polynomial and the solutions Z_i to $Z^N R(Z) = 0$ will be the same as the solutions of $R(Z) = 0$. Finding all roots of a polynomial is a standard though difficult task. Assuming this to have been done we may then check to see if the roots come in the pairs Z_i and $1/\bar{Z}_i$. If they do not, then $R(Z)$ was not really a spectrum. If they do, then for every zero inside the unit circle, we must have one outside. Refer to Fig. 3-1. Thus, if we decide to make $B(Z)$ be a minimum-phase wavelet with the spectrum $R(Z)$, we collect all of the roots outside the unit circle. Then we create $B(Z)$ with

$$B(Z) = b_N(Z - Z_1)(Z - Z_2)\ldots(Z - Z_N) \qquad (3\text{-}1\text{-}6)$$

This then summarizes the calculation of a minimum-phase wavelet from a given spectrum. When N is large, it is computationally very awkward compared to methods yet to be discussed. The value of the root method is that it shows certain basic principles.

1 Every spectrum has a minimum-phase wavelet which is unique within a complex scale factor of unit magnitude.
2 There are infinitely many time functions with any given spectrum.
3 Not all functions are possible autocorrelation functions.

The root method of spectral factorization was apparently developed by economists in the 1920s and 1930s. A number of early references may be found in Wold's book, *Stationary Time Series* [Ref. 10].

EXERCISES

1 How can you find the scale factor b_N in (3-1-6)?
2 Compute the autocorrelation of each of the four wavelets $(4, 0, -1)$, $(2, 3, -2)$, $(-2, 3, 2)$, $(1, 0, -4)$.
3 A power spectrum is observed to fit the form $P(\omega) = 38 + 10 \cos \omega - 12 \cos 2\omega$. What are some wavelets with this spectrum? Which is minimum phase? [HINT: $\cos 2\omega = 2 \cos^2 \omega - 1$; $2 \cos \omega = Z + 1/Z$; use quadratic formula.]
4 Show that if a wavelet $b_t = (b_0, b_1, \ldots, b_n)$ is real, the roots of the spectrum R come in the quadruplets Z_0, $1/Z_0$, $\bar{Z}_0$, and $1/\bar{Z}_0$. Look into the case of roots exactly on the unit circle and on the real axis. What is the minimum multiplicity of such roots?

3-2 ROBINSON'S ENERGY DELAY THEOREM [Ref. 11]

We will now show that a minimum-phase wavelet has less energy delay than any other one-sided wavelet with the same spectrum. More precisely, we will show that the energy summed from zero to any time t for the minimum-phase wavelet is greater than or equal to that of any other wavelet with the same spectrum. Refer to Fig. 3-2.

We will compare two wavelets P_{in} and P_{out} which are identical except for one zero, which is outside the unit circled for P_{out} and inside for P_{in}. We may write this as

$$P_{out}(Z) = (b + sZ)P(Z)$$

$$P_{in}(Z) = (s + bZ)P(Z)$$

where b is bigger than s and P is arbitrary but of degree n. Next we tabulate the terms in question.

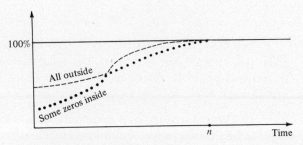

FIGURE 3-2
Percent of total energy in a filter between time 0 and time t.

t	P_{out}	P_{in}	$P_{\text{out}}^2 - P_{\text{in}}^2$	$\sum\limits_{k=0}^{t}(P_{\text{out}}^2 - P_{\text{in}}^2)$
0	bp_0	sp_0	$(b^2 - s^2)p_0{}^2$	$(b^2 - s^2)p_0{}^2$
1	$bp_1 + sp_0$	$sp_1 + bp_0$	$(b^2 - s^2)(p_1{}^2 - p_0{}^2)$	$(b^2 - s^2)p_1{}^2$
⋮				
k	$bp_k + sp_{k-1}$	$sp_k + bp_{k-1}$	$(b^2 - s^2)(p_k{}^2 - p_k{}^2{}_{-1})$	$(b^2 - s^2)p_k{}^2$
⋮				
$n+1$	sp_n	bp_n	$(b^2 - s^2)(-p_n{}^2)$	0

The difference, which is given in the right-hand column, is clearly always positive.

To prove that the miminum-phase wavelet delays energy the least, the preceding argument is repeated with each of the roots until they are all outside the unit circle.

EXERCISE

1 Do the foregoing minimum-energy-delay proof for complex-valued b, s, and P. [CAUTION: Does $P_{\text{in}} = (s + bZ)P$ or $P_{\text{in}} = (\bar{s} + \bar{b}Z)P$?]

3-3 THE TOEPLITZ METHOD

The Toeplitz method of spectral factorization is based on special properties of Toeplitz matrices [Ref. 12]. In this chapter we introduce the Toeplitz matrix to perform spectral factorization. In later chapters we will refer back several times to the algebra described here. When one desires to predict a time series, one can do this with a so-called *prediction filter*. This filter is found as the solution to Toeplitz simultaneous equations. Norman Levinson, in his explanatory appendix of Norbert Wiener's *Time Series*, first introduced the Toeplitz matrix to engineers; however, it had been widely known and used previously in the field of econometrics. It is only natural that it should appear first in economics because there the data are observed at discrete time points, whereas in engineering the idea of discretized time was rather artificial until the advent of digital computers. The need for prediction in economics is obvious. In seismology, it is not the prediction itself but the error in prediction which is of interest. Reflection seismograms are used in petroleum exploration. Ideally, the situation is like radar where the delay time is in direct proportion to physical distance. This is the case for the so-called *primary* reflections. A serious practical complication arises in shallow seas where large acoustic waves bounce back and forth between the sea surface and the sea floor. These are called *multiple* reflections. A mechanism for separation of the primary waves from the multiple reflections is provided by prediction. A multiple reflection is predictable from earlier echoes, but a primary reflection is not predictable from earlier echoes. Thus, the useful information is carried in the part of the seismogram which is *not* predictable. An oil company computer devoted to interpreting seismic exploration data typically solves about 100,000 sets of Toeplitz simultaneous equations in a day.

Another important application of the algebra associated with Toeplitz matrices is in high-resolution spectral analysis. This is where a power spectrum is to be estimated from a sample of data which is short (in time or space). The conventional statistical and engineering knowledge in this subject is based on assumptions which are frequently inappropriate in geophysics. The situation was fully recognized by John P. Burg who utilized some of the special properties of Toeplitz matrices to develop his maximum-entropy spectral estimation procedure described in a later chapter.

Another place where Toeplitz matrices play a key role is in the mathematical physics which describes layered materials. Geophysicists often model the earth by a stack of plane layers or by concentric spherical shells where each shell or layer is homogeneous. Surprisingly enough, many mathematical physics books do not mention Toeplitz matrices. This is because they are preoccupied with *forward* problems; that is, they wish to calculate the waves (or potentials) observed in a known configuration of materials. In geophysics, we are interested in both *forward* problems and in *inverse* problems where we observe waves on the surface of the earth and we wish to deduce material configurations inside the earth. A later chapter contains a description of how Toeplitz matrices play a central role in such inverse problems.

We start with a time function x_t which may or may not be minimum phase. Its spectrum is computed by $R(Z) = \overline{X}(1/Z)X(Z)$. As we saw in the preceding sections, given $R(Z)$ alone there is no way of knowing whether it was computed from a minimum-phase function or a nonminimum-phase function. We may suppose that there exists a minimum phase $B(Z)$ of the given spectrum, that is, $R(Z) = \overline{B}(1/Z) B(Z)$. Since $B(Z)$ is by hypothesis minimum phase, it has an inverse $A(Z) = 1/B(Z)$. We can solve for the inverse $A(Z)$ in the following way:

$$R(Z) = \overline{B}\left(\frac{1}{Z}\right) B(Z) = \frac{\overline{B}(1/Z)}{A(Z)} \qquad (3\text{-}3\text{-}1)$$

$$R(Z)A(Z) = \overline{B}\left(\frac{1}{Z}\right) = \overline{b}_0 + \frac{\overline{b}_1}{Z} + \cdots \qquad (3\text{-}3\text{-}2)$$

To solve for $A(Z)$, we identify coefficients of powers of Z. For the case where, for example, $A(Z)$ is the quadratic $a_0 + a_1 Z + a_2 Z^2$, the coefficient of Z^0 in (3-3-2) is

$$r_0 a_0 + r_{-1} a_1 + r_{-2} a_2 = \overline{b}_0 \qquad (3\text{-}3\text{-}3a)$$

The coefficient of Z^1 is

$$r_1 a_0 + r_0 a_1 + r_{-1} a_2 = 0 \qquad (3\text{-}3\text{-}3b)$$

and the coefficient of Z^2 is

$$r_2 a_0 + r_1 a_1 + r_0 a_2 = 0 \qquad (3\text{-}3\text{-}3c)$$

Bringing these together we have the simultaneous equations

$$\begin{bmatrix} r_0 & r_{-1} & r_{-2} \\ r_1 & r_0 & r_{-1} \\ r_2 & r_1 & r_0 \end{bmatrix} \begin{bmatrix} a_0 \\ a_1 \\ a_2 \end{bmatrix} = \begin{bmatrix} \bar{b}_0 \\ 0 \\ 0 \end{bmatrix} \qquad (3\text{-}3\text{-}4)$$

It should be clear how to generalize this to a set of simultaneous equations of arbitrary size. The main diagonal of the matrix contains r_0 in every position. The diagonal just below the main one contains r_1 everywhere. Likewise, the whole matrix is filled. Such a matrix is called a Toeplitz matrix. Let us define $a'_k = a_k/a_0$. Recall by the polynomial division algorithm that $\bar{b}_0 = 1/\bar{a}_0$. Define a positive number $v = 1/a_0 \bar{a}_0$. Now, dividing the vector on each side of (3-3-4) by a_0, we get the most popular form of the equations

$$\begin{bmatrix} r_0 & r_{-1} & r_{-2} \\ r_1 & r_0 & r_{-1} \\ r_2 & r_1 & r_0 \end{bmatrix} \begin{bmatrix} 1 \\ a'_1 \\ a'_2 \end{bmatrix} = \begin{bmatrix} v \\ 0 \\ 0 \end{bmatrix} \qquad (3\text{-}3\text{-}5)$$

This gives three equations for the three unknowns a'_1, a'_2, and v. To put (3-3-5) in a form where standard simultaneous equations programs could be used one would divide the vectors on both sides by v. After solving the equations, we get a_0 by noting that it has magnitude $1/\sqrt{v}$ and its phase is arbitrary, as with the root method of spectral factorization.

At this point, a pessimist might interject that the polynomial $A(Z) = a_0 + a_1 Z + a_2 Z^2$ determined from solving the set of simultaneous equations might not turn out to be minimum phase, so that we could not necessarily compute $B(Z)$ by $B(Z) = 1/A(Z)$. The pessimist might argue that the difficulty would be especially likely to occur if the size of the set (3-3-5) was not taken to be large enough. Actually experimentalists have known for a long time that the pessimists were wrong. A proof can now be performed rather easily, along with a description of a computer algorithm which may be used to solve (3-3-5).

The standard computer algorithms for solving simultaneous equations require time proportional to n^3 and computer memory proportional to n^2. The Levinson computer algorithm [Ref. 13] for Toeplitz matrices requires time proportional to n^2 and memory proportional to n. First notice that the Toeplitz matrix contains many identical elements. Levinson utilized this special Toeplitz symmetry to develop his fast method.

The method proceeds by the approach called recursion. That is, given the solution to the $k \times k$ set of equations, we show how to calculate the solution to the $(k + 1) \times (k + 1)$ set. One must first get the solution for $k = 1$; then one repeatedly (recursively) applies a set of formulas increasing k by one at each stage. We will show how the recursion works for real-time functions ($r_k = r_{-k}$) going from the 3×3 set of equations to the 4×4 set, and leave it to the reader to work out the general case.

Given the 3×3 simultaneous equations and their solution a_i

$$\begin{bmatrix} r_0 & r_1 & r_2 \\ r_1 & r_0 & r_1 \\ r_2 & r_1 & r_0 \end{bmatrix} \begin{bmatrix} 1 \\ a_1 \\ a_2 \end{bmatrix} = \begin{bmatrix} v \\ 0 \\ 0 \end{bmatrix} \qquad (3\text{-}3\text{-}6)$$

then the following construction defines a quantity e given r_3 (or r_3 given e)

$$\begin{bmatrix} r_0 & r_1 & r_2 & r_3 \\ r_1 & r_0 & r_1 & r_2 \\ r_2 & r_1 & r_0 & r_1 \\ r_3 & r_2 & r_1 & r_0 \end{bmatrix} \begin{bmatrix} 1 \\ a_1 \\ a_2 \\ 0 \end{bmatrix} = \begin{bmatrix} v \\ 0 \\ 0 \\ e \end{bmatrix} \qquad (3\text{-}3\text{-}7)$$

The first three rows in (3-3-7) are the same as (3-3-6); the last row is the new definition of e. The Levinson recursion shows how to calculate the solution a' to the 4×4 simultaneous equations which is like (3-3-6) but larger in size.

$$\begin{bmatrix} r_0 & r_1 & r_2 & r_3 \\ r_1 & r_0 & r_1 & r_2 \\ r_2 & r_1 & r_0 & r_1 \\ r_3 & r_2 & r_1 & r_0 \end{bmatrix} \begin{bmatrix} 1 \\ a_1' \\ a_2' \\ a_3' \end{bmatrix} = \begin{bmatrix} v' \\ 0 \\ 0 \\ 0 \end{bmatrix} \qquad (3\text{-}3\text{-}8)$$

The important trick is that from (3-3-7) one can write a "reversed" system of equations. (If you have trouble with the matrix manipulation, merely write out (3-3-8) as simultaneous equations, then reverse the order of the unknowns, and then reverse the order of the equations.)

$$\begin{bmatrix} r_0 & r_1 & r_2 & r_3 \\ r_1 & r_0 & r_1 & r_2 \\ r_2 & r_1 & r_0 & r_1 \\ r_3 & r_2 & r_1 & r_0 \end{bmatrix} \begin{bmatrix} 0 \\ a_2 \\ a_1 \\ 1 \end{bmatrix} = \begin{bmatrix} e \\ 0 \\ 0 \\ v \end{bmatrix} \qquad (3\text{-}3\text{-}9)$$

The Levinson recursion consists of subtracting a yet unknown portion c_3 of (3-3-9) from (3-3-7) so as to get the result (3-3-8). That is

$$\begin{bmatrix} r_0 & r_1 & r_2 & r_3 \\ r_1 & r_0 & r_1 & r_2 \\ r_2 & r_1 & r_0 & r_1 \\ r_3 & r_2 & r_1 & r_0 \end{bmatrix} \left(\begin{bmatrix} 1 \\ a_1 \\ a_2 \\ 0 \end{bmatrix} - c_3 \begin{bmatrix} 0 \\ a_2 \\ a_1 \\ 1 \end{bmatrix} \right) = \left(\begin{bmatrix} v \\ 0 \\ 0 \\ e \end{bmatrix} - c_3 \begin{bmatrix} e \\ 0 \\ 0 \\ v \end{bmatrix} \right) \qquad (3\text{-}3\text{-}10)$$

To make the right-hand side of (3-3-10) look like the right-hand side of (3-3-8), we have to get the bottom element to vanish, so we must choose $c_3 = e/v$. This implies that $v' = v - c_3 e = v - e^2/v = v[1 - (e/v)^2]$. Thus, the solution to the 4×4 system is derived from the 3×3 by

$$e \leftarrow \sum_{i=0}^{2} a_i r_{3-i} \qquad (3\text{-}3\text{-}11)$$

$$\begin{bmatrix} 1 \\ a_1' \\ a_2' \\ a_3' \end{bmatrix} \leftarrow \begin{bmatrix} 1 \\ a_1 \\ a_2 \\ 0 \end{bmatrix} - \frac{e}{v} \begin{bmatrix} 0 \\ a_2 \\ a_1 \\ 1 \end{bmatrix} \qquad (3\text{-}3\text{-}12)$$

$$v' \leftarrow v[1 - (e/v)^2] \qquad (3\text{-}3\text{-}13)$$

We have shown how to calculate the solution of the 4×4 Toeplitz equations from the solution of the 3×3 Toeplitz equations. The Levinson recursion consists of doing this type of step, starting from 1×1 and working up to $n \times n$.

Let us reexamine the calculation to see why $A(Z)$ turns out to be minimum

```
COMPLEX R,A,C,E,BOT,CONJG
        C(1)=-1.; R(1)=1.; A(1)=1.; V(1)=1.
200 DO 220 J=2,N
        A(J)=0.
        E=0.
        DO 210 I=2,J
210 E=E+R(I)*A(J-I+1)
        C(J)=E/V(J-1)
        V(J)=V(J-1)-E*CONJG(C(J))
        JH=(J+1)/2
        DO 220 I=1,JH
        BOT=A(J-I+1)-C(J)*CONJG(A(I))
        A(I)=A(I)-C(J)*CONJG(A(J-I+1))
220 A(J-I+1)=BOT
```

FIGURE 3-3
A computer program to do the Levinson recursion. It is assumed that the input r_k have been normalized by division by r_0. The complex arithmetic is optional.

phase. First, we notice that $v = 1/\bar{a}_0 a_0$ and $v' = 1/\bar{a}_0' a_0'$ are always positive. Then from (3-3-13) we see that $-1 < e/v < +1$. (The fact that $c = e/v$ is bounded by unity will later be shown to correspond to the fact that reflection coefficients for waves are so bounded.) Next, (3-3-12) may be written in polynomial form as

$$A'(Z) = A(Z) - (e/v)Z^3 A(1/Z) \qquad (3\text{-}3\text{-}14)$$

We know that Z^3 has unit magnitude on the unit circle. Likewise (for real time series), the spectrum of $A(Z)$ equals that of $A(1/Z)$. Thus (by the theorem of adding garbage to a minimum-phase wavelet) if $A(Z)$ is minimum phase, then $A'(Z)$ will also be minimum phase. In summary, the following three statements are equivalent:

1 $R(Z)$ is of the form $\bar{X}\left(\dfrac{1}{Z}\right)X(Z)$.
2 $|c_k| < 1$.
3 $A(Z)$ is minimum phase.

If any one of the above three is false, then they are all false. A program for the calculation of a_k and c_k from r_k is given in Fig. 3-3. In Chap. 8, on wave propagation in layers, programs are given to compute r_k from a_k or c_k.

EXERCISES

1 The top row of a 4 × 4 Toeplitz set of simultaneous equations like (3-3-8) is $(1, \tfrac{1}{4}, \tfrac{1}{16}, \tfrac{1}{4})$. What is the solution a_k?

2 How must the Levinson recursion be altered if time functions are complex? Specifically, where do complex conjugates occur in (3-3-11), (3-3-12), and (3-3-13)?

3 Let $A_m(Z)$ denote a polynomial whose coefficients are the solution to an $m \times m$ set of Toeplitz equations. Show that

$$v_n \, \delta_{nm} = \frac{1}{2\pi} \int_0^{2\pi} R(Z)A_m(Z)Z^{-n} \, d\omega \qquad n \leq m$$

which means that the polynomial $A_m(Z)$ is orthogonal to polynomial Z^n over the unit circle under the positive weighting function R. Utilizing this result, state why A_m is orthogonal to A_n, that is,

$$v_n \, \delta_{nm} = \frac{1}{2\pi} \int_0^{2\pi} R(Z)A_m(Z)\bar{A}_n\left(\frac{1}{Z}\right) d\omega$$

(HINT: First consider $n \leq m$, then all n.)

Toeplitz matrices are found in the mathematical literature under the topic of polynomials orthogonal on the unit circle. The author especially recommends Atkinson's book (Ref. 14).

3-4 WHITTLE'S EXP-LOG METHOD [Ref. 15]

In this method of spectral factorization we substitute power series into other power series. Thus, like the root method, it is good for learning but not good for computing. We start with some given autocorrelation r_t where

$$R(Z) = \cdots + r_{-1}Z^{-1} + r_0 + r_1 Z + r_2 Z^2 + \cdots$$

If $|R| > 2$ on the unit circle then a scale factor should be divided out. Insert this power series into the power series for logarithms.

$$U(Z) = \ln R(Z)$$

$$= (R - 1) - \frac{(R-1)^2}{2} + \frac{(R-1)^3}{3} - \cdots \qquad 0 < R \le 2$$

$$= \cdots + u_{-1}Z^{-1} + u_0 + u_1 Z + u_2 Z^2 + \cdots$$

Of course, in practice this would be a lot of effort, but it could be done in a systematic fashion with a computer program. Now define $U_t{}^+$ by dropping negative powers of Z from $U(Z)$

$$U^+(Z) = \frac{u_0}{2} + u_1 Z + u_2 Z^2 + \cdots$$

Insert this into the power series for the exponential

$$B(Z) = e^{U^+(Z)} = 1 + U^+ + \frac{(U^+)^2}{2!} + \frac{(U^+)^3}{3!} + \cdots$$

The desired minimum-phase wavelet is $B(Z)$; its spectrum is $R(Z)$. To see why this is so, consider the following identities.

$$R(Z) = e^{\ln R(Z)}$$

$$= \exp\left(\frac{u_0}{2} + \sum_{-\infty}^{-1} u_k Z^k + \frac{u_0}{2} + \sum_{+1}^{\infty} u_k Z^k\right)$$

$$= \exp\left(\frac{u_0}{2} + \sum_{-\infty}^{-1} u_k Z^k\right) \exp\left(\frac{u_0}{2} + \sum_{1}^{\infty} u_k Z^k\right)$$

$$= \exp\left[\bar{U}^+\left(\frac{1}{Z}\right)\right] \exp[U^+(Z)]$$

$$= \bar{B}\left(\frac{1}{Z}\right) B(Z)$$

Thus we have factored $R(Z)$ into the desired conjugate parts. Finally, let us see why $B(Z) = e^{U^+(Z)}$ is minimum phase. All we need to know about $U^+(Z)$ is that it is finite and does not contain powers of Z^{-1}. There are two proofs.

The first proof goes by observing that the imaginary part of $U(Z)$ on the unit circle is the phase angle of $B(Z)$. To be minimum phase, the phase of $B(Z)$ must not be augmented by multiples of 2π as Z goes around the unit circle. For minimum phase, the phase should be periodic with period 2π. The phase $u_1 \sin \omega + u_2 \sin 2\omega + \cdots$ obviously satisfies this condition. The second proof is more abstract. In the second proof we note that the only way for $B(Z_0) = e^{U^+(Z_0)}$ to be zero for some Z_0 would be if $U^+(Z_0)$ were equal to $-\infty$; in other words, if U^+ were nonconvergent. This is impossible inside the unit circle because we took the log series for $U(Z)$ to be absolutely convergent on the unit circle; this means $U^+(Z)$ is convergent (finite) inside the circle. Since $B(Z)$ cannot have zeros inside the unit circle, it must be minimum phase.

EXERCISES

1 How can you get $B(Z)$ if $|R| > 2$ on the unit circle?
2 Suppose $U^+(Z) = aZ$. For values of $a = -1, 0, 1, 100$, and i sketch the time function associated with $B(Z)$. Also sketch their spectra. Find one which resembles a gaussian.
3 Seismograms may be *spectrally balanced* by dividing the F.T. of each seismogram by its magnitude and multiplying by the average magnitude. This amounts to convolution with a symmetrical filter. How can spectral balancing be defined with causal filters?
4 Examine the coefficients of the Z derivative of $\ln B(Z) = U(Z)$ to find a recurrence for u_k given b_k and for b_k given u_k.

3-5 THE KOLMOGOROFF METHOD [Ref. 16]

If in a computer we have the coefficients $(x_0, x_1, x_2, \ldots)$ of a polynomial $X(Z)$, we say we are "working in the time domain." If we evaluate the polynomial $X(Z)$ at a number of positions on the unit circle, we have numbers, say $X(e^{i\omega_0})$, $X(e^{i\omega_1})$, $\ldots, X(e^{i\omega_n})$, which we call a frequency-domain representation of the polynomial $X(Z)$. We have seen that the fast Fourier transform is a very cheap way of going from time domain to frequency domain and back. This makes it very worthwhile to look at Whittle's factorization method in the frequency domain. Furthermore, we will understand spectral factorization from yet another point of view.

We may begin with a time function or Z transform $X(Z) = x_0 + x_1 Z + \cdots$. Let us denote by X_k the transform of the time function, that is, $X(Z)$ evaluated at numerous $(k = 0, 1, \ldots, n)$ places on the unit circle. Consider the identities

$$R_k = \bar{X}_k X_k$$
$$= e^{\ln(R_k)} = e^{U_k}$$

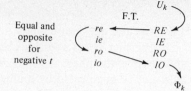

FIGURE 3-4
Determination of the phase function.

Now we add and subtract a still arbitrary function Φ_k to the exponential

$$R_k = \exp[\tfrac{1}{2}(U_k - i\Phi_k)] \exp [\tfrac{1}{2}(U_k + i\Phi_k)]$$
$$= \bar{B}_k B_k$$

Now the big question is what Φ_k should be used to guarantee that B_k transforms to a minimum-phase, one-sided time function? By looking at Whittle's method, we note that the only significant properties of $U^+(Z)$ are that it is finite and that the time function u_t vanishes before $t = 0$. Thus we expect that Φ_k should be chosen so that when $U_k + i\Phi_k$ is transformed into the time domain the resulting time function u_t^+ should vanish for negative time. This may be done as depicted in Fig. 3-4.

To see how easy it really is to get the imaginary odd part IO, we fetch the integration filter from Sec. 2-8 (on bilinear transformation) and display it in Fig. 3-5.

To get Φ_k we take U_k into the time domain, getting u_t. Then we multiply by the real step function of time in Fig. 3-5, obtaining $u_t^+ = u_0, u_1, \ldots,$. This implies that in the frequency domain U_k has been convolved with $\delta_{k=0} + i \cdot (90°$ phase shift filter). Thus, Φ has been generated.

Let us reconsider the operation of dropping all of the negative powers of Z in $U(Z)$ as we did in the previous section to get $U^+(Z)$. For simplicity, consider the case r_t real; then u_t is real.

$$U(Z) = \cdots + u_1 Z^{-1} + u_0 + u_1 Z + u_2 Z^2 + \cdots$$
$$U = u_0(\cos 0) + 2u_1 \cos \omega + 2u_2 \cos 2\omega + \cdots$$

Now let us make up a new function Φ by replacing cosine by sine in the foregoing expression

$$\Phi = 2u_1 \sin \omega + 2u_2 \sin 2\omega + \cdots$$

We now see that combining U with $i\Phi$ we get U^+.

$$\tfrac{1}{2}(U + i\Phi) = \tfrac{1}{2}u_0 + u_1 Z + u_2 Z^2 + \cdots$$
$$= U^+(Z)$$

Notice that the operation of changing $\cos t$ to $\sin t$ would be called 90° phase shift filtering. Here we have changed $\cos \omega$ to $\sin \omega$ with the result that $U^+(Z)$ has only positive coefficients of Z.

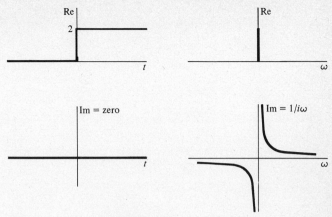

FIGURE 3-5
The transform pair used in the Hilbert transform.

The Kolmogoroff method of spectral factorization is very fast in a computer because fast Fourier transforms may be used. Its principle disadvantage is that summation around the unit circle is always slightly different than integration about the circle. When the spectrum is simple but poles are very close to the unit circle, then the Toeplitz method may prove more satisfactory. A simple program to do spectral factorization is given in Fig. 3-6.

EXERCISES

1 Insert the additional arrows in Fig. 3-4 which are required when dealing with complex time functions.

2 What is the meaning of *minimum-phase waveform* if the roles of time domain and frequency domain are interchanged?

3 Show how to do the inverse Hilbert transform, given ϕ find u. What is the interpretation of the fact that you cannot get u_0?

4 Consider a model of a portion of the earth where x is the north coordinate, $+z$ represents altitude above the earth, and magnetic bodies are distributed in the earth so as to create no magnetic field component in the east-west direction. One may show that the magnetic field h above the earth is represented by

$$\begin{bmatrix} h_x(x, z) \\ h_z(x, z) \end{bmatrix} = \int_{-\infty}^{+\infty} F(k) \begin{bmatrix} -ik \\ |k| \end{bmatrix} e^{ikx - |k|z} \, dk$$

Here $F(k)$ is some spatial frequency spectrum.

(a) By using Fourier transforms, how does one compute $h_x(x, 0)$ from $h_z(x, 0)$ and vice versa?

(b) Given $h_z(x, 0)$, how does one compute $h_z(x, z)$?

(c) Notice that at $z = 0$

$$w(x) = h_z(x) + ih_x(x) = \int_{-\infty}^{+\infty} e^{ikx} F(k)(|k| + k) \, dk$$

```
         LX=4                              LX MUST BE A POWER OF 2.
         WRITE(6,FMT) (CX(I),I=1,LX)       PRINT THE TEST EXAMPLE.
         CALL FORK(LX,CX,1.)               FOURIER TRANSFORM.
         DO 10 I=1,LX
   10    CX(I)=.5*CLOG(CX(I)*CONJG(CX(I)))  FORM THE LOG OF THE SPECTRUM
         CALL FORK(LX,CX,-1.)              BEGIN THE HILBERT TRANSFORM
         K=LX/2                            LX MUST BE EVEN
         DO 20 J=2,K                       LEAVE  T=0  ALONE
         CX(J)=CX(J)+CX(J)                 DOUBLE VALUES AT POSITIVE TIME
   20    CX(LX+2-J)=0.                     ZERO VALUES AT NEGATIVE TIME
         CALL FORK(LX,CX,+1.)              END HILBERT TRANSFORM
         DO 30 I=1,LX
   30    CX(I)=CEXP(CX(I))                 EXPONENTIATE
         CALL FORK(LX,CX,-1.)              INVERSE FOURIER TRANSFORM
         WRITE(6,FMT)(CX(I),I=1,LX)        PRINT MINIMUM PHASE WAVELET.
```

IN: (0.0, 0.0),   (0.0, 8.0),   (16.0, 0.0),   (0.0, 0.0),

OUT: (15.6,-0.0),   (0.0,-8.7),   (-0.4,-0.0),   (0.0,-0.7),

FIGURE 3-6
A program to do spectral factorization by means of fast Fourier transform and Hilbert transform. Complex arithmetic is mandatory. Results are approximate since integration around the unit circle has been approximated by summation over four points.

and that $F(k)(|k| + k)$ is a one-sided function of k. With a total field magnetometer one observes

$$h_x^2(x) + h_z^2(x) = w(x)\bar{w}(x)$$

What can you say about getting $F(k)$ from this?
(d) How unique are $h_x(x)$ and $h_z(x)$ if $w(x)\bar{w}(x)$ is given?

3-6 CAUSALITY AND WAVE PROPAGATION

The principle of causality, i.e., no response before a stimulus, places certain restraints upon the nature of wave propagation. Causality often seems to be violated because of

1 Approximations in a theory which are made to simplify it.
2 Fitting inappropriate curves to experimental observations.
3 Approximations made in converting differential equations to difference equations.

Violations of causality often result from seemingly inconsequential approximations at spectral frequencies outside the range of practical interest. The seemingly inconsequential approximations often take on significance because of the very weak convergence in the Hilbert transform. With regard to *1* and *2*, the violation of causality may turn out to be a low price to pay compared with the cost of a more precise analysis. With regard to *3* or computation in general, the violation of

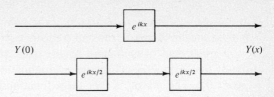

FIGURE 3-7
Modeling wave propagation with filters.

causality usually has disastrous consequences. Even though difference equations may have proper behavior at frequencies of interest, exponential growth at other frequencies is almost always so severe as to completely obliterate the desired solution.

Most wave-propagation theories are worked out in the frequency domain in homogeneous materials. If they are further done in cartesian geometry, the end result is that propagation along the x axis is given by

$$Y(x) = Y(0)\, e^{ikx}. \qquad (3\text{-}6\text{-}1)$$

What the wave theory provides is an expression for k in terms of frequency and all the physical parameters such as velocity, viscosity, etc. We will now take an " after-the-derivation " look at $k(\omega)$ and see if it satisfies causality. In other words, (3-6-1) is like the filter situation depicted in Fig. 3-7. $Y(0)$ is the filter input, $Y(x)$ is like the output, and e^{ikx} is a filter which models the effect of propagation. The question is whether the filters are one-sided. This depends upon the detailed form of $k(\omega)$.

First, consider the simple case when $k(\omega)$ is real for all real ω. Then e^{ikx} is an all-pass filter. It is realizable, as shown on p. 42, if and only if $k(\omega)x$ is a monotonically increasing function of ω.

Next, suppose $k(\omega)$ is complex. This means that attenuation will occur. Clearly energy conservation requires that the real part of ikx be negative.

Define $P(Z)$ to be a realizable all-pass filter and $B(Z)$ to be a realizable minimum-phase wavelet. Any realizable function can be represented in the form $B(Z)P(Z)$. We will require e^{ikx} to be so represented.

$$B(Z)P(Z) = e^{ikx} = \exp\left[i(k_r + ik_i)x\right] \qquad (3\text{-}6\text{-}2)$$

Taking logs and using ϕ to denote phase, we have

$$\ln|B| + i\phi_B + i\phi_P = -k_i x + ik_r x \qquad (3\text{-}6\text{-}3)$$

Splitting into real and imaginary parts we obtain

$$\ln|B| = -k_i x$$

$$\phi_P = k_r x - \phi_B$$

First of all, from the attenuation $-k_i x = \ln|B|$ we can by Hilbert transform compute ϕ_B. When it is subtracted from $k_r x$ we are left with the phase shift ϕ_P of the all-pass filter. Recall from Sec. 2-6 (on all-pass filters) that if and only if ϕ_P is monotonically increasing, we have a causal all-pass filter. In conclusion, the test of causality for a function $k(\omega)$ lies in computing $\phi_P(\omega)$ and seeing whether it is monotonically increasing for all frequencies.

A great deal of confusion results from attenuation laws which are thought to be of the form $k_i \simeq \omega^2$ (gaussian attenuation) or $k_i \simeq |\omega|$ (sometimes called *constant Q*). The reason is that these functions are not integrable so they do not have convergent Hilbert transforms; that is, the ϕ_B turns out to be infinite. These functions correspond to time pulses like the gaussian time function e^{-t^2} which do not have a time before which they are zero. Such functions cannot strictly be considered to be associated with any causal process; however, they may give excellent practical approximations. In reality, physical dissipation is generally associated with a relaxation phenomenon having a finite relaxation time. For frequency values below the relaxation value, the attenuation may appear to be increasing indefinitely with frequency, but, in fact, the attenuation decreases above the characteristic relaxation frequency. The attenuation also provides a phase velocity aberration because of the addition of ϕ_B to ϕ_P but this is generally considered to be too small to measure.

Now let us take up an example of computer modeling of wave propagation. The simplest example is propagation without dissipation. Then $k = \omega/v$ where v is the velocity. Rather than attempt to construct a filter which will carry waves a long distance we will only attempt to carry them a distance Δx. The filter may be used N times to propagate a distance $x = N \, \Delta x$. We will use the bilinear truncation of the power series for exponential

$$e^u \approx \frac{1 + \dfrac{u}{2}}{1 - \dfrac{u}{2}} \qquad (3\text{-}6\text{-}4)$$

and the bilinear representation for $i\omega \, \Delta t$, namely

$$i\omega \, \Delta t \approx 2\left(\frac{Z - 1}{Z + 1}\right) \qquad (3\text{-}6\text{-}5)$$

With these and $k = \omega/v$ and the definition $a = \Delta x/v \, \Delta t$ we obtain

$$e^{ik \, \Delta x} \approx e^{ik \, \Delta x} = \frac{1 + a\left(\dfrac{Z - 1}{Z + 1}\right)}{1 - a\left(\dfrac{Z - 1}{Z + 1}\right)}$$

$$= \frac{(Z + 1) + a(Z - 1)}{(Z + 1) - a(Z - 1)} = \frac{(1 - a) + (1 + a)Z}{(1 + a) + (1 - a)Z} \qquad (3\text{-}6\text{-}6)$$

That this is the general form of an all-pass filter may be seen by taking $Z_0 = -(1 + a)/(1 - a)$. Then (3-6-6) becomes

$$e^{ik \, \Delta x} = \frac{-1/Z_0 + Z}{1 - Z/Z_0} \qquad (3\text{-}6\text{-}7)$$

The denominator will be minimum phase and the all-pass filter will be realizable if $Z_0 > 1$ or $\Delta x / v \, \Delta t > 0$. This means that present and past values of the inputs will determine the output if the wave is being projected in the $+x$ direction. If it is desired to project the wave backwards in space (Δx negative), then Δt must be taken negative, so that present and future input values determine the present output.

EXERCISE

1 What is the velocity error as a function of ω for the filter of (3-6-6) for values of $a = .1$, 1., and 2. HINTS: Let $\hat{k} = \omega / \hat{v}$ so $v / \hat{v} = \hat{k} v / \omega$. Then note that

$$i\hat{k} \, \Delta x = \ln e^{i\hat{k} \, \Delta x} = \ln Z + \ln \left(\frac{-1}{Z_0 Z} + 1 \right) - \ln \left(1 - \frac{Z}{Z_0} \right)$$

$$= + i\omega \, \Delta t - 2 \ln \left(1 - \frac{Z}{Z_0} \right)$$

4

RESOLUTION

In locating an earthquake or a petroleum drilling site there will be an uncertainty in location, say (Δx, Δy, Δz) caused by measurement errors and the physical size of the target. In measuring a voltage there will be a measuring accuracy Δv. The frequency of useful seismic waves will have a bandwidth $\Delta \omega$. The time at which an earthquake occurs will have an uncertainty given by the duration of shaking Δt. A telescope of diameter Δd has at best a resolving power measured by a certain angular range $\Delta \theta$. It is often desirable to make measurements in such a way as to reduce the quantities Δx, Δy, Δz, Δv, $\Delta \omega$, Δt, Δd, and $\Delta \theta$ to values as small as possible. These measures of resolution (which are called variances, tolerances, uncertainties, bandwidths, durations, spreads, spans, etc.) sometimes intereact with one another in such a way that any experimental modification which reduces one must necessarily increase another or some combination of the others. The purpose of this chapter is to discuss some of the commonly occurring situations where such conflicting interactions occur.

In this chapter we use Δt to denote the time duration of a signal. We use τ to denote the amount of time which passes between sample points. In other chapters, Δt is synonymous with τ, the sample interval.

4-1 TIME-FREQUENCY RESOLUTION

The famous "uncertainty principle" of quantum mechanics resulted from observations that subatomic particles behave like waves with wave frequency proportional to particle momentum. The classical laws of mechanics enable prediction of the future of a mechanical system by extrapolation from presently known position and momentum. But because of the wave nature of matter with momentum proportional to frequency, such prediction requires simultaneous knowledge of both the location and the frequency of a wave. A sinusoidal wave has a perfectly clearly determined frequency, but it is spread over the infinitely long time axis. At the other extreme is a delta function, which is nicely compressed to a point on the time axis but contains a mixture of all frequencies. A mathematical analysis of the uncertainty principle is thus an analysis relating functions to their Fourier transforms.

Such an analysis begins by definitions of time duration and spectral bandwidth. The time duration of a damped exponential function is infinite if by duration you mean the span of nonzero function values. However, for nearly all practical purposes the time span is chosen as the time required for the amplitude to decay to e^{-1} of its original value. For many functions the span is defined by the span between points on the time or frequency axis where the curve (or its envelope) drop to half of the maximum value. The main idea is that the time span Δt or the frequency span $\Delta \omega$ should be able to include most of the total energy but need not contain all of it. The precise definition of Δt and $\Delta \omega$ is somewhat arbitrary and may be chosen to simplify analysis. The general statement is that for any function the time duration Δt and the spectral bandwidth $\Delta \omega$ are related by

$$\Delta \omega \, \Delta t \geq 2\pi \qquad (4\text{-}1\text{-}1)$$

Although it is easy to verify (4-1-1) in many special cases, it is not very easy to deduce (4-1-1) as a general principle. This has, however, been done by D. Gabor [Ref. 17]. He chose to define Δt and $\Delta \omega$ by second moments.

A similar and perhaps more basic concept than the product of time and frequency spreads is the relationship between spectral bandwidth and *rise time* of a system response function. The *rise time* Δt of a system response is also defined somewhat arbitrarily, often as the time span between the time of excitation and the time at which the system response is half its ultimate value. In principle, a broad frequency response can result from a rapid decay time as well as from a rapid rise time. *Tightness* in the inequality (4-1-1) may be associated with situations in which a certain rise time is quicky followed by an equal decay time. *Slackness* in the inequality (4-1-1) may be associated with increasing inequality between rise time and decay time. Slackness could also result from other combinations of rises and falls such as random combinations. Many systems respond very rapidly compared to the rate at which they subsequently decay. Focusing our attention on such systems, we can now seek to derive the inequality (4-1-1) applied to rise time and bandwidth. The first step is to choose a definition for rise time. The choice is determined not only for clarity and usefulness but also by the need to ensure tractability of the subsequent analysis. I have found a reasonable defintiion of rise time to be

$$\Delta t = \frac{\int_0^\infty b(t)^2 \, dt}{\int_0^\infty \frac{1}{t} b(t)^2 \, dt} \qquad (4\text{-}1\text{-}2)$$

where $b(t)$ is the response function under consideration. The numerator is just a normalizing factor. The denominator says we have defined Δt by the first negative moment. For example, if $b(t)$ is a step function, then the denominator integral diverges, giving the desired $\Delta t = 0$ rise time. If $b(t)^2$ grows linearly from zero to t_0 and then vanishes, the rise time Δt is $t_0/2$, again a reasonable definition.

Although the Z transform method is a great aid in studying situations where divergence (as $1/t$) plays a key role, it does have the disadvantage that it destroys the formal identity between the time domain and the frequency domain. Presumably this disadvantage is not fundamental since we can always go to a limiting process in which the discretized time domain tends to a continuum. In order to utilize the analytic simplicity of the Z transform we now consider the dual to the rise-time problem. Instead of a time function whose square vanishes identically at negative time we now consider a spectrum $\bar{B}(1/Z)B(Z)$ which vanishes at negative frequencies. We measure how fast this spectrum can rise after $\omega = 0$. We will find this to be related to the time duration Δt of the complex time function b_t. More precisely, we will now define the lowest significant frequency component $\omega = \Delta\omega$ in the spectrum analogously to (4-1-2) to be

$$\Delta\omega = \frac{\int_{-\infty}^\infty \bar{B}B \, d\omega}{\int_{-\infty}^\infty \frac{1}{\omega} \bar{B}B \, d\omega} \qquad (4\text{-}1\text{-}3)$$

Without loss of generality we can assume that the spectrum has been normalized so that the numerator integral is unity. In other words, the zero lag of the autocorrelation of b_t is $+1$. Then

$$\frac{1}{\Delta\omega} = \int_{-\infty}^\infty \frac{1}{\omega} \bar{B}B \, d\omega \qquad (4\text{-}1\text{-}4)$$

Now we recall the bilinear transform which gives us various Z transform expressions for $(-i\omega)^{-1}$. The one we ordinarily use is the integral $(\ldots 0, 0, 0.5, 1., 1., \ldots)$. We could also use $-(\ldots 1, 1, 1, 0.5, 0, 0, 0 \ldots)$. The pole right on the unit circle at $Z = 1$ causes some nonuniqueness. Because $1/i\omega$ is an imaginary odd frequency function we will take the desired expansion to be the odd function of time given by

$$(-i\omega)^{-1} = \frac{(\cdots - Z^{-2} - Z^{-1} + 0 + Z + Z^2 + \cdots)}{2} \qquad (4\text{-}1\text{-}5)$$

Converting (4-1-4) to an integral on the unit circle in Z transform notation we have

$$\frac{1}{\Delta\omega} = \int_{-\infty}^{+\infty} - i\frac{1}{-i\omega}\,\bar{B}B\,d\omega$$

$$= \frac{-i}{2\pi}\int_{-\pi}^{+\pi}\frac{1}{2}(\cdots - Z^{-2} - Z^{-1} + Z + Z^2 + \cdots)\bar{B}\left(\frac{1}{Z}\right)B(Z)\,d\omega \qquad (4\text{-}1\text{-}6)$$

But since this integral selects the coefficient of Z^0 of its argument we have

$$\frac{1}{\Delta\omega} = \frac{-i}{4\pi}\left[(r_{-1} - r_1) + (r_{-2} - r_2) + \cdots\right] \qquad (4\text{-}1\text{-}7)$$

where r_t is the autocorrelation function of b_t. This may be further expressed as

$$\frac{1}{\Delta\omega} = \sum_{t=1}^{\infty}\operatorname{Im}(-r_t) \leq \sum_{t=1}^{\infty}|r_t| \qquad (4\text{-}1\text{-}8)$$

The sum in (4-1-8) is like an integral representing area under the $|r_t|$ function. Imagine the $|r_t|$ function replaced by a rectangle function of equal area. This would define a Δt_{auto} for the $|r_t|$ function. Any autocorrelation function satisfies $|r_t| < r_0$ and we have normalized $r_0 = 1$. Thus, we extend the inequality (4-1-8) by

$$\frac{1}{\Delta\omega} \leq \sum_{t=1}^{\infty}|r_t| < \Delta t_{\text{auto}} \qquad (4\text{-}1\text{-}9)$$

Finally, we must relate the duration of a time function Δt to the duration of its autocorrelation Δt_{auto}. Generally speaking, it is easy to find a long time function which has short autocorrelation. Just take an arbitrary short-time function and convolve it by a long and tortuous all-pass filter. The new function is long, but its autocorrelation is short. If a time function has n nonzero points, then its autocorrelation has only $2n - 1$ nonzero points. It is obviously impossible to get a long autocorrelation function out of a short time function. It is not even fair to say that the autocorrelation is twice as long as the original time function because the autocorrelation must lie under some tapering function. To construct a time function with as long an autocorrelation as possible, the best thing to do is to concentrate the energy in two lumps, one at each end of the time function. Even from this extreme example, we see that it is not unreasonable to assert that

$$\Delta t \geq \Delta t_{\text{auto}} \qquad (4\text{-}1\text{-}10)$$

inserting into (4-1-9) we have the uncertainty relation

$$\Delta t\,\Delta\omega \geq 1 \qquad (4\text{-}1\text{-}11)$$

The more usual form of the uncertainty principle uses the frequency variable $f = 2\omega$ and a different definition of Δt, namely *time duration* rather than *rise time*. It is

$$\Delta t\,\Delta f \geq 1 \qquad (\Delta t \text{ is duration}) \qquad (4\text{-}1\text{-}12)$$

The choice of a 2π scaling factor to convert *rise time* to *duration* is indicative of the approximate nature of the inequalities.

EXERCISES

1 Consider $B(Z) = [1 - (Z/Z_0)^n]/(1 - Z/Z_0)$ in the limit Z_0 goes to the unit circle. Sketch the time function and its squared amplitude. Sketch the frequency function and its squared amplitude. Choose Δf and Δt.

2 A time series made up of two frequencies may be written as

$$b_t = A \cos \omega_1 t + B \sin \omega_1 t + C \cos \omega_2 t + D \sin \omega_2 t$$

Given $\omega_1, \omega_2, b_0, b_1, b_2, b_3$ show how to calculate the amplitude and phase angles of the two sinusoidal components.

3 Consider the frequency function graphed below.

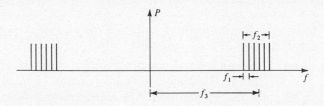

FIGURE E 4-1-3

Describe the time function in rough terms indicating the times corresponding to $1/f_1$, $1/f_2$, and $1/f_3$. Try to avoid algebraic calculation. Sketch an approximate result.

PROBLEM FOR RESEARCH

Can you find a method of defining $\Delta \omega$ and Δt of one-sided wavelets in such a way that for minimum-phase wavelets only the uncertainty principle takes on the equality sign?

4-2 TIME-STATISTICAL RESOLUTION

If you flipped a coin 100 times, it is possible that you would get exactly 50 "heads" and 50 "tails." More likely it would be something between 60–40 and 40–60. Typically, how much deviation from 50 would you expect to see? The average (mean) value should be 50, but some other value is almost always obtained from a random sample. The other value is called the *sample mean*. We would like to know how much difference to expect between the sample mean and the true mean. The average squared difference is called the *variance of the sample mean*. For a very large sample, the sample mean should be proportionately much closer to the true mean than for a smaller sample. This idea will lead to an uncertainty relation between the probable error in the estimated mean and the size of the sample. Let us be more precise.

The "true value" of the mean could be defined by flipping the coin n times and conceiving of n going to infinity. A more convenient definition of "true value" is that the experiment could be conceived of as having been done separately under identical conditions by an infinite number of people (an ensemble). Such an artifice will enable us to define a time-variable mean for coins which change with time.

The utility of the concept of an ensemble is often subjected to serious attack both from the point of view of the theoretical foundations of statistics and from the point of view of experimentalists applying the techniques of statistics. Nonetheless a great body of geophysical literature uses the artifice of assuming the existence of an unobservable ensemble. The advocates of using ensembles (the Gibbsians) have the advantage over their adversaries (the Bayesians) in that their mathematics is more tractable (and more explainable). So, let us begin!

A conceptual average over the ensemble, called an expectation, is denoted by the symbol E. The index for summation over the ensemble is never shown explicitly; every random variable is presumed to have one. Thus, the true mean at time t may be defined as

$$m_t = E(x_t) \qquad (4\text{-}2\text{-}1)$$

If the mean does not vary with time, we may write

$$m = E(x_t) \qquad \text{(all } t) \qquad (4\text{-}2\text{-}2)$$

Likewise, we may be interested in a property of x_t called its *variance* which is a measure of variability about the mean defined by

$$\sigma_t{}^2 = E[(x_t - m_t)^2] \qquad (4\text{-}2\text{-}3)$$

The x_t random numbers could be defined in such a way that σ or m or both is either time-variable or constant. If both are constant, we have

$$\sigma^2 = E[(x_t - m)^2] \qquad (4\text{-}2\text{-}4)$$

When manipulating algebraic expressions the symbol E behaves like a summation sign, namely

$$E = (\lim n \to \infty) \frac{1}{n} \sum_1^n \qquad (4\text{-}2\text{-}5)$$

Notice that the summation index is not given, since the sum is over the ensemble, not time.

Now let x_t be a time series made up from (identically distributed, independently chosen) random numbers in such a way that m and σ do not depend on time. Suppose we have a sample of n points of x_t and are trying to determine the value of m. We could make an estimate $\hat{m}$ of the mean m with the formula

$$\hat{m} = \frac{1}{n} \sum_{t=1}^n x_t \qquad (4\text{-}2\text{-}6)$$

A somewhat more elaborate method of estimating the mean would be to take a weighted average. Let w_t define a set of weights normalized so that

$$\sum w_t = 1 \qquad (4\text{-}2\text{-}7)$$

With these weights the more elaborate estimate $\hat{m}$ of the mean is

$$\hat{m} = \sum w_t x_t \qquad (4\text{-}2\text{-}8)$$

Actually (4-2-6) is just a special case of (4-2-8) where the weights are $w_t = 1/n$; $t = 1, 2, \ldots, n$.

Our objective in this section is to determine how far the estimated mean $\hat{m}$ is likely to be from the true mean m for a sample of length n. One possible definition of this excursion Δm is

$$(\Delta m)^2 = E[(\hat{m} - m)^2] \qquad (4\text{-}2\text{-}9)$$

$$= E\{[(\sum w_t x_t) - m]^2\} \qquad (4\text{-}2\text{-}10)$$

Now utilize the fact that $m = m \sum w_t = \sum w_t m$

$$(\Delta m)^2 = E\left\{\left[\sum_t w_t(x_t - m)\right]^2\right\} \qquad (4\text{-}2\text{-}11)$$

$$= E\left\{\left[\sum_t w_t(x_t - m)\right]\left[\sum_s w_s(x_s - m)\right]\right\} \qquad (4\text{-}2\text{-}12)$$

$$= E\left[\sum_t \sum_s w_t w_s(x_t - m)(x_s - m)\right] \qquad (4\text{-}2\text{-}13)$$

Now the expectation symbol E may be regarded as a summation sign and brought inside the sums on t and s.

$$(\Delta m)^2 = \sum_t \sum_s w_t w_s E[(x_t - m)(x_s - m)] \qquad (4\text{-}2\text{-}14)$$

By the randomness of x_t and x_s the expectation on the right, that is, the sum over the ensemble, gives zero unless $s = t$. If $s = t$, then the expectation is the variance defined by (4-2-4). Thus we have

$$(\Delta m)^2 = \sum_t \sum_s w_t w_s \sigma^2 \delta_{ts} \qquad (4\text{-}2\text{-}15)$$

$$= \sum_t w_t^2 \sigma^2 \qquad (4\text{-}2\text{-}16)$$

or

$$\Delta m = \sigma\left(\sum_t w_t^2\right)^{1/2} \qquad (4\text{-}2\text{-}17)$$

Now let us examine this final result for n weights each of size $1/n$. For this case, we get

$$\Delta m = \sigma\left[\sum_1^n \left(\frac{1}{n}\right)^2\right]^{1/2} = \frac{\sigma}{(n)^{1/2}} \qquad (4\text{-}2\text{-}18)$$

This is the most important property of random numbers which is not intuitively obvious. For a zero mean situation it may be expressed in words: "n random numbers of unit magnitude add up to a magnitude of about the square root of n."

When one is trying to estimate the mean of a random series which has a time-variable mean, one faces a basic dilemma. If one includes a lot of numbers in the sum to get Δm small, then m may be changing while one is trying to measure it. In contrast, $\hat{m}$ measured from a short sample of the series might deviate greatly from the true m (defined by an infinite sum over the ensemble at any point in time). This is the basic dilemma faced by a stockbroker when a client tells him, "Since the market fluctuates a lot I'd like you to sell my stock sometime when the price is above the mean selling price."

If we imagine that a time series is sampled every τ seconds and we let $\Delta t = n\tau$ denote the length of the sample then (4-2-18) may be written as

$$(\Delta m)^2 \, \Delta t = \sigma^2 \tau \qquad (4\text{-}2\text{-}19)$$

It is clearly desirable to have both Δm and Δt as small as possible. If the original random numbers x_t were correlated with one another, for example, if x_t were an approximation to a continuous function, then the sum of the n numbers would not cancel to root n. This is expressed by the inequality

$$(\Delta m)^2 \, \Delta t \geq \sigma^2 \tau \qquad (4\text{-}2\text{-}20)$$

The inequality (4-2-20) may be called an uncertainty relation between accuracy and time resolution.

In considering other sets of weights one may take a definition of Δt which is more physically sensible than τ times the number of weights. For example, if the weights w_t are given by a sampled gaussian function as shown in Fig. 4-1, then Δt could be taken as the separation of half-amplitude points, $1/e$ points, the time span which includes 95 percent of the area, or it could be given many other "sensible" interpretations. Given a little slop in the definition of Δm and Δt, it is clear that the inequality of (4-2-20) is not to be strictly applied.

Given a sample of a zero mean random time series x_t, we may define another series y_t by $y_t = x_t^2$. The problem of estimating the variance $\sigma^2 = p$ of x_t is identical to the problem of estimating the mean m of y_t. If the sample is short, we may expect an error Δp in our estimate of the variance. Thus, in a scientific paper one would like to write for the mean

$$m = \hat{m} \pm \Delta m \qquad (4\text{-}2\text{-}21)$$

$$= \hat{m} \pm \sigma/\sqrt{n} \qquad (4\text{-}2\text{-}22)$$

but since the variance σ^2 often is not known either, it is necessary to use the estimated $\hat{\sigma}$, that is

$$m = \hat{m} \pm \hat{\sigma}/\sqrt{n} \qquad (4\text{-}2\text{-}23)$$

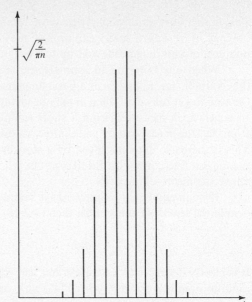

FIGURE 4-1
Binomial coefficients tend to the gaussian function. Plotted are the coefficients of Z^t in $(.5 + .5Z)^{20}$.

Of course (4-2-23) really is not right because we really should add something to indicate additional uncertainty due to error in $\hat{\sigma}$. This estimated error would again have an error, ad infinitum. To really express the result properly, it is necessary to have a probability density function to calculate all the $E(x^n)$ which are required. The probability function can be either estimated from the data or chosen theoretically. In practice, for a reason given in a later section, the gaussian function often occurs. In the exercises it is shown that

$$\Delta p = p\sqrt{\frac{2}{n}} \quad (4\text{-}2\text{-}24)$$

Since $\Delta t = n\tau$, by squaring we have

$$\left(\frac{\Delta p}{p}\right)^2 \frac{\Delta t}{2} \geq 1 \quad (4\text{-}2\text{-}25)$$

The inequality applies if the random numbers x_t are not totally unpredictable random numbers. If x_t is an approximation to a continuous function, then it is highly predictable and there will be a lot of slack in the inequality.

Correlation is a concept similar to cosine. A cosine measures the angle between two vectors. It is given by the dot product of the two vectors divided by their magnitudes

$$c = \frac{(\mathbf{x} \cdot \mathbf{y})}{[(\mathbf{x} \cdot \mathbf{x})(\mathbf{y} \cdot \mathbf{y})]^{1/2}}$$

Correlation is the same sort of thing, except x and y are scalar random variables, so instead of having a vector subscript their subscript is the implicit ensemble subscript. Correlation is defined

$$c = \frac{E(xy)}{[(E(x^2)E(y^2)]^{1/2}}$$

In practice one never has an ensemble. There is a practical problem when the ensemble average is simulated by averaging over a sample. The problem arises with small samples and is most dramatically illustrated for a sample with only one element. Then the sample correlation is

$$\hat{c} = \frac{xy}{|x|\,|y|} = \pm 1$$

regardless of what value the random number x or the random number y should take. In fact, it turns out that the sample correlation $\hat{c}$ will always scatter away from zero.

No doubt this accounts for many false "discoveries." The topic of bias and variance of coherency estimates is a complicated one, but a rule of thumb seems to be to expect bias and variance of $\hat{c}$ on the order of $1/\sqrt{n}$ for samples of size n.

EXERCISES

1 Suppose the mean of a sample of random numbers is estimated by a triangle weighting function, i.e.,

$$\hat{m} = s \sum_{i=0}^{n} (n - i)x_i$$

Find the scale factor s so that $E(\hat{m}) = m$. Calculate Δm. Define a reasonable Δt. Examine the uncertainty relation.

2 A random series x_t with a possibly time-variable mean may have the mean estimated by the feedback equation

$$\hat{m}_t = (1 - \varepsilon)\hat{m}_{t-1} + bx_t$$

(a) Express $\hat{m}_t$ as a function of x_t, $x_{t-1}, \ldots$, and not $\hat{m}_{t-1}$.
(b) What is Δt, the effective averaging time?
(c) Find the scale factor b so that if $m_t = m$, then $E(\hat{m}_t) = m$.
(d) Compute the random error $\Delta m = [E(\hat{m} - m)^2]^{1/2}$ [answer goes to $\sigma(\varepsilon/2)^{1/2}$ as ε goes to zero].
(e) What is $(\Delta m)^2 \, \Delta t$ in this case?

3 Show that

$$(\Delta P)^2 = \frac{1}{n} [E(x^4) - \sigma^4]$$

4 Define the behavior of an independent zero-mean-time series x_t by defining the probabilities that various amplitudes will be attained. Calculate $E(x_i)$, $E(x_i^2)$, $(\Delta P)^2$.

If you have taken a course in probability theory, use a gaussian probability density function for x_t. HINT:

$$P(x) = \frac{1}{\sigma\sqrt{2\pi}} e^{-x^2/2\sigma^2}$$

and

$$\int_0^\infty x^{2n} e^{-ax^2} \, dx = \frac{1 \cdot 3 \cdot 5 \cdots (2n-1)}{2^{n+1}a^n} \sqrt{\frac{\pi}{a}}$$

4-3 FREQUENCY-STATISTICAL RESOLUTION

Observations of sea level for a long period of time can be summarized in terms of a few statistical averages such as the mean height m and the variance σ^2. Another important kind of statistical average for use on such geophysical time series is the *power spectrum*. Some mathematical models explain only statistical averages of data and not the data themselves. In order to recognize certain pitfalls and understand certain fundamental limitations on work with power spectra, we first consider an idealized example.

Let x_t be a time series made up of independently chosen random numbers. Suppose we have n of these numbers. We can then define the data sample polynomial $X(Z)$

$$X(Z) = x_0 + x_1 Z + x_2 Z^2 + \cdots + x_{n-1} Z^{n-1} \qquad (4\text{-}3\text{-}1)$$

We can now make up a power spectral estimate $\hat{R}(Z)$ from this sample of random numbers by

$$\hat{R}(Z) = \frac{1}{n} \overline{X}\left(\frac{1}{Z}\right) X(Z) \qquad (4\text{-}3\text{-}2)$$

The difference between this and our earlier definition of spectrum is that a *power* spectrum has the divisor n to keep the expected result from increasing linearly with the somewhat arbitrary sample size n.

The definition of power spectrum is the expected value of $\hat{R}$, namely

$$R(Z) = E[\hat{R}(Z)] \qquad (4\text{-}3\text{-}3)$$

It might seem that a practical definition would be to let n tend to infinity in (4-3-2) Such a definition would lead us into a pitfall which is the main topic of the present section. Specifically, from Fig. 4-2 we conclude that $\hat{R}(Z)$ is a much fuzzier function than $R(Z)$, so that

$$R(Z) \neq \lim_{n \to \infty} \hat{R}(Z) \qquad (4\text{-}3\text{-}4)$$

To understand why this is so, we identify coefficients of like powers of Z in (4-3-2). We have

$$\hat{r}_k = \frac{1}{n} \sum_{t=0}^{n-k-1} \overline{x}_t x_{t+k} \qquad k = -n+1 \text{ to } n-1 \qquad (4\text{-}3\text{-}5)$$

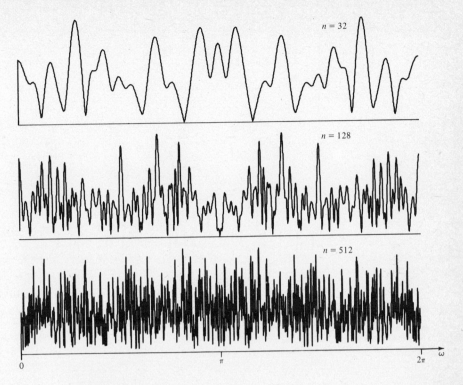

FIGURE 4-2
Amplitude spectra $[\hat{R}(Z)]^{1/2}$ of samples of n random numbers. These functions seem to oscillate over about the same range for $n = 512$ as they do for $n = 32$. As n tends to infinity we expect infinitely rapid oscillation.

enabling us to write (4-3-2) for real time series $x_t = \bar{x}_t$ as

$$\hat{R} = \hat{r}_0 + 2 \sum_{k=1}^{n-1} \hat{r}_k \cos k\omega \qquad (4\text{-}3\text{-}6)$$

Let us examine (4-3-6) for large n. To do this, we will need to know some of the statistical properties of the random numbers. Let them have zero mean $m = E(x_t) = 0$ and let them have known constant variance $\sigma^2 = E(x_t^2)$ and recall our assumption of independence which means that $E(x_t x_{t+s}) = 0$ if $t \neq s$. Because of random fluctuations, we have learned to expect that $\hat{r}_0$ will come out to be σ^2 plus a random fluctuation component which decreases with sample size as $1/\sqrt{n}$, namely

$$\hat{r}_0 = \sigma^2 \pm \frac{\sigma^2}{\sqrt{n}} \qquad (4\text{-}3\text{-}7a)$$

Likewise, $\hat{r}_1$ should come out to be zero but the definition (4-3-5) leads us to expect a fluctuation component

$$\hat{r}_1 = \pm \frac{n-1}{n} \frac{\sigma^2}{\sqrt{n}} \qquad (4\text{-}3\text{-}7b)$$

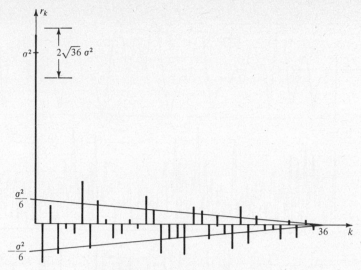

FIGURE 4-3
Positive lags of autocorrelation of 36 random numbers.

For the kth correlation value $k > 1$ we expect a fluctuation of order

$$\hat{r}_k = \pm \frac{n-k}{n} \frac{\sigma^2}{\sqrt{n}} \qquad (4\text{-}3\text{-}7c)$$

Equation (4-3-7) for a particular set of random numbers is displayed in Fig. 4-3. Now one might imagine that as n goes to infinity the fluctuation terms vanish and (4-3-2) takes the limiting form $\hat{R} = \sigma^2$. Such a conclusion is false. The reason is that although the individual fluctuation terms go as $1/\sqrt{n}$ the summation in (4-3-6) contains n such terms. Luckily, these terms are randomly canceling one another so the sum does not diverge as $\sqrt{n}$. We recall that the sum of n random signed numbers of unit magnitude is expected to add up to a random number in the range $\pm\sqrt{n}$. Thus the sum (4-3-6) adds up to

$$\hat{R} \approx \left(1 \pm \frac{\sqrt{n}}{\sqrt{n}}\right)\sigma^2 = (1 \pm 1)\sigma^2 \qquad (4\text{-}3\text{-}8)$$

This is the basic result that a power spectrum estimated from the energy density of a sample of random numbers has a fluctuation from frequency to frequency and from sample to sample which is as large as the expected spectrum.

It should be clear that letting n go to infinity does not take us to the theoretical result $\hat{R} = \sigma^2$. The problem is that, as we increase n, we increase the frequency resolution but not the statistical resolution. To increase the statistical resolution we need to simulate ensemble averaging. There are two ways to do this: (1) Take the sample of n points and break it into k equal-length segments of n/k points each.

Compute an $R(\omega)$ for each segment and then add all k of the $R(\omega)$ together, or (2) form $R(\omega)$ from the n-point sample. Of the $n/2$ independent amplitudes, replace each one by an average over its k nearest neighbors. Whichever method, (1) or (2), is used it will be found that $\Delta f = 0.5k/n\tau$ and $(\Delta p/p)^2 =$ inverse of number of degrees of freedom averaged over $= 1/k$. Thus, we have

$$\Delta f \left(\frac{\Delta p}{p}\right)^2 = \frac{.5}{n\tau}$$

If some of the data are not used, or are not used effectively, we get the usual inequality

$$\Delta f \left(\frac{\Delta p}{p}\right)^2 \geq \frac{0.5}{n\tau}$$

Thus we see that, if there are enough data available (n large enough), we can get as good resolution as we like. Otherwise, improved statistical resolution is at the cost of frequency resolution and vice versa.

We are right on the verge of recognizing a resolution tradeoff, not only between Δf and Δp but also with $\Delta t = n\tau$, the time duration of the data sample. Recognizing now that the time duration of our data sample is given by $\Delta t = n\tau$, we obtain the inequality

$$\Delta f \Delta t \left(\frac{\Delta p}{p}\right)^2 > \frac{1}{2} \qquad (4\text{-}3\text{-}9)$$

This inequality will be further interpreted and rederived from a somewhat different point of view in the next section.

In time-series analysis we have the concept of coherency which is analogous to the concept of correlation defined in Sec. 4-2. There we had for two random variables x and y that

$$c = \frac{E(xy)}{[E(x^2)E(y^2)]^{1/2}}$$

Now if x_t and y_t are time series, they may have a relationship between them which depends on time-delay, scaling, or even filtering. For example, perhaps $Y(Z) = F(Z)X(Z) + N(Z)$ where $F(Z)$ is a filter and n_t is unrelated noise. The generalization of the correlation concept is to define *coherency* by

$$C = \frac{E\left[X\left(\frac{1}{Z}\right)Y(Z)\right]}{[E(\bar{X}X)E(\bar{Y}Y)]^{1/2}}$$

Correlation is a real scalar. Coherency is complex and expresses the frequency dependence of correlation. In forming an estimate of coherency it is always essential to simulate some ensemble averaging. Note that if the ensemble averaging were to be omitted, the coherency (squared) calculation would give

$$|C|^2 = \bar{C}C = \frac{(\overline{X}Y)(\overline{X}Y)}{(\overline{X}X)(\overline{Y}Y)} = +1$$

FIGURE 4-4
Model of random time series generation.

which states that the coherency squared is $+1$ independent of the data. Because correlation scatters away from zero we find that coherency squared is biased away from zero.

4-4 TIME-FREQUENCY-STATISTICAL RESOLUTION

Many time functions are not completely random from point to point but become more random when viewed over a longer time scale. A popular mathematical model embodying this concept is to make a so-called *stationary time series* by putting random numbers into a filter as depicted in Fig. 4-4. The input x_t may be *independent random numbers* or *white light*. [The two terms mean nearly the same thing in practice but the first term is the stronger; it means that x_t is in no way related to x_s if $t \neq s$, whereas white light means that $E(x_t x_s) = 0$ if $t \neq s$.] The output random time series y_t may vary rather slowly from point to point if f_t is a low-pass filter. This is the usual case when we are modeling continuous time functions. The random time series may be called a *stationary* random time series if neither the filter nor any property of the random numbers (such as m or σ) vary with time. Stationarity is often assumed even where it cannot be strictly true.

This model will be useful later when we consider the problem of predicting a future point on y_t from knowledge of past values. Now we will use the model to examine the estimation of the spectrum of y_t given a sample of n points of y_t. To begin with, we have a very precise meaning for the spectrum of y_t. We have

$$Y(Z) = F(Z)X(Z) \qquad (4\text{-}4\text{-}1)$$

and its conjugate

$$\bar{Y}\left(\frac{1}{Z}\right) = \bar{F}\left(\frac{1}{Z}\right)\bar{X}\left(\frac{1}{Z}\right) \qquad (4\text{-}4\text{-}2)$$

Multiplying (4-4-1) by (4-4-2) we get

$$\bar{Y}\left(\frac{1}{Z}\right)Y(Z) = \bar{X}\left(\frac{1}{Z}\right)X(Z)\bar{F}\left(\frac{1}{Z}\right)F(Z) \qquad (4\text{-}4\text{-}3)$$

but, from the previous section, we learned that $E(\bar{X}X) = \sigma^2$. Considering σ^2 to be unity, we see that the expected power spectrum of the output $\bar{Y}$ is the energy spectrum of the filter F. The overall situation is depicted in Fig. 4-5. The interest-

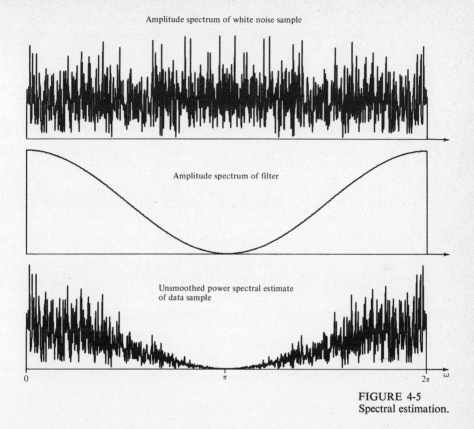

Amplitude spectrum of white noise sample

Amplitude spectrum of filter

Unsmoothed power spectral estimate
of data sample

0 $\qquad$ π $\qquad$ 2π $\quad \omega$

FIGURE 4-5
Spectral estimation.

ing question is how well can we estimate the spectrum when we start with an
n-point sample of y_t. We will describe three computationally different methods,
all having the same fundamental limitations.

The first method uses a bank of filters as shown in Fig. 4-6. When random
numbers excite the narrowband filter, the output is somewhat like a sine wave. It
differs in one important respect. A sine wave has constant amplitude, but the output of a narrowband filter has an amplitude which swings over a range. This is
illustrated in Fig. 4-7. If the bandwidth is narrow, the amplitude changes slowly.
If the impulse response of the filter has duration Δt_{filter}, then the output amplitude
at time t will be randomly related to the amplitude at time $t + \Delta t_{\text{filter}}$. Thus,

FIGURE 4-6
Spectral estimate of a random series.

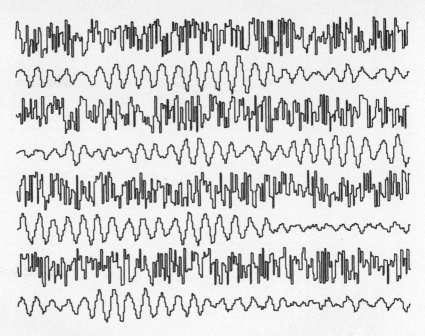

FIGURE 4-7
1,024 random numbers before and after narrowband filtering. The filter was $(1-Z)/[(1-Z/Z_0)(1-Z/\bar{Z}_0)]$ where $Z_0 = 1.02\, e^{i\pi/5}$.

in statistical averaging, it is not the number of time points but the number of intervals Δt_{filter} which enhance the reliability of the average. Consequently, the decay time of the integrator $\Delta t_{\text{integrator}}$ will generally be chosen to be greater than $\Delta t_{\text{filter}} = 1/\Delta f$. The variability Δp of the output p decreases as $\Delta t_{\text{integrator}}$ increases. Since v_t has independent values over time spans of about $\Delta t_{\text{filter}} = 1/\Delta f$, then the "degrees of freedom" smoothed over can be written $\Delta t_{\text{integrator}}/\Delta t_{\text{filter}} = \Delta f\,\Delta t_{\text{integrator}}$. The variability $\Delta p/p$ is proportional to the inverse square root of the number of degrees of freedom, and so we get

$$\left(\frac{\Delta p}{p}\right)^2 = \frac{1}{\Delta f\,\Delta t_{\text{integrator}}}$$

or, introducing the usual inequality,

$$\Delta t\,\Delta f\left(\frac{\Delta p}{p}\right)^2 > 1 \qquad (4\text{-}4\text{-}4)$$

The inequality (4-4-4) indicates the three-parameter uncertainty which is fundamental to estimating power spectra of random functions. Two other methods of estimating the spectrum of y_t from a sample of length n are exactly the same as the methods described in Sec. 4-3 as ways of estimating the spectrum of white light. In fact, (4-4-4) turns out to be the same as (4-3-9).

The usual interpretation is that to attain a frequency resolution of Δf and a relative accuracy of $\Delta p/p$ a time sample of duration at least $\Delta t \geq 1/[\Delta f(\Delta p/p)^2]$ will be required. Although this sort of interpretation is generally correct, it will break down for highly resonant series recorded for a short time. Then the data sample may be predictable an appreciable distance off its ends so that the effective Δt is somewhat (perhaps appreciably) larger than the sample length.

EXERCISES

1 It is popular to taper the ends of a data sample so that the data go smoothly to zero at the ends of the sample. Choose a weighting function and discuss in a semiquantitative fashion its effect on Δt, Δf, and $(\Delta p/p)^2$.

2 Answer the question of Exercise 1, where the autocorrelation function is tapered rather than the data sample.

4-5 THE CENTRAL-LIMIT THEOREM

The central-limit theorem of probability and statistics is perhaps the most important theorem in the fields of probability and statistics. A derivation of the central limit theorem explains why the gaussian probability function is so frequently encountered in nature; not just in physics but also in the biological and social sciences. No experimental scientist should be unaware of the basic ideas behind this theorem. Although the result is very deep and is even today the topic of active research, we can get to the basic idea quite easily.

One way to obtain random integers from a known probability function is to write integers on slips of paper and place them in a hat. Draw one slip at a time. After each drawing replace the slip in the hat. The probability of drawing the integer i is given by the ratio a_i of the number of slips containing the integer i divided by the total number of slips. Obviously the sum over i of a_i must be unity. Another way to get random integers is to throw one of a pair of dice. Then all a_i equal zero except $a_1 = a_2 = a_3 = a_4 = a_5 = a_6 = \frac{1}{6}$. The probability that the integer i will occur on the first drawing and the integer j will occur on the second drawing is $a_i a_j$. If you draw two slips or throw a pair of dice, then the probability that the sum of i and j equals k is readily seen to be

$$c_k = \sum_i a_i a_{k-i} \qquad (4\text{-}5\text{-}1)$$

Since this equation is a convolution, we may look into the meaning of the Z transform

$$A(Z) = \cdots a_{-1}Z^{-1} + a_0 + a_1 Z + a_2 Z^2 + \cdots \qquad (4\text{-}5\text{-}2)$$

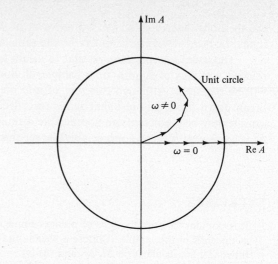

FIGURE 4-8
The complex numbers $a_k e^{i\omega k}$ added together.

In terms of Z transforms the probability that i plus j equals k is simply the coefficient of Z^k in

$$C(Z) = A(Z)A(Z) \qquad (4\text{-}5\text{-}3)$$

Obviously, if we add n of the random numbers, the probability that the sum of them equals k is given by the coefficient of Z^k in

$$G(Z) = [A(Z)]^n \qquad (4\text{-}5\text{-}4)$$

The central-limit theorem of probability says that as n goes to infinity the polynomial $G(Z)$ goes to a special form, almost regardless of the specific polynomial $A(Z)$. The specific form is such that a graph of the coefficients of $G(Z)$ comes closer and closer to fitting under the envelope of the bell-shaped gaussian function. Let us see why this happens. Our development will lack a mathematical rigor because the theorem is not always true. There are pathological A functions which do not result in G tending to gaussian. Despite the fact that some of the pathological functions sometimes turn up in applications, we will not take the time here to look at such instances.

   Consider the size of $A(Z)$ for real ω. If $\omega = 0$, the sum of the terms of $A(Z)$ may be visualized in the complex plane as a sum of vectors $a_k e^{i\omega k}$ all pointing in the positive real direction. If $\omega \neq 0$ the vectors point in different directions. This is shown in Fig. 4-8.

   In raising $A(e^{i\omega})$ to the nth power, the values of ω of greatest concern are those near $\omega = 0$ where A is largest—because in any region where A is small A^n will be extremely small. Near $\omega = 0$ or $Z = 1$ we may expand $A(Z)$ in a power series in ω

$$A(e^{i\omega}) = A\bigg|_0 + \frac{\partial}{\partial \omega} A\bigg|_0 \omega + \frac{\partial^2}{\partial \omega^2} A\bigg|_0 \frac{\omega^2}{2!} + \cdots \qquad (4\text{-}5\text{-}5)$$

Note that the coefficients of this power series are proportional to the moments m_i of the probability function; that is

$$A(e^{i\omega}) = \sum_k a_k e^{ik\omega} \qquad (4\text{-}5\text{-}6)$$

$$A(1) = \sum_k a_k = m_0 = 1 \qquad (4\text{-}5\text{-}7)$$

$$\frac{\partial}{\partial\omega} A = \sum_k ik a_k e^{ik\omega} \qquad (4\text{-}5\text{-}8)$$

$$\left.\frac{\partial}{\partial\omega} A\right|_0 = \sum_k ik a_k = im_1 \qquad (4\text{-}5\text{-}9)$$

$$\left.\frac{\partial^2}{\partial\omega^2} A\right|_0 = -\sum k^2 a_k = -m_2 \qquad (4\text{-}5\text{-}10)$$

When we raise A(Z) to the nth power we will make the conjecture that only the given first three terms of the power series expansion will be important. (This assumption clearly fails if any of the moments of the probability function are infinite.) Thus, we are saying that as far as G is concerned the only important things about A are its mean value $m = m_1$ and its second moment m_2. If this is really so, we may calculate G by replacing A with any function B having the same mean and same second moment as A. We may use the simplest function we can find. A good choice is the so called binomial probability function given by

$$B = \frac{Z^m(Z^\sigma + Z^{-\sigma})}{2} \qquad (4\text{-}5\text{-}11)$$

$$= \frac{e^{i(m+\sigma)\omega} + e^{i(m-\sigma)\omega}}{2} \qquad (4\text{-}5\text{-}12)$$

Let us verify its first moment

$$\frac{\partial B}{\partial\omega} = \frac{i[(m+\sigma)e^{i(m+\sigma)\omega} + (m-\sigma)e^{i(m-\sigma)\omega}]}{2} \qquad (4\text{-}5\text{-}13)$$

$$\left.\frac{\partial B}{\partial\omega}\right|_0 = im \qquad (4\text{-}5\text{-}14)$$

Now let us verify its second moment

$$\left.\frac{\partial^2 B}{\partial\omega^2}\right|_0 = -\frac{(m+\sigma)^2 + (m-\sigma)^2}{2} \qquad (4\text{-}5\text{-}15)$$

$$\left.\frac{\partial^2 B}{\partial\omega^2}\right|_0 = -(m^2 + \sigma^2) \qquad (4\text{-}5\text{-}16)$$

Hence, σ should be chosen so that

$$m_2 = m^2 + \sigma^2 \qquad (4\text{-}5\text{-}17)$$

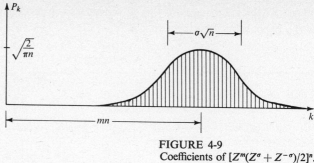

FIGURE 4-9
Coefficients of $[Z^m(Z^\sigma + Z^{-\sigma})/2]^n$.

Of course, we cannot expect that m and σ will necessarily turn out to be integers; therefore (4-5-11) will not necessarily be a Z transform in the usual sense. It does not really matter; we simply interpret (4-5-11) as saying:

1   The probability of drawing the number $m + \sigma$ is one-half.
2   The probability of $m - \sigma$ is one-half.
3   The probability of any other number is zero.

Now, raising $(Z^\sigma + Z^{-\sigma})$ to the nth power gives a series in powers of Z^σ whose coefficients are symmetrically distributed about Z to the zero power and whose magnitudes are given by the binomial coefficients. A sketch of the coefficients of $B(Z)^n$ is given in Fig. 4-9.

We will now see how, for large n, the binomial coefficients asymptotically approach a gaussian. Approaching this limit is a bit tricky. Obviously, the sum of n random integers will diverge as $\sqrt{n}$. Likewise the coefficients of powers of Z in $(\frac{1}{2} + Z/2)^n$ individually get smaller while the number of coefficients gets larger. We recall that in time series analysis we used the substitution $Z = e^{i\omega\,\Delta t}$. We commonly chose $\Delta t = 1$, which had the meaning that data points were given at integral points on the time axis. In the present probability theory application of Z transforms, the choice $\Delta t = 1$ arises from our original statement that the numbers chosen randomly from the slips of paper were integers. Now we wish to add n of these random numbers together; and so, it makes sense to rescale the integers to be integers divided by $\sqrt{n}$. Then we can make the substitution $Z = e^{i\omega\,\Delta t} = e^{i\omega/\sqrt{n}}$. The coefficient of Z^k now refers to the probability of drawing the number $k/\sqrt{n}$. Raising $(Z^\sigma + Z^{-\sigma})/2$ to the nth power to find the probability distribution for the sum of n independently chosen numbers, we get

$$[B(Z)]^n = \left(\frac{Z^\sigma + Z^{-\sigma}}{2}\right)^n$$

$$= \left(\cos\frac{\sigma\omega}{\sqrt{n}}\right)^n$$

Using the first term of the series expansion for cosine we have

$$[B(Z)]^n \approx \left(1 - \frac{\sigma^2 \omega^2}{2n}\right)^n$$

Using the well-known fact that $(1 + x/n)^n \to e^x$, we have for large n

$$[B(Z)]^n \approx e^{-\sigma^2 \omega^2/2} \qquad (4\text{-}5\text{-}18)$$

The probability that the number t will result from the sum is now found by inverse Fourier transformation of (4-5-18). The Fourier transform of the gaussian (4-5-18) may be looked up in a table of integrals. It is found to be the gaussian

$$p(t) = \frac{1}{\sigma \sqrt{2\pi}} e^{-t^2/2\sigma^2}$$

4-6 CONFIDENCE INTERVALS

It is always important to have some idea of the size and influence of random errors. It is often important to be able to communicate this idea to others in the form of a statement such as

$$m = \hat{m} \pm \sigma/\sqrt{n}$$

In a matter of any controversy you may be called upon to define a probability that the true mean lies in your stated interval; in other words, what is your *confidence* that m lies in the *interval*

$$\hat{m} - \Delta m < m < \hat{m} + \Delta m$$

Before you can answer questions about probability, it is necessary to make some assumptions and assertions about the probability functions which control your random errors. The assertion that errors are independent of one another is your most immediate hazard. If they are not, as is often the case, you may be able to readjust the numerical value of n to be an estimate of the number of independent errors. We did something like this in time series analysis when we took n to be not the number of points on the time series but the number of intervals of length Δt_{filter}. The second big hazard in trying to state a confidence interval is the common assumption that, because of the central-limit theorem and for lack of better information, the errors follow a gaussian probability function. If in fact the data errors include blunders which arise from human errors or blunders from transient electronic equipment difficulties, then the gaussian assumption can be very wrong and can lead you into serious errors in geophysical interpretation. Some useful help is found in the field of nonparametric statistics (see, for example, Ref. [18]).

To begin with, it is helpful to rephrase the original question into one involving the median rather than the mean. The median m_1 is defined as that value which is expected to be less than half of the population and greater than the other half. In many—if not most—applications the median is a ready, practical substitute for the arithmetic mean. The median is insensitive to a data point, which, by some blunder, is near infinity. In fact, median and mean are equal when the probability function is symmetrical. For a sample of n numbers $(x_i, i = 1, 2, \ldots, n)$, the median m_1 may be estimated by reordering the numbers from smallest to largest and then selecting the number in the middle as the estimate of the median $\hat{m}_1$. Specifically, let the reordered x_i be denoted by x_i' where $x_i' \leq x_{i+1}'$. Then we have $\hat{m}_1 = x_{n/2}'$. Now it turns out that without knowledge of the probability density function for the random variables x_i we will still be able to compute the probability that the true median m_1 is contained in the interval

$$x_{n/2-\alpha\sqrt{n}}' < m_1 < x_{n/2+\alpha\sqrt{n}}' \qquad (4\text{-}6\text{-}1)$$

For example, set $\alpha = 1$ and $N = 100$, the assertion is that we can now calculate the probability that the true median m_1 lies between the 40th and the 60th percentile of our data. The trick is this: Define a new random variable

$$y = \text{step}\,(x - m_1) \qquad (4\text{-}6\text{-}2)$$

The step function equals $+1$ if $x > m_1$ and equals 0 if $x < m_1$. The new random variable y takes on only values of zero and one with equal probability; thus we know its probability function even though we may not know the probability function for the random variable x. Now define a third random variable s as

$$s = \sum_{i=1}^{n} y_i$$

Since each y_i is zero or one, then s must be an integer between zero and n. Furthermore, the probability that s takes the value j is given by the coefficient of Z^j of $(\frac{1}{2} + Z/2)^n$. Now the probability that s lies in the interval $n/2 - \alpha\sqrt{n} < s < n/2 + \alpha\sqrt{n}$ is readily determined by adding the required coefficients of Z^j, and this probability is by definition equal to the probability that the median m_1 lies in the interval (4-6-1). For $\alpha = 1$ and large n this probability works out to about 95 percent.

5

MATRICES AND MULTICHANNEL TIME SERIES

Familiarity with matrices is essential to computer modeling in both physical and social sciences. As this is a big subject covered by many excellent texts at all levels, our review will be a quick one. We focus on those properties required in the succeeding chapters. We avoid proofs, and although constructions given should be useful in most situations, there will be occasional matrices (which we will dismiss as pathological cases) in which our constructions will fail. In practice, the user should always check computed results. Unfortunately, the so-called pathological cases arise in practice far more often than might be expected. When matrix difficulties arise, the first tendency of the scientist is to use a higher-precision arithmetic. In the author's experience, physically meaningful calculations rarely require high precision. When higher precision seems to be needed, it is often because something is happening physically which shows that the problem being solved is a poorly posed problem. If a slight change in the problem should not make a drastic change in the answer, then it may happen that a different organization of the calculations will obviate the need for high precision. Anyway, our discussion here will focus on the nonpathological cases, but the reader is warned that pathological cases will certainly be encountered in practice and when they are they will be a stern test of the reader's mathematical knowledge and physical insight.

5-1 REVIEW OF MATRICES

A set of simultaneous equations may be written as

$$\mathbf{Ax} = \mathbf{b} \qquad (5\text{-}1\text{-}1a)$$

where A is a square matrix (nonsquare matrices are taken up in Chap. 6 on least squares) and $\mathbf{x}$ and $\mathbf{b}$ are column vectors. In a 2×2 case, $(5\text{-}1\text{-}1a)$ becomes

$$\begin{bmatrix} a_{11} & a_{12} \\ a_{21} & a_{22} \end{bmatrix} \begin{bmatrix} x_1 \\ x_2 \end{bmatrix} = \begin{bmatrix} b_1 \\ b_2 \end{bmatrix} = x_1 \begin{bmatrix} a_{11} \\ a_{21} \end{bmatrix} + x_2 \begin{bmatrix} a_{12} \\ a_{22} \end{bmatrix} \qquad (5\text{-}1\text{-}1b)$$

Equation $(5\text{-}1\text{-}1b)$ has a simple geometric interpretation. The two columns of the matrix and the column $\mathbf{b}$ are regarded as vectors in a plane. Equation $(5\text{-}1\text{-}1b)$ says x_1 times the first column vector plus x_2 times the second column vector equals the $\mathbf{b}$ column vector. Difficulty arises when the two column vectors of the matrix point in the same direction. Unless $\mathbf{b}$ just happens to be that direction, no solution x_1, x_2 is possible. The same thing may be said about the general case. A solution $\mathbf{x}$ to equation $(5\text{-}1\text{-}1a)$ exists if $\mathbf{b}$ lies in the space spanned by the columns of $\mathbf{A}$. In most practical situations the matrix $\mathbf{A}$ and the column $\mathbf{b}$ arise from independent considerations so that it is often reasonable to require the columns of $\mathbf{A}$ to span a space which will contain an arbitrary $\mathbf{b}$ vector. If $\mathbf{A}$ is an $n \times n$ matrix, then its columns are required to span an n-dimensional space. In particular, the n-dimensional parallelopiped with edges given by the columns of $\mathbf{A}$ should not have a zero volume. Such a volume is given by the determinant of $\mathbf{A}$.

Another set of simultaneous equations which arises frequently in practice is the so-called homogeneous equations

$$\mathbf{Ax} = \mathbf{0} \qquad (5\text{-}1\text{-}2)$$

This set always has the solution $\mathbf{x} = \mathbf{0}$ which is often called the trivial solution. For $(5\text{-}1\text{-}2)$ to have nontrivial solution values for $\mathbf{x}$ the determinant of A should vanish, meaning that the columns of A do not span an n-dimensional space. We will return later to the subject of actually solving sets of simultaneous equations.

A most useful feature of matrices is that their elements may be not only numbers but that they may be other matrices. Viewed differently, a big matrix may be partitioned into smaller submatrices. A surprising thing is that the product of two matrices is the same whether there are partitions or not. Study the identity

$$\left[\begin{array}{cc|c} a & b & c \\ d & e & f \end{array}\right] \left[\begin{array}{cc} g & h \\ i & j \\ \hline k & l \end{array}\right] = \begin{bmatrix} a & b \\ d & e \end{bmatrix} \begin{bmatrix} g & h \\ i & j \end{bmatrix} + \begin{bmatrix} c \\ f \end{bmatrix} [k \quad l] \qquad (5\text{-}1\text{-}3)$$

In terms of summation notation, the left-hand side of $(5\text{-}1\text{-}3)$ means

$$C_{ik} = \sum_{j=1}^{3} A_{ij} B_{jk} \qquad (5\text{-}1\text{-}4)$$

whereas the right-hand side means

$$C_{ik} = \sum_{j=1}^{2} A_{ij} B_{jk} + \sum_{j=3}^{3} A_{ij} B_{jk} \qquad (5\text{-}1\text{-}5)$$

Equations (5-1-4) and (5-1-5) are obviously the same; this shows that this partitioning of a matrix product is merely rearranging the terms. Partitioning does not really do anything at all from a mathematical point of view, but it is extremely important from the point of view of computation or discussion.

We now utilize matrix partitioning to develop the bordering method of matrix inversion. The bordering method is not the fastest or the most accurate method but it is quite simple, even for nonsymmetric complex-valued matrices, and it also gives the determinant and works for homogeneous equations. The bordering method proceeds by recursion. Given the inverse to a $k \times k$ matrix, the method shows how to find the inverse of a $(k+1) \times (k+1)$ matrix, which is the same old $k \times k$ matrix with an additional row and column attached to its borders. Specifically, $\mathbf{A}$, $\mathbf{e}$, $\mathbf{f}$, g, and $\mathbf{A}^{-1}$ are taken to be known in (5-1-6). The task is to find $\mathbf{W}$, $\mathbf{x}$, $\mathbf{y}$, and z.

$$\begin{bmatrix} \mathbf{A} & \mathbf{f} \\ \hline \mathbf{e} & g \end{bmatrix} \begin{bmatrix} \mathbf{W} & \mathbf{y} \\ \hline \mathbf{x} & z \end{bmatrix} = \begin{bmatrix} \mathbf{I} & 0 \\ \hline 0 & 1 \end{bmatrix} \qquad (5\text{-}1\text{-}6)$$

The first thing to do is multiply the partitions in (5-1-6) together. For the first column of the product we obtain

$$\mathbf{AW} + \mathbf{fx} = \mathbf{I} \qquad (5\text{-}1\text{-}7)$$

$$\mathbf{eW} + g\mathbf{x} = 0 \qquad (5\text{-}1\text{-}8)$$

A choice of $\mathbf{W}$ of

$$\mathbf{W} = \mathbf{A}^{-1}(\mathbf{I} - \mathbf{fx}) \qquad (5\text{-}1\text{-}9)$$

leads to (5-1-7) being satisfied identically. This leaves $\mathbf{x}$ still unknown, but we may find it by substituting (5-1-9) into (5-1-8)

$$\mathbf{x} = \frac{\mathbf{eA}^{-1}}{\mathbf{eA}^{-1}\mathbf{f} - g} \qquad (5\text{-}1\text{-}10)$$

Now, to get the column unknowns $\mathbf{y}$ and z, we compute the second column of the product (5-1-6)

$$\mathbf{Ay} + \mathbf{f}z = 0 \qquad (5\text{-}1\text{-}11)$$

$$\mathbf{e}\mathbf{y} + gz = 1 \qquad (5\text{-}1\text{-}12)$$

Multiply (5-1-11) by $\mathbf{A}^{-1}$

$$\mathbf{y} = -\mathbf{A}^{-1}\mathbf{f}z \qquad (5\text{-}1\text{-}13)$$

This gives the column vector $\mathbf{y}$ within a scale factor z. To get the scale factor, we insert (5-1-13) into (5-1-12)

$$-\mathbf{eA}^{-1}\mathbf{f}z + gz = 1$$

$$z = \frac{1}{g - \mathbf{eA}^{-1}\mathbf{f}} \qquad (5\text{-}1\text{-}14)$$

```
                              SUBROUTINE CMAINE(N,B,A)
                        C     A=MATRIX INVERSE OF B
                              COMPLEX B,A,C,R,DEL
                              DIMENSION A(N,N),B(N,N),R(100),C(100)
                              DO 10 I=1,N
                              DO 10 J=1,N
                         10   A(I,J)=0.
                              DO 40 L=1,N
                              DEL=B(L,L)
                              DO 30 I=1,L
                              C(I)=0.
                              R(I)=0.
                              DO 20 J=1,L
                              C(I)=C(I)+A(I,J)*B(J,L)
                         20   R(I)=R(I)+B(L,J)*A(J,I)
                         30   DEL=DEL-B(L,I)*C(I)
                              C(L)=-1.
                              R(L)=-1.
                              DO 40 I=1,L
                              C(I)=C(I)/DEL
                              DO 40 J=1,L
                         40   A(I,J)=A(I,J)+C(I)*R(J)
                              RETURN
                              END
```

FIGURE 5-1
A Fortran computer program for matrix inversion based on the bordering method.

It may, in fact, be shown that the determinant of the matrix being inverted is given by the product over all the bordering steps of the denominator of (5-1-14). Thus, if at any time during the recursion the denominator of (5-1-14) goes to zero, the matrix is singular and the calculation cannot proceed.

Let us summarize the recursion: One begins with the upper left-hand corner of a matrix. The corner is a scalar and its inverse is trivial. Then it is considered to be bordered by a row and a column as shown in (5-1-6). Next, we find the inverse of this 2×2 matrix. The process is continued as long as one likes. A typical step is first compute z by (5-1-14) and then compute A^{-1} of one larger size by

$$A^{-1} \leftarrow \left[\begin{array}{c} A^{-1} \\ \hline \text{zeros} \end{array}\right] + z \left[\begin{array}{c} A^{-1}f \\ \hline -1 \end{array}\right] [eA^{-1} \mid -1] \quad (5\text{-}1\text{-}15)$$

where (5-1-15) was made up from (5-1-9), (5-1-10), and (5-1-13). A Fortran computer program to achieve this is shown in Fig. 5-1.

It is instructive to see what becomes of A^{-1} if A is perturbed steadily in such a way that the determinant of A becomes singular. If the element g in the matrix of (5-1-6) is moved closer and closer to $eA^{-1}f$, then we see from (5-1-14) that z tends to infinity. What is interesting is that the second term in (5-1-15) comes to dominate the first, and the inverse tends to infinity times the product of a column c with a row r.

The usual expressions $AA^{-1} = I$ or $A^{-1}A = I$ in the limit of small z^{-1} tend to

$$Acr = z^{-1}I \quad (5\text{-}1\text{-}16)$$

or

$$crA = z^{-1}I \quad (5\text{-}1\text{-}17)$$

In the usual case (rank $\mathbf{A} = n - 1$, not rank $\mathbf{A} < n - 1$) where neither $\mathbf{c}$ nor $\mathbf{r}$ vanish identically, (5-1-16) and (5-1-17) in the limit $z^{-1} = 0$ become

$$\mathbf{Ac} = \mathbf{0} \quad \text{(5-1-18)}$$

$$\mathbf{rA} = \mathbf{0} \quad \text{(5-1-19)}$$

In summary, then, to solve an ordinary set of simultaneous equations like (5-1-1), one may compute the matrix inverse of $\mathbf{A}$ by the bordering method and then multiply (5-1-1) by $\mathbf{A}^{-1}$ obtaining

$$\mathbf{x} = \mathbf{A}^{-1}\mathbf{b} \quad \text{(5-1-20)}$$

In the event $\mathbf{b}$ vanishes, we are seeking the solution to homogeneous equations and we expect that z will explode in the last step of the bordering process. (If it happens earlier, one should be able to rearrange things.) The solution is then given by the column $\mathbf{c}$ in (5-1-18).

The row homogeneous equations of (5-1-19) was introduced because such a set arises naturally for the solution to the row eigenvectors of a nonsymmetric matrix. In the next section, we will go into some detailed properties of eigenvectors. A column eigenvector $\mathbf{c}$ of a matrix $\mathbf{A}$ is defined by the solution to

$$\mathbf{Ac} = \lambda\mathbf{c} \quad \text{(5-1-21)}$$

where λ is the so-called eigenvalue. At the same time, one also considers a row eigenvector equation

$$\mathbf{rA} = \lambda\mathbf{r} \quad \text{(5-1-22)}$$

To have a solution for (5-1-21) or (5-1-22), one must have $\det(\mathbf{A} - \lambda\mathbf{I}) = 0$. After finding the roots λ_j of the polynomial $\det(\mathbf{A} - \lambda\mathbf{I})$, one may form a new matrix $\mathbf{A}'$ for each λ_j by

$$\mathbf{A}' = \mathbf{A} - \lambda_j\mathbf{I} \quad \text{(5-1-23)}$$

then the solution to

$$\mathbf{A}'\mathbf{x} = \mathbf{0} \quad \text{(5-1-24)}$$

arises from the column $\mathbf{c}$ at the last step of the bordering. It is the column eigenvector. Likewise, the row eigenvector is the row in the last step of the bordering algorithm.

EXERCISES

1 Indicate the sizes of all the matrices in equations (5-1-7) to (5-1-14)

2 Show how (5-1-15) follows from (5-1-9), (5-1-10), (5-1-13), and (5-1-14).

5-2 SYLVESTER'S MATRIX THEOREM

Sylvester's theorem provides a rapid way to calculate functions of a matrix. Some simple functions of a matrix of frequent occurrence are A^{-1} and A^N(for N large). Two more matrix functions which are very important in wave propagation are e^A and $A^{1/2}$. Before going into the somewhat abstract proof of Sylvester's theorem, we will take up a mumerical example. Consider the matrix

$$A = \begin{bmatrix} 3 & -2 \\ 1 & 0 \end{bmatrix} \qquad (5\text{-}2\text{-}1)$$

It will be necessary to have the column eigenvectors and the eigenvalues of this matrix; they are given by

$$\begin{bmatrix} 3 & -2 \\ 1 & 0 \end{bmatrix} \begin{bmatrix} 1 \\ 1 \end{bmatrix} = 1 \begin{bmatrix} 1 \\ 1 \end{bmatrix} \qquad (5\text{-}2\text{-}2)$$

$$\begin{bmatrix} 3 & -2 \\ 1 & 0 \end{bmatrix} \begin{bmatrix} 2 \\ 1 \end{bmatrix} = 2 \begin{bmatrix} 2 \\ 1 \end{bmatrix} \qquad (5\text{-}2\text{-}3)$$

Since the matrix A is not symmetric, it has row eigenvectors which differ from the column vectors. These are

$$[-1 \quad 2] \begin{bmatrix} 3 & -2 \\ 1 & 0 \end{bmatrix} = 1[-1 \quad 2] \qquad (5\text{-}2\text{-}4)$$

$$[1 \quad -1] \begin{bmatrix} 3 & -2 \\ 1 & 0 \end{bmatrix} = 2[1 \quad -1] \qquad (5\text{-}2\text{-}5)$$

We may abbreviate equations (5-2-2) through (5-2-5) by

$$\begin{aligned} A \quad c_1 &= \lambda_1 \quad c_1 \\ A \quad c_2 &= \lambda_2 \quad c_2 \\ r_1 \quad A &= \lambda_1 \quad r_1 \\ r_2 \quad A &= \lambda_2 \quad r_2 \end{aligned} \qquad (5\text{-}2\text{-}6)$$

The reader will observe that r or c could be multiplied by an arbitrary scale factor and (5-2-6) would still be valid. The eigenvectors are said to be normalized if scale factors have been chosen so that $r_1 \cdot c_1 = 1$ and $r_2 \cdot c_2 = 1$. It will be observed that $r_1 \cdot c_2 = 0$ and $r_2 \cdot c_1 = 0$, a general result to be established in the exercises.

Let us consider the behavior of the matrix $c_1 r_1$.

$$c_1 r_1 = \begin{bmatrix} 1 \\ 1 \end{bmatrix} [-1 \quad 2] = \begin{bmatrix} -1 & 2 \\ -1 & 2 \end{bmatrix}$$

Any power of this matrix is the matrix itself, for example its square.

$$\begin{bmatrix} -1 & 2 \\ -1 & 2 \end{bmatrix} \begin{bmatrix} -1 & 2 \\ -1 & 2 \end{bmatrix} = \begin{bmatrix} -1 & 2 \\ -1 & 2 \end{bmatrix}$$

This property is called idempotence (Latin for self-power). It arises because $(c_1 r_1)(c_1 r_1) = c_1 (r_1 \cdot c_1) r_1 = c_1 r_1$. The same thing is of course true of $c_2 r_2$. Now notice that the matrix $c_1 r_1$ is "perpendicular" to the matrix $c_2 r_2$, that is

$$\begin{bmatrix} 2 & -2 \\ 1 & -1 \end{bmatrix} \begin{bmatrix} -1 & 2 \\ -1 & 2 \end{bmatrix} = \begin{bmatrix} 0 & 0 \\ 0 & 0 \end{bmatrix}$$

since r_2 and c_2 are perpendicular.

Sylvester's theorem says that any function f of the matrix A may be written

$$f(A) = f(\lambda_1) c_1 r_1 + f(\lambda_2) c_2 r_2$$

The simplest example is $f(A) = A$

$$A = \lambda_1 c_1 r_1 + \lambda_2 c_2 r_2$$

$$= 1 \begin{bmatrix} -1 & 2 \\ -1 & 2 \end{bmatrix} + 2 \begin{bmatrix} 2 & -2 \\ 1 & -1 \end{bmatrix} = \begin{bmatrix} 3 & -2 \\ 1 & 0 \end{bmatrix} \qquad (5\text{-}2\text{-}7)$$

Another example is

$$A^2 = 1^2 \begin{bmatrix} -1 & 2 \\ -1 & 2 \end{bmatrix} + 2^2 \begin{bmatrix} 2 & -2 \\ 1 & -1 \end{bmatrix} = \begin{bmatrix} 7 & -6 \\ 3 & -2 \end{bmatrix}$$

The inverse is

$$A^{-1} = 1^{-1} \begin{bmatrix} -1 & 2 \\ -1 & 2 \end{bmatrix} + 2^{-1} \begin{bmatrix} 2 & -2 \\ 1 & -1 \end{bmatrix} = \tfrac{1}{2} \begin{bmatrix} 0 & 2 \\ -1 & 3 \end{bmatrix}$$

The identity matrix may be expanded in terms of the eigenvectors of the matrix A.

$$A^0 = I = 1^0 \begin{bmatrix} -1 & 2 \\ -1 & 2 \end{bmatrix} + 2^0 \begin{bmatrix} 2 & -2 \\ 1 & -1 \end{bmatrix} = \begin{bmatrix} 1 & 0 \\ 0 & 1 \end{bmatrix}$$

Before illustrating some more complicated functions let us see what it takes to prove Sylvester's theorem. We will need one basic result which is in all the books on matrix theory, namely, that most matrices (see exercises) can be diagonalized. In terms of our 2×2 example this takes the form

$$\begin{bmatrix} r_1 \\ r_2 \end{bmatrix} A [c_1 \mid c_2] = \begin{bmatrix} \lambda_1 & 0 \\ 0 & \lambda_2 \end{bmatrix} \qquad (5\text{-}2\text{-}8)$$

where

$$\begin{bmatrix} r_1 \\ r_2 \end{bmatrix} [c_1 \mid c_2] = \begin{bmatrix} 1 & 0 \\ 0 & 1 \end{bmatrix} \qquad (5\text{-}2\text{-}9)$$

Since a matrix commutes with its inverse, (5-2-9) implies

$$[c_1 \mid c_2] \begin{bmatrix} r_1 \\ r_2 \end{bmatrix} = \begin{bmatrix} 1 & 0 \\ 0 & 1 \end{bmatrix} \qquad (5\text{-}2\text{-}10)$$

Postmultiply (5-2-8) by the row matrix and premultiply by the column matrix. Using (5-2-10), we get

$$A = [c_1 \mid c_2] \begin{bmatrix} \lambda_1 & 0 \\ 0 & \lambda_2 \end{bmatrix} \begin{bmatrix} r_1 \\ r_2 \end{bmatrix} \qquad (5\text{-}2\text{-}11)$$

Equation (5-2-11) is (5-2-7) in disguise, as we can see by writing (5-2-11) as

$$A = [c_1 \mid c_2] \left\{ \begin{bmatrix} \lambda_1 & 0 \\ 0 & 0 \end{bmatrix} + \begin{bmatrix} 0 & 0 \\ 0 & \lambda_2 \end{bmatrix} \right\} \begin{bmatrix} r_1 \\ r_2 \end{bmatrix}$$

$$= [c_1 \mid c_2] \begin{bmatrix} \lambda_1 & 0 \\ 0 & 0 \end{bmatrix} \begin{bmatrix} r_1 \\ r_2 \end{bmatrix} + [c_1 \mid c_2] \begin{bmatrix} 0 & 0 \\ 0 & \lambda_2 \end{bmatrix} \begin{bmatrix} r_1 \\ r_2 \end{bmatrix}$$

$$= \lambda_1 c_1 r_1 + \lambda_2 c_2 r_2$$

Now to get A^2 we have

$$A^2 = (\lambda_1 c_1 r_1 + \lambda_2 c_2 r_2)(\lambda_1 c_1 r_1 + \lambda_2 c_2 r_2)$$

Using the orthonormality of $c_1 r_1$ and $c_2 r_2$ this reduces to

$$A^2 = \lambda_1{}^2 c_1 r_1 + \lambda_2{}^2 c_2 r_2$$

It is clear how (5-2-11) can be used to prove Sylvester's theorem for any polynomial function of A. Clearly, there is nothing peculiar about 2×2 matrices either. This works for $n \times n$. Likewise, one may consider infinite series functions in A. Since almost any function can be made up of infinite series, we can consider also transcendental functions like sine, cosine, exponential.

Exponentials arise naturally as the solutions to differential equations. Consider the matrix differential equation

$$\frac{d}{dx} E = AE \qquad (5\text{-}2\text{-}12)$$

One may readily verify the power series solution

$$E = I + Ax + \frac{A^2 x^2}{2!} + \cdots$$

This is the power series definition of an exponential function. If the matrix A is one of that vast majority which can be diagonalized, then the exponential can be more simply expressed by Sylvester's theorem. For the numerical example we have been considering, we have

$$E = e^x \begin{bmatrix} -1 & 2 \\ -1 & 2 \end{bmatrix} + e^{2x} \begin{bmatrix} 2 & -2 \\ 1 & -1 \end{bmatrix}$$

The exponential matrix is a solution to the differential equation (5-2-12) without regard to boundaries. It frequently happens that physics gives one a differential equation

$$\frac{d}{dx} \begin{bmatrix} y_1 \\ y_2 \end{bmatrix} = A \begin{bmatrix} y_1 \\ y_2 \end{bmatrix} \qquad (5\text{-}2\text{-}13)$$

subject to two boundary conditions on either of y_1 or y_2 or a combination. One may verify that

$$\begin{bmatrix} y_1 \\ y_2 \end{bmatrix} = e^{\mathbf{A}x} \begin{bmatrix} k_1 \\ k_2 \end{bmatrix}$$

is the solution to (5-2-13) for arbitrary constants k_1 and k_2. Boundary conditions are then used to determine the numerical values of k_1 and k_2. Note that k_1 and k_2 are just $y_1(x = 0)$ and $y_2(x = 0)$.

An interesting situation arises with the square root of a matrix. A 2×2 matrix like $\mathbf{A}$ will have four square roots because there are four possible combinations for choice of plus or minus signs on $\sqrt{\lambda_1}$ and $\sqrt{\lambda_2}$. In general, an $n \times n$ matrix has 2^n square roots. An important application arises in a later chapter, where we will deal with the differential operator $(k^2 + \partial^2/\partial x^2)^{1/2}$. The square root of an operator is explained in very few books and few people even know what it means. The best way to visualize the square root of this differential operator is to relate it to the square root of the matrix $\mathbf{M}$ where

$$\mathbf{M} = k^2 \begin{bmatrix} 1 & & & \\ & 1 & & \\ & & 1 & \\ & & & 1 \end{bmatrix} + \frac{1}{(\Delta x)^2} \begin{bmatrix} ? & ? & & \\ 1 & -2 & 1 & \\ & 1 & -2 & 1 \\ & & 1 & -2 & 1 \\ & & & ? & ? \end{bmatrix}$$

The right-hand matrix is a second difference approximation to a second partial derivative. Let us define

$$\mathbf{M} = k^2 \mathbf{I} + \mathbf{T}$$

Clearly we wish to consider $\mathbf{M}$ generalized to a very large size so that the end effects may be minimized. In concept, we can make $\mathbf{M}$ as large as we like and for any size we can get $2^{\mathbf{M}}$ square roots. In practice there will be only two square roots of interest, one with the plus roots of all the eigenvalues and the other with all the minus roots. How can we find these "principal value" square roots? An important case of interest is where we can use the binomial theorem so that

$$(k^2 \mathbf{I} + \mathbf{T})^{1/2} = \pm k \left(\mathbf{I} + \frac{\mathbf{T}}{k^2} \right)^{1/2}$$

$$= \pm k \left(\mathbf{I} + \frac{\mathbf{T}}{2k^2} - \frac{\mathbf{T}^2}{8k^4} + \cdots \right)$$

The result is justified by merely squaring the assumed square root. Alternatively, it may be justified by means of Sylvester's theorem. It should be noted that on squaring the assumed square root one utilizes the fact that $\mathbf{I}$ and $\mathbf{T}$ commute. We are led to the idea that the square root of the differential operator may be interpreted as

$$\left(k^2 + \frac{\partial^2}{\partial x^2} \right)^{1/2} = k + \frac{1}{2k} \frac{\partial^2}{\partial x^2} + \cdots$$

provided that k is not a function of x. If k is a function of x, the square root of the differential operator still has meaning but is not so simply computed with the binomial theorem.

EXERCISES

1 Premultiply (5-2-6b) by r_1 and postmultiply (5-2-6c) by c_2, then subtract. Is $\lambda_1 \neq \lambda_2$ a necessary condition for r_1 and c_2 to be perpendicular? Is it a sufficient condition?
2 Show the Cayley-Hamilton theorem, that is, if

$$0 = f(\lambda) = \det(A - \lambda I) = p_0 + p_1 \lambda + p_2 \lambda^2 + \cdots p_n \lambda^n$$

then

$$f(A) = p_0 + p_1 A + p_2 A^2 + \cdots + p_n A^n = 0$$

3 Verify that, for a general 2×2 matrix A, for which

$$\lambda_1 \neq \lambda_2,$$

$$c_1 r_1 = (\lambda_2 I - A)/(\lambda_2 - \lambda_1)$$

where λ_1 and λ_2 are eigenvalues of A. What is the general form for $c_2 r_2$?
4 For a symmetric matrix it can be shown that there is always a complete set of eigenvectors. A problem sometimes arises with nonsymmetric matrices. Study the matrix

$$\begin{bmatrix} 1 & 1 - \varepsilon^2 \\ -1 & 3 \end{bmatrix}$$

as $\varepsilon \to 0$ to see why one eigenvector is lost. This is called a defective matrix. (This example is from T. R. Madden.)
5 A wide variety of wave-propagation problems in a stratified medium reduce to the equation

$$\frac{d}{dx} \begin{bmatrix} y_1 \\ y_2 \end{bmatrix} = \begin{bmatrix} 0 & b \\ a & 0 \end{bmatrix} \begin{bmatrix} y_1 \\ y_2 \end{bmatrix}$$

What is the x dependence of the solution when ab is positive? When ab is negative? Assume a and b are independent of x. Use Sylvester's theorem. What would it take to get a defective matrix? What are the solutions in the case of a defective matrix?
6 Consider a matrix of the form $I + vv^T$ where v is a column vector and v^T is its transpose. Find $(I + vv^T)^{-1}$ in terms of a power series in vv^T. [Note that $(vv^T)^N$ collapses to vv^T times a scaling factor, so the power series reduces considerably.]
7 The following "cross-product" matrix often arises in electrodynamics. Let $B = (B_x, B_y, B_z)$

$$U = \frac{1}{\sqrt{B \cdot B}} \begin{bmatrix} 0 & -B_z & B_y \\ B_z & 0 & -B_x \\ -B_y & B_x & 0 \end{bmatrix}$$

(a) Write out elements of $I + U^2$.
(b) Show that $U(I + U^2) = 0$ or $U^3 = -U$.

(c) Let **v** be an arbitrary vector. In what geometrical directions do $\mathbf{Uv}$, $\mathbf{U}^2\mathbf{v}$, and $(\mathbf{I} + \mathbf{U}^2)\mathbf{v}$ point?

(d) What are the eigenvalues of $\mathbf{U}$. [HINT: Use part (b).]

(e) Why cannot $\mathbf{U}$ be canceled from $\mathbf{U}^3 = -\mathbf{U}$?

(f) Verify that the idempotent matrices of $\mathbf{U}$ are

$$\mathbf{c}_1\mathbf{r}_1 = (\mathbf{I} + \mathbf{U}^2)$$

$$\mathbf{c}_2\mathbf{r}_2 = \frac{1}{2(i\mathbf{U} - \mathbf{U}^2)}$$

$$\mathbf{c}_3\mathbf{r}_3 = \frac{1}{2(-i\mathbf{U} - \mathbf{U}^2)}$$

5-3 MATRIX FILTERS, SPECTRA, AND FACTORING

Two time series can be much more interesting than one because of the possibility of interactions between them. The general linear model for two series is depicted in Fig. 5-2

The filtering operation in the figure can be expressed as a matrix times vector operation, where the elements of the matrix and vectors are Z transform polynomials. That is,

$$\begin{bmatrix} Y_1(Z) \\ Y_2(Z) \end{bmatrix} = \begin{bmatrix} B_{11}(Z) & B_{12}(Z) \\ B_{21}(Z) & B_{22}(Z) \end{bmatrix} \begin{bmatrix} X_1(Z) \\ X_2(Z) \end{bmatrix} \qquad (5\text{-}3\text{-}1)$$

One fact which is obvious but unfamiliar is that a matrix with polynomial elements is exactly the same thing as a polynomial with matrix coefficients. This is illustrated by the example:

$$\begin{bmatrix} 1 + Z + 2Z^2 & Z \\ 1 & 1 + Z^2 \end{bmatrix} = \begin{bmatrix} 1 & 0 \\ 1 & 1 \end{bmatrix} + \begin{bmatrix} 1 & 1 \\ 0 & 0 \end{bmatrix}Z + \begin{bmatrix} 2 & 0 \\ 0 & 1 \end{bmatrix}Z^2$$

Now we can address ourselves to the inverse problem; given a filter $\mathbf{B}$ and the outputs $\mathbf{Y}$ how can we find the inputs $\mathbf{X}$? The solution is analogous to that of single time series. Let us regard $\mathbf{B}(Z)$ as a matrix of polynomials. One knows, for example, that the inverse of any 2×2 matrix

$$\begin{bmatrix} a & b \\ c & d \end{bmatrix} \quad \text{is} \quad \frac{\begin{bmatrix} d & -b \\ -c & a \end{bmatrix}}{ad - bc}$$

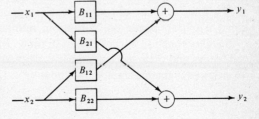

FIGURE 5-2
Two time series x_1 and x_2 input to a matrix of four filters illustrates the general linear model of multichannel filtering.

Consequently $\mathbf{Y} = \mathbf{BX}$ may be solved for $\mathbf{X}$ as $\mathbf{X} = \mathbf{B}^{-1}\mathbf{Y}$ where

$$\mathbf{B}^{-1} = \begin{bmatrix} B_{22}(Z) & -B_{12}(Z) \\ -B_{21}(Z) & B_{11}(Z) \end{bmatrix} \frac{1}{B_{11}(Z)B_{22}(Z) - B_{12}(Z)B_{21}(Z)}$$

The denominator is a scalar. We have treated scalar denominators before. If all the zeros lie outside the unit circle, we can use an ordinary power series for the inverse; otherwise, it is not minimum-phase and we use a Laurent series.

When one generalizes to many time series, the numerator matrix is the so-called adjoint matrix and the denominator is the determinant. The adjoint matrix can be formed without the use of any division operations. In other words, elements in the adjoint matrix are in the form of sums of products. For this reason, we may say that the criterion for a minimum-phase matrix wavelet is that the determinant of its Z transform has no zeros inside the unit circle.

Equation (5-3-1) is a useful description of Fig. 5-2 in most applications. However in some applications (where the filter is an unknown to be determined), a transposed form of (5-3-1) is more useful. If b_{12} was interchanged with b_{21} in Fig. 5-2, we could use the "row data" expression

$$[Y_1(Z) \quad Y_2(Z)] = [X_1(Z) \quad X_2(Z)]\begin{bmatrix} B_{11} & B_{12} \\ B_{21} & B_{22} \end{bmatrix} \quad (5\text{-}3\text{-}2)$$

Now that we have generalized the concept of filtering from scalar-valued time series to vector-valued series, it is natural to generalize the idea of spectrum. For vector-valued time functions, the spectrum is a matrix called the *spectral matrix* and it is given by

$$\mathbf{R}(\omega) = \begin{bmatrix} r_{11} & r_{12} \\ r_{21} & r_{22} \end{bmatrix} = \begin{bmatrix} \overline{Y}_1\left(\dfrac{1}{Z}\right) \\ \overline{Y}_2\left(\dfrac{1}{Z}\right) \end{bmatrix} [Y_1(Z) \quad Y_2(Z)] \quad (5\text{-}3\text{-}3)$$

$$= \begin{bmatrix} \overline{Y}_1 Y_1 & \overline{Y}_1 Y_2 \\ \overline{Y}_2 Y_1 & \overline{Y}_2 Y_2 \end{bmatrix}$$

It will be noticed that the vector times vector operation defining (5-3-3) is an "outer product" rather than the more usually occurring "inner product." The diagonals of the spectral matrix $\mathbf{R}$ contain the usual auto-spectrum of each channel. Off-diagonals contain the cross spectrum. Because (5-3-3) is an outer product, the matrix is singular. Now, instead of taking $[Y_1(Z) \quad Y_2(Z)]$ to have a time function with a finite amount of energy, let us suppose the filter inputs to (5-3-2), namely $(x_1(t), x_2(t))$ are made up of random numbers, independently drawn from some probability function at every point in time. In this case, $y_1(t)$ and $y_2(t)$ are random time series and their spectral matrix is defined like (5-3-3) but taking an expectation (average over the ensemble). We have

$$\mathbf{R}(\omega) = E\begin{bmatrix} \overline{Y}_1 \\ \overline{Y}_2 \end{bmatrix} [Y_1 \quad Y_2]$$

substituting from (5-3-2)

$$\mathbf{R}(\omega) = E \begin{bmatrix} \bar{B}_{11} & \bar{B}_{21} \\ \bar{B}_{12} & \bar{B}_{22} \end{bmatrix} \begin{bmatrix} \bar{X}_1 \\ \bar{X}_2 \end{bmatrix} [X_1 \quad X_2] \begin{bmatrix} B_{11} & B_{12} \\ B_{21} & B_{22} \end{bmatrix}$$

Now, grouping the ensemble summation with the random variables, we get

$$\mathbf{R} = \begin{bmatrix} \bar{B}_{11} & \bar{B}_{21} \\ \bar{B}_{12} & \bar{B}_{22} \end{bmatrix} \left\{ E \begin{bmatrix} \bar{X}_1 \\ \bar{X}_2 \end{bmatrix} [X_1 \quad X_2] \right\} \begin{bmatrix} B_{11} & B_{12} \\ B_{21} & B_{22} \end{bmatrix} \qquad (5\text{-}3\text{-}4)$$

Next, we explicitly introduce the assumption that the random numbers $x_1(t)$ are drawn independently of $x_2(t)$, thus $E(\bar{X}_2(1/z)X_1(z)) = 0$ and the assumption that $x_i(t)$ is white $E[x_i(t)x_i(t+s)] = 0$ if $s \neq 0$ and of unit variance $E[x_i(t)^2] = 1$. Thus (5-3-4) becomes

$$\mathbf{R} = \begin{bmatrix} \bar{B}_{11} & \bar{B}_{21} \\ \bar{B}_{12} & \bar{B}_{22} \end{bmatrix} \begin{bmatrix} 1 & 0 \\ 0 & 1 \end{bmatrix} \begin{bmatrix} B_{11} & B_{12} \\ B_{21} & B_{22} \end{bmatrix}$$

$$= \begin{bmatrix} \bar{B}_{11} & \bar{B}_{21} \\ \bar{B}_{12} & \bar{B}_{22} \end{bmatrix} \begin{bmatrix} B_{11} & B_{12} \\ B_{21} & B_{22} \end{bmatrix} \qquad (5\text{-}3\text{-}5)$$

Of course, in practice the spectral matrix must be estimated, say $\hat{\mathbf{R}}$, from finite samples of data. This means that ensemble summation must be simulated. If the ensemble sum in (5-3-4) is simulated by summation over one point (no summation), then (5-3-4) is a singular matrix like (5-3-3). As discussed earlier, the accuracy of the elements of the spectral matrix improves with the square root of the number of ensemble elements summed over.

Single-channel spectral factorization gives insight into numerous important problems in mathematical physics. We have seen that the concepts of filter and spectrum extend in quite a useful fashion to multichannel data. It was only natural that a great deal of effort should have gone to spectral factorization of multichannel data. This effort has been successful. However, in retrospect, from the point of view of computer modeling and interpretation of observed waves, it must be admitted that multichannel spectral factorization has not been especially useful. Nevertheless a brief summary of results will be given.

The root method The author extended the single-channel root method to the multichannel case [Ref. 19]. The method is even more cumbersome in the multi-channel case. A most surprising thing about the solution is that it includes a much broader result: that a polynomial with matrix coefficients may be factored. For example,

$$\begin{bmatrix} 1 & 0 \\ 0 & 1 \end{bmatrix} + Z \begin{bmatrix} -3 & -1 \\ 14 & -11 \end{bmatrix} + Z^2 \begin{bmatrix} -4 & 4 \\ -58 & 28 \end{bmatrix}$$

factors 6 ways to

$$\left\{ I + Z \begin{bmatrix} 2 & -1 \\ 20 & -7 \end{bmatrix} \right\} \left\{ I + Z \begin{bmatrix} -5 & 0 \\ -6 & -4 \end{bmatrix} \right\} \qquad \left\{ I + Z \begin{bmatrix} -4 & 0 \\ -10 & -2 \end{bmatrix} \right\} \left\{ I + Z \begin{bmatrix} 1 & -1 \\ 24 & -9 \end{bmatrix} \right\}$$

$$\left\{ I + Z \begin{bmatrix} 0 & -1 \\ 10 & -7 \end{bmatrix} \right\} \left\{ I + Z \begin{bmatrix} -3 & 0 \\ 4 & -4 \end{bmatrix} \right\} \qquad \left\{ I + Z \begin{bmatrix} -4 & 0 \\ -4 & -3 \end{bmatrix} \right\} \left\{ I + Z \begin{bmatrix} 1 & -1 \\ 18 & -8 \end{bmatrix} \right\}$$

$$\left\{ I + Z \begin{bmatrix} -1 & -1 \\ 8 & -7 \end{bmatrix} \right\} \left\{ I + Z \begin{bmatrix} -2 & 0 \\ 6 & -4 \end{bmatrix} \right\} \qquad \left\{ I + Z \begin{bmatrix} -4 & 0 \\ 2 & -5 \end{bmatrix} \right\} \left\{ I + Z \begin{bmatrix} 1 & -1 \\ 12 & -6 \end{bmatrix} \right\}$$

The Toeplitz method The only really practical method for finding an invertible matrix wavelet with a given spectrum is the multichannel Toeplitz method. The necessary algebra is developed in a later section on multichannel time series prediction.

The exp-log and Hilbert transform methods A number of famous mathematicians including Norbert Wiener have worked on the problem from the point of view of extending the exp-log or the Hilbert transform method. The principal stumbling block is that $\exp(A + B)$ does not equal $\exp(A) \exp(B)$ unless A and B happen to commute, that is, $AB = BA$. This is usually not the case. Although many difficult papers have appeared on the subject (some stating that they solved the problem), the author is unaware of anyone who ever wrote a computer program which works at fast Fourier transform speeds as does the single-channel Hilbert transform method.

EXERCISES

1 Think up a matrix filter where the two outputs $y_1(t)$ and $y_2(t)$ are the same but for a scale factor. Clearly **X** cannot be recovered from **Y**. Show that the determinant of the filter vanishes. Find another example in which the determinant is zero at one frequency but nonzero elsewhere. Explain in the time domain in what sense the input cannot be recovered from the output.

2 Given a thermometer which measures temperature plus α times pressure and a pressure gage which measures pressure plus β times the time rate of change of the temperature, find the matrix filter which converts the observed series to temperature and pressure. [HINT: Use either the time derivative approximation $1 - Z$ or $2(1 - Z)/(1 + Z)$.]

3 Let

$$\mathbf{B}(Z) = \begin{bmatrix} 1 & 0 \\ 3 & 2 \end{bmatrix} + Z \begin{bmatrix} 2 & 1 \\ 0 & 0 \end{bmatrix}$$

Identify coefficients of powers of Z in $\mathbf{B}(Z)\mathbf{A}(Z) = \mathbf{I}$, to recursively develop the coefficients of $\mathbf{A}(Z) = [\mathbf{B}(Z)]^{-1}$.

4 Express the inverse of

$$\begin{bmatrix} 1 + 2Z & Z \\ 3 & 2 \end{bmatrix}$$

in a Taylor or Laurent series as is necessary.

5 The determinant of a polynomial with matrix coficients may be independent of Z. Applied to matrix filters, this may mean that an inverse filter may have only a finite number of powers in Z instead of the infinite series one always has with scalar filters. What is the most nontrivial example you can find?

5-4 MARKOV PROCESSES

A Markov process is another mathematical model for a time series. Until now it has found little use in geophysics, but we will include it anyway because it might become useful and it is easily explained with the methods previously developed.

Suppose that x_t could take on only integer values. A given value is called a state. As time proceeds, transitions are made from the jth state to the ith state according to a probability matrix p_{ij}. The system has no memory. The next state is probabilistically dependent on the current state but independent of the previous states. The classic example is of a frog in a lily pond. As time goes by, the frog jumps from one lily pad to another. He may be more likely to jump to a near one than to a far one. He may prefer big to small pads, and he doesn't remember the last pad he was on. The state of the system is the number of the pad the frog currently occupies. The transitions are his jumps.

To begin with, one defines a state probability $\pi_i(k)$, the probability that the system will occupy state i after k transitions if its state is known at $k = 0$. We also define the transition matrix P_{ij}. Then

$$\pi(k + 1) = \mathbf{P}\pi(k) \qquad (5\text{-}4\text{-}1)$$

The initial-state probability vector is $\pi(0)$. Since the initial state is known, then $\pi(0)$ is all zeros except for a one (1) in the position corresponding to the initial state. For example, see the state-transition diagram of Fig. 5-3.

The diagram corresponds to the probability matrix

$$\begin{bmatrix} \pi_1 \\ \pi_2 \\ \pi_3 \\ \pi_4 \end{bmatrix}_{k+1} = \begin{bmatrix} 0 & \frac{1}{2} & 0 & 0 \\ 0 & 0 & \frac{1}{2} & 0 \\ 1 & \frac{1}{2} & 0 & 0 \\ 0 & 0 & \frac{1}{2} & 1 \end{bmatrix} \begin{bmatrix} \pi_1 \\ \pi_2 \\ \pi_3 \\ \pi_4 \end{bmatrix}_k$$

Since at each time a transition must occur, we have that the sum of the elements in a column must be unity. In other words, the row vector $[1 \quad 1 \quad 1 \quad 1]$ is an eigenvector of $\mathbf{P}$ with unit eigenvalue. Let us define the Z transform of the probability vector as

$$\mathbf{\Pi}(Z) = \pi(0) + Z\pi(1) + Z^2\pi(2) + \cdots \qquad (5\text{-}4\text{-}2)$$

In terms of Z transforms (5-4-1) becomes

$$[\pi(1) + Z\pi(2) + \cdots] = \mathbf{P}[\pi(0) + Z\pi(1) + \cdots]$$
$$Z^{-1}[\mathbf{\Pi}(Z) - \pi(0)] = \mathbf{P}\mathbf{\Pi}(Z)$$
$$(\mathbf{I} - Z\mathbf{P})\mathbf{\Pi}(Z) = \pi(0)$$
$$\mathbf{\Pi}(Z) = (\mathbf{I} - Z\mathbf{P})^{-1}\pi(0) \qquad (5\text{-}4\text{-}3)$$

Thus we have expressed the general solution to the problem as a function of the matrix $\mathbf{P}$ times an initial-state vector. There will be values of Z for which the

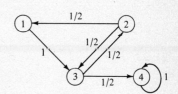

FIGURE 5-3
An example of a state-transition diagram.

inverse matrix to $(\mathbf{I} - \mathbf{ZP})$ does not exist. These values Z_j are given by $\det(\mathbf{I} - Z_j\mathbf{P})$ $= 0$ or $\det(\mathbf{P} - Z_j^{-1}\mathbf{I}) = 0$. Clearly the Z_j^{-1} are the eigenvalues of $\mathbf{P}$. Utilizing Sylvester's theorem, then, we have

$$(\mathbf{I} - \mathbf{ZP})^{-1} = \sum_j \frac{\mathbf{c}_j\mathbf{r}_j}{1 - \dfrac{Z}{Z_j}} \qquad (5\text{-}4\text{-}4)$$

Some modification to (5-4-4) is required if there are repeated eigenvalues. Equation (5-4-4) is essentially a partial fraction expansion. A typical term has the form

$$\frac{1}{1 - \dfrac{Z}{Z_j}} = 1 + \frac{Z}{Z_j} + \left(\frac{Z}{Z_j}\right)^2 + \cdots$$

Thus coefficients at successive powers of Z decline with time in the form $(Z_j^{-1})^t$. It is clear that, if probabilities are to be bounded, the roots $1/Z_j$ must be inside the unit circle (recall minimum phase). We have already shown that one of the roots Z_1 is always unity. This leads to the "steady-state" solution $1^t = 1$. In our particular example, one can see by inspection that the steady-state probability vector is $[0 \quad 0 \quad 0 \quad 1]^T$ so the general solution is of the form

$$\pi(t) = \left\{ \begin{bmatrix} 0 \\ 0 \\ 0 \\ 1 \end{bmatrix} [1 \quad 1 \quad 1 \quad 1] + \sum_{j=2}^{4} Z_j^{-t}\mathbf{c}_j\mathbf{r}_j \right\} \pi(0)$$

Finally, a word of caution must be added. Occasionally defective matrices arise (incomplete set of eigenvectors) and for these the Sylvester theorem does not apply. In such cases, the solutions turn out to contain not only terms like Z_j^{-t} but also terms like tZ^{-t} and t^2Z^{-t}. It is the same situation as that applying to ordinary differential equations with constant coefficients. Ordinarily, the solutions are of the form $(r_i)^t$ where r_i is the ith root of the *indicial* equation but the presence of repeated roots gives rise to solutions like tr_i^t. A mathematical survey of the subject is given by Seneta [Ref. 20].

6

DATA MODELING BY LEAST SQUARES

The reconciliation of theory and data is the essence of science. A ubiquitous tool in this task is the method of least-squares fitting. Elementary calculus books generally consider the fitting of a straight line to scattered data points. Such an elementary application gives scant hint of the variety of practical problems which can be solved by the method of least squares. Some geophysical examples which we will consider include locating earthquakes, analyzing tides, expanding the earth's gravity and magnetic fields in spherical harmonics, and doing interesting things with time series. When the past of a time series is available, one may find that least squares can be used to determine a filter which predicts some future values of the time series. When a time series which has been highly predictable for a long stretch of time suddenly becomes much less predictable an "event" is said to have occurred. A filter which emphasizes such events is called a *prediction-error filter*. If one is searching for a particular dispersed wavelet in a time series, it may help to design a filter which compresses the wavelet into some more recognizable shape, an impulse for example. Such a wave-shaping filter may be designed by least squares. With multiple time series which arise from several sensors detecting waves in space, least squares may be used to find filters which respond only to certain directions and wave speeds.

Before we begin with the general theory, let us take up a simple example in the subject of time series analysis. Given the input, say $\mathbf{x} = (2, 1)$ to some filter, say $\mathbf{f} = (f_0, f_1)$ then the output is necessarily $\mathbf{c} = (2f_0, f_0 + 2f_1, f_1)$. To design an inverse filter we would wish to have $\mathbf{c}$ come out as close as possible to $(1, 0, 0)$. In order to minimize the difference between the actual and the desired outputs we minimize

$$E(f_0, f_1) = (2f_0 - 1)^2 + (f_0 + 2f_1)^2 + (f_1)^2$$

The sum E of the squared errors will attain a minimum if f_0 and f_1 are chosen so that

$$0 = \frac{\partial E}{\partial f_0} = 2(2f_0 - 1)2 + 2(f_0 + 2f_1)$$

$$0 = \frac{\partial E}{\partial f_1} = 2(f_0 + 2f_1)2 + 2f_1$$

Cancelling a 2 and arranging this into the standard form for simultaneous equations, we get

$$\begin{bmatrix} 5 & 2 \\ 2 & 5 \end{bmatrix} \begin{bmatrix} f_0 \\ f_1 \end{bmatrix} = \begin{bmatrix} 2 \\ 0 \end{bmatrix}$$

and the solution is

$$\begin{bmatrix} f_0 \\ f_1 \end{bmatrix} = \frac{1}{21} \begin{bmatrix} 5 & -2 \\ -2 & 5 \end{bmatrix} \begin{bmatrix} 2 \\ 0 \end{bmatrix} = \begin{bmatrix} \frac{10}{21} \\ -\frac{4}{21} \end{bmatrix}$$

The actual $\mathbf{c}$ which comes out of this filter is $(\frac{20}{21}, +\frac{2}{21}, -\frac{4}{21})$ which is not a bad approximation to $(1, 0, 0)$.

6-1 MORE EQUATIONS THAN UNKNOWNS

When there are more linear equations than unknowns, it is usually impossible to find a solution which satisfies all the equations. Then one often looks for a solution which approximately satisfies all the equations. Let $\mathbf{a}$ and $\mathbf{c}$ be known and $\mathbf{x}$ be unknown in the following set of equations where there are more equations than unknowns.

$$\begin{bmatrix} a_{11} & a_{12} & \cdots & a_{1m} \\ a_{21} & a_{22} & & \\ a_{31} & & & \\ \vdots & & & \\ a_{n1} & & \cdots & a_{nm} \end{bmatrix} \begin{bmatrix} x_1 \\ x_2 \\ \vdots \\ x_m \end{bmatrix} \approx \begin{bmatrix} c_1 \\ c_2 \\ \vdots \\ c_n \end{bmatrix} \quad (6\text{-}1\text{-}1)$$

Usually there will be no set of x_i which exactly satisfies (6-1-1). Let us define an error vector e_j by

$$\begin{bmatrix} a_{11} & a_{12} & \cdots & a_{1m} \\ a_{21} & a_{22} & & \\ a_{31} & & & \\ \vdots & & & \\ a_{n1} & & \cdots & a_{nm} \end{bmatrix} \begin{bmatrix} x_1 \\ x_2 \\ \vdots \\ x_m \end{bmatrix} - \begin{bmatrix} c_1 \\ c_2 \\ \vdots \\ c_n \end{bmatrix} = \begin{bmatrix} e_1 \\ e_2 \\ \vdots \\ e_n \end{bmatrix} \quad (6\text{-}1\text{-}2)$$

It simplifies the development to rewrite this equation as follows (a trick I learned from John P. Burg).

$$
\begin{bmatrix}
-c_1 & a_{11} & \cdots & a_{1m} \\
-c_2 & a_{21} & & \\
-c_3 & & & \\
\vdots & & & \\
-c_n & a_{n1} & \cdots & a_{nm}
\end{bmatrix}
\begin{bmatrix}
1 \\
x_1 \\
\vdots \\
x_m
\end{bmatrix}
=
\begin{bmatrix}
e_1 \\
e_2 \\
\vdots \\
e_n
\end{bmatrix}
\qquad (6\text{-}1\text{-}3)
$$

We may abbreviate this equation as

$$
\mathbf{Bx} = \mathbf{e} \qquad (6\text{-}1\text{-}4)
$$

where $\mathbf{B}$ is the matrix containing $\mathbf{c}$ and $\mathbf{a}$. The ith error may be written as a dot product and either vector may be written as the column

$$
e_i = [b_{i1} \quad b_{i2} \quad \cdots]
\begin{bmatrix} 1 \\ x_1 \\ \vdots \end{bmatrix}
= [1 \quad x_1 \quad \cdots]
\begin{bmatrix} b_{i1} \\ b_{i2} \\ \vdots \end{bmatrix}
$$

Now we will minimize the sum squared error E defined as $\sum e_i^2$

$$
E = \sum_i [1 \quad x_1 \quad \cdots]
\begin{bmatrix} b_{i1} \\ b_{i2} \\ \vdots \end{bmatrix}
[b_{i1} \quad b_{i2} \quad \cdots]
\begin{bmatrix} 1 \\ x_1 \\ \vdots \end{bmatrix}
\qquad (6\text{-}1\text{-}5)
$$

The summation may be brought inside the constants

$$
E = [1 \quad x_1 \quad x_2 \quad \cdots]
\left\{ \sum_{i=1}^{n}
\begin{bmatrix} -c_i \\ a_{i1} \\ a_{i2} \\ \vdots \end{bmatrix}
[-c_i \quad a_{i1} \quad a_{i2} \quad \cdots]
\right\}
\begin{bmatrix} 1 \\ x_1 \\ x_2 \\ \vdots \end{bmatrix}
\qquad (6\text{-}1\text{-}6)
$$

The matrix in the center, call it r_{ij}, is symmetrical. It is a positive (more strictly, nonnegative) definite matrix because you will never be able to find an $\mathbf{x}$ for which E is negative, since E is a sum of squared e_i. We find the $\mathbf{x}$ with minimum E by requiring $\partial E/\partial x_1 = 0$, $\partial E/\partial x_2 = 0$, $\ldots$, $\partial E/\partial x_m = 0$. Notice that this will give us exactly one equation for each unknown. In order to clarify the presentation we will specialize (6-1-6) to two unknowns.

$$
E = [1 \quad x_1 \quad x_2]
\begin{bmatrix}
r_{00} & r_{01} & r_{02} \\
r_{10} & r_{11} & r_{12} \\
r_{20} & r_{21} & r_{22}
\end{bmatrix}
\begin{bmatrix} 1 \\ x_1 \\ x_2 \end{bmatrix}
\qquad (6\text{-}1\text{-}7)
$$

Setting to zero the derivative with respect to x_1, we get

$$
0 = \frac{\partial E}{\partial x_1} = [0 \quad 1 \quad 0]R
\begin{bmatrix} 1 \\ x_1 \\ x_2 \end{bmatrix}
+ [1 \quad x_1 \quad x_2]R
\begin{bmatrix} 0 \\ 1 \\ 0 \end{bmatrix}
\qquad (6\text{-}1\text{-}8)
$$

Since $r_{ij} = r_{ji}$, both terms on the right are equal. Thus (6-1-8) may be written

$$0 = \frac{\partial E}{\partial x_1} = 2[r_{10} \quad r_{11} \quad r_{12}]\begin{bmatrix} 1 \\ x_1 \\ x_2 \end{bmatrix} \qquad (6\text{-}1\text{-}9)$$

Likewise, differentiating with respect to x_2 gives

$$0 = \frac{\partial E}{\partial x_2} = 2[r_{20} \quad r_{21} \quad r_{22}]\begin{bmatrix} 1 \\ x_1 \\ x_2 \end{bmatrix} \qquad (6\text{-}1\text{-}10)$$

Equations (6-1-9) and (6-1-10) may be combined

$$\begin{bmatrix} 0 \\ 0 \end{bmatrix} = \begin{bmatrix} r_{10} & r_{11} & r_{12} \\ r_{20} & r_{21} & r_{22} \end{bmatrix}\begin{bmatrix} 1 \\ x_1 \\ x_2 \end{bmatrix} \qquad (6\text{-}1\text{-}11)$$

This form is two equations in two unknowns. One might write it in the more conventional form

$$\begin{bmatrix} r_{11} & r_{12} \\ r_{21} & r_{22} \end{bmatrix}\begin{bmatrix} x_1 \\ x_2 \end{bmatrix} = -\begin{bmatrix} r_{10} \\ r_{20} \end{bmatrix} \qquad (6\text{-}1\text{-}12)$$

The matrix of (6-1-11) lacks only a top row to be equal to the matrix of (6-1-7). To give it that row, we may augment (6-1-11) by

$$v = r_{00} + r_{01}x_1 + r_{02}x_2 \qquad (6\text{-}1\text{-}13)$$

where (6-1-13) may be regarded as a definition of a new variable v. Putting (6-1-13) on top of (6-1-11) we get

$$\begin{bmatrix} v \\ 0 \\ 0 \end{bmatrix} = \begin{bmatrix} r_{00} & r_{01} & r_{02} \\ r_{10} & r_{11} & r_{12} \\ r_{20} & r_{21} & r_{22} \end{bmatrix}\begin{bmatrix} 1 \\ x_1 \\ x_2 \end{bmatrix} \qquad (6\text{-}1\text{-}14)$$

The solution $\mathbf{x}$ of (6-1-12) or (6-1-14) is that set of x_k for which E is a minimum. To get an interpretation of v, we may multiply both sides by $[1 \quad x_1 \quad x_2]$, getting

$$v = [1 \quad x_1 \quad x_2]\begin{bmatrix} v \\ 0 \\ 0 \end{bmatrix} = [1 \quad x_1 \quad x_2]R\begin{bmatrix} 1 \\ x_1 \\ x_2 \end{bmatrix} \qquad (6\text{-}1\text{-}15)$$

Comparing (6-1-15) with (6-1-7), we see that v is the minimum value of E.

Occasionally, it is more convenient to have the essential equations in partitioned matrix form. In partitioned matrix form, we have for the error (6-1-6)

$$E = [1 \mid \mathbf{x}]^T\left[\begin{array}{c|c} -\mathbf{c}^T \\ \hline \mathbf{A}^T \end{array}\right]\left[-\mathbf{c} \mid \mathbf{A}\right]\begin{bmatrix} 1 \\ \hline \mathbf{x} \end{bmatrix} \qquad (6\text{-}1\text{-}16)$$

The final equation (6-1-14) splits into

$$V = \mathbf{c}^T\mathbf{c} - \mathbf{c}^T\mathbf{A}\mathbf{x} \qquad (6\text{-}1\text{-}17)$$

$$0 = -\mathbf{A}^T\mathbf{c} + \mathbf{A}^T\mathbf{A}\mathbf{x} \qquad (6\text{-}1\text{-}18)$$

where (6-1-18) represents simultaneous equations to be solved for $\mathbf{x}$. Equation (6-1-18) is what you have to set up in a computer. It is easily remembered by a quick and dirty (very dirty) derivation. That is, we began with the overdetermined equations $\mathbf{A}\mathbf{x} \approx \mathbf{c}$; premultiplying by $\mathbf{A}^T$ gives $(\mathbf{A}^T\mathbf{A})\mathbf{x} = \mathbf{A}^T\mathbf{c}$ which is (6-1-18).

In physical science applications, the variable z_j is frequently a complex variable, say $z_j = x_j + iy_j$. It is always possible to go through the foregoing analyses, treating the problem as though x_i and y_i were real independent variables. There is a considerable gain in simplicity and a saving in computational effort by treating z_j as a single complex variable. The error E may be regarded as a function of either x_j and y_j or z_j and $\bar{z}_j$. In general $j = 1, 2, \ldots, N$, but we will treat the case $N = 1$ here and leave the general case for the Exercises. The minimum is found where

$$0 = \frac{dE}{dx} = \frac{\partial E}{\partial z}\frac{dz}{dx} + \frac{\partial E}{\partial \bar{z}}\frac{d\bar{z}}{dx} = \frac{\partial E}{\partial z} + \frac{\partial E}{\partial \bar{z}} \qquad (6\text{-}1\text{-}19)$$

$$0 = \frac{dE}{dy} = \frac{\partial E}{\partial z}\frac{dz}{dy} + \frac{\partial E}{\partial \bar{z}}\frac{d\bar{z}}{dy} = i\left(\frac{\partial E}{\partial z} - \frac{\partial E}{\partial \bar{z}}\right) \qquad (6\text{-}1\text{-}20)$$

Multiplying (6-1-20) by i and adding and subtracting these equations, we may express the minimum condition more simply as

$$0 = \frac{\partial E}{\partial z} \qquad (6\text{-}1\text{-}21)$$

$$0 = \frac{\partial E}{\partial \bar{z}} \qquad (6\text{-}1\text{-}22)$$

However, the usual case is that E is a positive real quadratic function of z and $\bar{z}$ and that $\partial E/\partial z$ is merely the complex conjugate of $\partial E/\partial \bar{z}$. Then the two conditions (6-1-21) and (6-1-22) may be replaced by either one of them. Usually, when working with complex variables we are minimizing a positive quadratic form like

$$E(z^*, z) = |\mathbf{A}z - \mathbf{c}|^2 = (z^*\mathbf{A}^* - \mathbf{c}^*)(\mathbf{A}z - \mathbf{c}) \qquad (6\text{-}1\text{-}23)$$

where $*$ denotes complex-conjugate transpose. Now (6-1-22) gives

$$0 = \frac{\partial E}{\partial z^*} = \mathbf{A}^*(\mathbf{A}z - \mathbf{c}) \qquad (6\text{-}1\text{-}24)$$

which is just the complex form of (6-1-18).

Let us consider an example. Suppose a set of wave arrival times t_i is measured at sensors located on the x axis at points x_i. Suppose the wavefront is to be fitted to a parabola $t_i \approx a + bx_i + cx_i^2$. Here, the x_i are knowns and a, b, and c are unknowns. For each sensor i we have an equation

$$[-t_i \quad 1 \quad x_i \quad x_i^2] \begin{bmatrix} 1 \\ a \\ b \\ c \end{bmatrix} \approx 0 \qquad (6\text{-}1\text{-}25)$$

When i has greater range than 3 we have more equations than unknowns. In this example, (6-1-14) takes the form

$$\left\{ \sum_{i=1}^{n} \begin{bmatrix} -t_i \\ 1 \\ x_i \\ x_i^2 \end{bmatrix} [-t_i \quad 1 \quad x_i \quad x_i^2] \right\} \begin{bmatrix} 1 \\ a \\ b \\ c \end{bmatrix} = \begin{bmatrix} v \\ 0 \\ 0 \\ 0 \end{bmatrix} \qquad (6\text{-}1\text{-}26)$$

This may be solved by standard methods for a, b, and c.

The last three rows of (6-1-26) may be written

$$\sum_{i=1}^{n} \begin{bmatrix} 1 \\ x_i \\ x_i^2 \end{bmatrix} e_i = \begin{bmatrix} 0 \\ 0 \\ 0 \end{bmatrix} \qquad (6\text{-}1\text{-}27)$$

This says the error vector e_i is perpendicular (or normal) to the functions 1, x, and x^2, which we are fitting to the data. For that reason these equations are often called normal equations.

EXERCISES

1 Extend (6-1-24) by fitting waves observed in the x, y plane to a two-dimensional quadratic.

2 Let $y(t)$ constitute a complex-valued function at successive integer values of t. Fit $y(t)$ to a least-squares straight line $y(t) \approx \alpha + \beta t$ where $\alpha = \alpha_r + i\alpha_i$ and $\beta = \beta_r + i\beta_i$. Do it two ways: (a) Assume α_r, α_i, β_i, and β_r are four independent variables, and (b) Assume α, $\bar{\alpha}$, β, and $\bar{\beta}$ are independent variables. (Leave answer in terms of $s_n = \sum_i t^n$.)

3 Equation (6-1-14) has assumed all quantities are real. Generalize equation (6-1-14) to all complex quantities. Verify that the matrix is Hermitian.

4 At the jth seismic observatory (latitude x_j, longitude y_j) earthquake waves are observed to arrive at time t_j. It has been conjectured that the earthquake has an origin time t, latitude x, and longitude y. The theoretical travel time may be looked up in a travel time table $T(\Delta)$ where T is the travel time and Δ is the great circle angle. One has

$$\cos \Delta = \sin y \sin y_i + \cos y \cos y_i \cos (x - x_i)$$

The time residual at the jth station, supposing that the earthquake occurred at (x, y, t), is

$$e_j = t + T(\Delta_j) - t_j$$

The time residual, supposing that the earthquake occurred at $(x + dx, y + dy, t + dt)$, is

$$e_J = t + dt + T(\Delta_J) + \left(\frac{\partial T}{\partial \Delta} \frac{\partial \Delta}{\partial x}\right)_J dx + \left(\frac{\partial T}{\partial \Delta} \frac{\partial \Delta}{\partial y}\right)_J dy - t_J$$

Find equations to solve for dx, dy, and dt which minimize the sum-squared time residuals.

5 Gravity g_J has been measured at N irregularly spaced points on the surface of the earth (colatitude x_J, longitude y_J, $j = 1, N$). Show that the matrix of the normal equation which fits the data to spherical harmonics may be written as a sum of a column times its transpose, as in the preceding problem. How would the matrix simplify if there were infinitely many uniformly spaced data points? (NOTE: Spherical harmonics S are the class of functions

$$S_n^m(x, y) = P_n^m(\cos x) \exp{(imy)}$$

for $(m = -n, \ldots, -1, 0, 1, \ldots, n)$ and $(n = 0, 1, \ldots, \infty)$ where P_n^m is an associated Legendre polynomial of degree n and order m.

6 Ocean tides fit sinusoidal functions of known frequencies quite accurately. Associated with the tide is an earth tilt. A complex time series may be made from the north-south tilt plus $\sqrt{-1}$ times the east-west tilt. The observed complex time series may be fitted to an analytical form $\sum_{j=1}^{N} A_j e^{i\omega_j t}$. Find a set of equations which may be solved for the A_j which gives the best fit of the formula to the data. Show that some elements of the normal equation matrix are sums which may be summed analytically.

7 The general solution to Laplace's equation in cylindrical coordinates (r, θ) for a potential field P which vanishes at $r = \infty$ is given by

$$P(r, \theta) = \text{Re} \sum_{m=0}^{\infty} A_m \frac{e^{im\theta}}{r^{m+1}}$$

Find the potential field surrounding a square object at the origin which is at unit potential. Do this by finding N of the coefficients A_m by minimizing the squared difference between $P(r, \theta)$ and unity integrated around the square. Give the answer in terms of an inverse matrix of integrals. Which coefficients A_m vanish exactly by symmetry?

6-2 WEIGHTS AND CONSTRAINTS

It often happens that some observations are considered more reliable than others. One may desire to weight the more reliable data more heavily in the calculation. In other words, we may multiply the ith equation by a weight $\sqrt{w_i}$

$$\sqrt{w_i}[-c_i \quad a_{i1} \quad a_{i2} \quad \cdots] \begin{bmatrix} 1 \\ x_1 \\ x_2 \\ \vdots \end{bmatrix} = \sqrt{w_i} e_i \qquad \text{(6-2-1)}$$

Now the weighted sum-squared error will be

$$E = \sum_i w_i e_i^2 \qquad \text{(6-2-2)}$$

Following the method of the last section, it is easy to show that the $\mathbf{x}$ which minimizes the weighted error E of (6-2-2) is the $\mathbf{x}$ which satisfies the simultaneous equations

$$\left\{ \sum_i w_i \begin{bmatrix} -c_i \\ a_{i1} \\ a_{i2} \\ \vdots \end{bmatrix} \begin{bmatrix} -c_i & a_{i1} & a_{i2} & \cdots \end{bmatrix} \right\} \begin{bmatrix} 1 \\ x_1 \\ x_2 \\ \vdots \end{bmatrix} = \begin{bmatrix} v \\ 0 \\ 0 \\ \vdots \end{bmatrix} \qquad (6\text{-}2\text{-}3)$$

Choice of a set of weights is often a rather subjective matter. However, if data are of uneven quality, it cannot be avoided. Omitting w is equivalent to choosing it equal to unity.

A case of common interest is where some equations should be solved exactly. Such equations are called constraint equations. Constraint equations often arise out of theoretical considerations so they may, in principle, not have any error. The rest of the equations often involve some measurement. Since the measurement can often be made many times, it is easy to get a lot more equations than unknowns Since measurement always involves error, we then use the method of least squares to minimize the average error. In order to be certain that the constraint equations are solved exactly, one could use the trick of applying very large weight factors to the constraint equations. A problem is that "very large" is not well defined. A weight equal 10^{10} might not be large enough to guarantee the constraint equation is satisfied with sufficient accuracy. On the other hand, 10^{10} might lead to disastrous round-off when solving the simultaneous equations in a computer with eight-digit accuracy. The best approach is to analyze the situation theoretically for $w \to \infty$.

An example of a constraint equation is that the sum of the x_i equals M. Another constraint would be $x_1 = x_2$. Arranged in a matrix, these two constraint equations are

$$\begin{bmatrix} -M & 1 & 1 & 1 & 1 & \cdots \\ 0 & 1 & -1 & 0 & 0 & \cdots \end{bmatrix} \begin{bmatrix} 1 \\ x_1 \\ x_2 \\ \vdots \end{bmatrix} = \begin{bmatrix} 0 \\ 0 \end{bmatrix} \qquad (6\text{-}2\text{-}4)$$

We write a general set of k constraint equations as

$$\mathbf{G} \begin{bmatrix} 1 \\ x_1 \\ x_2 \\ \vdots \end{bmatrix} = \begin{bmatrix} 0 \\ \vdots \end{bmatrix} \qquad (6\text{-}2\text{-}5)$$

Minimizing the error as $w \to \infty$ of the equations

$$\sqrt{w}\mathbf{G}\mathbf{x} \approx 0$$
$$\mathbf{B}\mathbf{x} \approx 0$$

is algebraically similar to minimizing the error of $\mathbf{B}\mathbf{x} \approx \mathbf{0}$. The rows of $\sqrt{w}\mathbf{G}$ are just like some extra rows for $\mathbf{B}$. The resulting equation for $\mathbf{x}$ is

$$\left\{ \sum_{i=1}^{n} \begin{bmatrix} -c_i \\ a_{i1} \\ \vdots \end{bmatrix} [-c_i \quad a_{i1} \quad \cdots] + \sum_{i=1}^{k} w_i \begin{bmatrix} g_{i0} \\ g_{i1} \\ \vdots \end{bmatrix} [g_{i0} \quad g_{i1} \quad \cdots] \right\} \begin{bmatrix} 1 \\ x_1 \\ \vdots \end{bmatrix} = \begin{bmatrix} v \\ 0 \\ \vdots \end{bmatrix} \qquad (6\text{-}2\text{-}6)$$

Now we will take all the w_i to equal $1/\varepsilon$ and we will let ε tend to zero. Also let

$$\mathbf{x} = \mathbf{x}^{(0)} + \varepsilon \mathbf{x}^{(1)} + \varepsilon^2 \mathbf{x}^{(2)} + \cdots \qquad (6\text{-}2\text{-}7a)$$

$$\mathbf{v} = \mathbf{v}^{(0)} + \varepsilon \mathbf{v}^{(1)} + \varepsilon^2 \mathbf{v}^{(2)} + \cdots \qquad (6\text{-}2\text{-}7b)$$

With this, (6-2-6) may be written

$$\left(\mathbf{B}^T \mathbf{B} + \frac{1}{\varepsilon} \mathbf{G}^T \mathbf{G} \right) (\mathbf{x}^{(0)} + \varepsilon \mathbf{x}^{(1)} + \cdots) = \mathbf{v}^{(0)} + \mathbf{v}^{(1)} \varepsilon + \cdots \qquad (6\text{-}2\text{-}8)$$

Identify coefficients of powers of ε

$$\varepsilon^{-1}: \qquad \mathbf{G}^T \mathbf{G} \mathbf{x}^{(0)} = 0 \qquad (6\text{-}2\text{-}9a)$$

$$\varepsilon^0: \qquad \mathbf{B}^T \mathbf{B} \mathbf{x}^{(0)} + \mathbf{G}^T \mathbf{G} \mathbf{x}^{(1)} = \mathbf{v}^{(0)} \qquad (6\text{-}2\text{-}9b)$$

$$\varepsilon^1, \varepsilon^2: \qquad \text{not required}$$

Equation (6-2-9a) is m equations in m unknowns. It will automatically be satisfied if the k equations in (6-2-5) are satisfied. Equation (6 2-9b) appears to involve the m unknowns in $\mathbf{x}^{(0)}$ plus m more unknowns in $\mathbf{x}^{(1)}$. In fact, we do not need $\mathbf{x}^{(1)}$; the k unknowns

$$\boldsymbol{\lambda} = \mathbf{G} \mathbf{x}^{(1)} \qquad (6\text{-}2\text{-}10)$$

will suffice.

Arranging (6-2-9b) and (6-2-5) together and dropping superscripts, we get a square matrix in $m + k$ unknowns.

$$\left[\begin{array}{c|c} \mathbf{B}^T \mathbf{B} & \mathbf{G}^T \\ \hline \mathbf{G} & 0 \end{array} \right] \left[\begin{array}{c} \mathbf{x} \\ \hline \boldsymbol{\lambda} \end{array} \right] = \begin{bmatrix} v \\ 0 \\ \vdots \end{bmatrix} \qquad (6\text{-}2\text{-}11)$$

Equation (6-2-11) is now a simultaneous set for the unknowns $\mathbf{x}$ and $\boldsymbol{\lambda}$. It might also be thought of as the solution to the problem of minimizing the quadratic form

$$E = [\mathbf{x}^T \quad \boldsymbol{\lambda}^T] \begin{bmatrix} \mathbf{B}^T \mathbf{B} & \mathbf{G}^T \\ \mathbf{G} & 0 \end{bmatrix} \begin{bmatrix} \mathbf{x} \\ \boldsymbol{\lambda} \end{bmatrix}$$

$$= \mathbf{x}^T \mathbf{B}^T \mathbf{B} \mathbf{x} + \boldsymbol{\lambda}^T \mathbf{G} \mathbf{x} + \mathbf{x}^T \mathbf{G}^T \boldsymbol{\lambda}$$

and since we can always transpose a scalar,

$$E = \mathbf{x}^T \mathbf{B}^T \mathbf{B} \mathbf{x} + 2 \boldsymbol{\lambda}^T \mathbf{G} \mathbf{x} \qquad (6\text{-}2\text{-}12)$$

According to the method of Lagrange multipliers, one may minimize a quadratic form subject to constraints by minimizing instead a sum of the quadratic form plus constraint terms where each constraint term is the product of a constraint equation multiplied by a Lagrange multiplier λ_i. This is precisely what we have in (6-2-12), and the solution is given by (6-2-11). Lagrange multipliers frequently arise in connection with integral equations. The concept is readily transformed to matrices merely by approximating integration by summation.

EXERCISE

1 In determining a density *vs.* depth profile of the earth one might minimize the squared difference between some theoretical quantities (say, the frequencies of free oscillation) and the observed quantities. By astronomical means, total mass and moment of inertia of the earth are very well known. If the earth is divided into arbitrarily thin shells of equal thickness, what are the two astronomical constraint equations on the layer densities ρ_i? If the least-squares problem is nonlinear (as it often is) it may be linearized by assuming that a given set of densities ρ_i is a good guess which satisfies the constraints and doing least squares for the perturbation $d\rho_i$. What are the constraint equations on $d\rho_i$?

6-3 FEWER EQUATIONS THAN UNKNOWNS

What is one to do when one has fewer equations than unknowns: give up? Certainly not, just apply the principle of simplicity. Let us find the simplest solution which satisfies all the equations. This situation often arises. Suppose, after having made a finite number of measurements one is trying to determine a continuous function, for example, the mass density $\rho(r)$ as a function of depth in the earth. Then, in a computer $\rho(r)$ would be represented by $\rho(r)$ sampled at N depths $r_i, i = 1, 2, \ldots, N$. Then merely by taking N large, one has more unknowns than equations.

One measure of simplicity is that the unknown function x_i has minimum wiggliness. In other words minimize

$$E = \sum (x_i - x_{i-1})^2 \qquad (6\text{-}3\text{-}1)$$

subject to satisfying exactly the observation or constraint equations

$$\mathbf{Gx} = \mathbf{0} \qquad (6\text{-}3\text{-}2)$$

Another more popular measure of simplicity (which does not imply an ordering of the variables x_i) is the minimization of

$$E = \sum_{i=1}^{n} x_i^2 \qquad (6\text{-}3\text{-}3)$$

If we set out to minimize (6-3-3) without any constraints, x would satisfy the simultaneous equations

$$\begin{bmatrix} 1 & & & \text{zeros} \\ & 1 & & \\ & & 1 & \\ \text{zeros} & & & \ddots \end{bmatrix} \begin{bmatrix} 1 \\ x_1 \\ x_2 \\ \vdots \end{bmatrix} = \begin{bmatrix} v \\ 0 \\ 0 \\ 0 \end{bmatrix}$$

By inspection one sees the obvious result that $x_i = 0$. Now let us include two constraint equations and, for definiteness, take three unknowns. The method of the previous section gives

$$\left[\begin{array}{cccc|cc} 1 & & & & -d_1 & -d_2 \\ & 1 & & & g_{11} & g_{21} \\ & & 1 & & g_{12} & g_{22} \\ & & & 1 & g_{13} & g_{23} \\ \hline -d_1 & g_{11} & g_{12} & g_{13} & 0 & 0 \\ -d_2 & g_{21} & g_{22} & g_{23} & 0 & 0 \end{array}\right] \left[\begin{array}{c} 1 \\ x_1 \\ x_2 \\ x_3 \\ \hline \lambda_1 \\ \lambda_2 \end{array}\right] = \left[\begin{array}{c} v \\ 0 \\ 0 \\ 0 \\ \hline 0 \\ 0 \end{array}\right] \qquad (6\text{-}3\text{-}4)$$

Equation (6-3-4) has a size equal to the number of variables plus the number of constraints. It may be solved numerically or it may be first reduced to a matrix whose size is given by the number of constraints. Let us split up (6-3-4) into two equations:

$$\left[\begin{array}{c} x_1 \\ x_2 \\ x_3 \end{array}\right] + \left[\begin{array}{cc} g_{11} & g_{21} \\ g_{12} & g_{22} \\ g_{13} & g_{23} \end{array}\right] \left[\begin{array}{c} \lambda_1 \\ \lambda_2 \end{array}\right] = \left[\begin{array}{c} 0 \\ 0 \\ 0 \end{array}\right] \qquad (6\text{-}3\text{-}5)$$

and

$$\left[\begin{array}{ccc} g_{11} & g_{12} & g_{13} \\ g_{21} & g_{22} & g_{23} \end{array}\right] \left[\begin{array}{c} x_1 \\ x_2 \\ x_3 \end{array}\right] = \left[\begin{array}{c} d_1 \\ d_2 \end{array}\right] \qquad (6\text{-}3\text{-}6)$$

We abbreviate these equations by $\mathbf{x} + \mathbf{G}^T\boldsymbol{\lambda} = 0$ and $\mathbf{G}\mathbf{x} = \mathbf{d}$. Premultiply (6-3-5) by $\mathbf{G}$,

$$\mathbf{G}\mathbf{x} + \mathbf{G}\mathbf{G}^T\boldsymbol{\lambda} = 0$$

insert (6-3-6)

$$\mathbf{d} + \mathbf{G}\mathbf{G}^T\boldsymbol{\lambda} = 0$$

solve for $\boldsymbol{\lambda}$

$$\boldsymbol{\lambda} = -(\mathbf{G}\mathbf{G}^T)^{-1}\mathbf{d}$$

put back into (6-3-5)

$$\mathbf{x} = \mathbf{G}^T(\mathbf{G}\mathbf{G}^T)^{-1}\mathbf{d}$$

Written out in full this is

$$\left[\begin{array}{c} x_1 \\ x_2 \\ x_3 \end{array}\right] = \left[\mathbf{G}^T\right]\left[\mathbf{G}\mathbf{G}^T\right]^{-1}\left[\mathbf{d}\right] \qquad (6\text{-}3\text{-}7)$$

This is the final result, a minimum wiggliness solution $\mathbf{x}$ which exactly satisfies an underdetermined set called the constraint equations.

EXERCISES

1 If wiggliness is defined by (6-3-1) instead of (6-3-3), what form does (6-3-7) take?
2 Given the mass and moment of intertia of the earth, calculate mass density as a function of depth utilizing the principle of minimum wiggliness (6-3-7). What criticism do you have of this procedure? (HINT: An elegant solution uses integrals instead of infinite sums.)
3 Use the techniques of this section on (6-2-11) to reduce the size of the matrices to be inverted.

6-4 HOUSEHOLDER TRANSFORMATIONS AND GOLUB'S METHOD [Ref. 21]

Our previous discussions of least squares always led us to matrices of the form $\mathbf{A}^T\mathbf{A}$ which then needed to be inverted. Golub's method of using Householder transformations works directly with the matrix $\mathbf{A}$ and has the advantage that it is considerably more accurate than methods which invert $\mathbf{A}^T\mathbf{A}$. It seems that about twice as much precision is required to invert $\mathbf{A}^T\mathbf{A}$ than is needed to deal directly with $\mathbf{A}$. Another reason for learning about Golub's method is that the calculation is organized in a completely different way; therefore, it will often turn out to have other advantages or disadvantages which differ from one application to the next.

A reflection transformation is a matrix of the form $\mathbf{R} = (\mathbf{I} - 2\mathbf{v}\mathbf{v}^T/\mathbf{v}^T\mathbf{v})$ where $\mathbf{v}$ is an arbitrary vector. Obviously $\mathbf{R}$ is symmetric, that is, $\mathbf{R} = \mathbf{R}^T$. It also turns out that the reflection transformation is its own inverse, that is, $\mathbf{R} = \mathbf{R}^{-1}$. To see this, we verify by substitution that $\mathbf{R}^2 = \mathbf{I}$.

$$\left(\mathbf{I} - \frac{2\mathbf{v}\mathbf{v}^T}{\mathbf{v}^T\mathbf{v}}\right)^2 = \mathbf{I} - \frac{4\mathbf{v}\mathbf{v}^T}{\mathbf{v}^T\mathbf{v}} + \frac{4\mathbf{v}(\mathbf{v}^T\mathbf{v})\mathbf{v}^T}{(\mathbf{v}^T\mathbf{v})^2} \quad (6\text{-}4\text{-}1)$$
$$= \mathbf{I}$$

A matrix transformation $\mathbf{M}$ is said to be *unitary* if $\mathbf{M}^T\mathbf{M} = \mathbf{I}$. When a matrix $\mathbf{M}$ is unitary it means that the vector $\mathbf{x}$ has the same length as the vector $\mathbf{M}\mathbf{x}$. These lengths are $\mathbf{x}^T\mathbf{x}$ and $(\mathbf{M}\mathbf{x})^T(\mathbf{M}\mathbf{x}) = \mathbf{x}^T\mathbf{M}^T\mathbf{M}\mathbf{x} = \mathbf{x}^T\mathbf{I}\mathbf{x} = \mathbf{x}^T\mathbf{x}$ which are the same. Reflection transformations are unitary because $\mathbf{R}^{-1} = \mathbf{R}^T$. They have a simple physical interpretation. Consider an orthogonal coordinate system in which one of the coordinate axes is aligned along the $\mathbf{v}$ vector. Reflection transformation reverses the sign of this coordinate axis vector (since $\mathbf{R}\mathbf{v} = -\mathbf{v}$) but it leaves unchanged all the other coordinate axis vectors. Thus it is obvious geometrically that reflection transformations preserve lengths and that applying the transformation twice returns any original vector to itself. Now, we seek a special reflection transformation called the Householder transformation which converts a matrix of the form on the left to the form on the right where a is an arbitrary element

$$\mathbf{H}\begin{bmatrix} a & a & a & a \\ 0 & a & a & a \\ 0 & 0 & a & a \\ 0 & 0 & a & a \\ 0 & 0 & a & a \end{bmatrix} = \begin{bmatrix} a & a & a & a \\ 0 & a & a & a \\ 0 & 0 & a & a \\ 0 & 0 & 0 & a \\ 0 & 0 & 0 & a \end{bmatrix} \quad (6\text{-}4\text{-}2)$$

Having determined the required transformation, we will know how to convert any matrix to an upper triangular form like

$$
\begin{bmatrix}
a & a & a & a \\
0 & a & a & a \\
0 & 0 & a & a \\
0 & 0 & 0 & a \\
0 & 0 & 0 & 0 \\
0 & 0 & 0 & 0
\end{bmatrix}
\qquad (6\text{-}4\text{-}3)
$$

by a succession of Householder transforms. Golub recognized the value of this technique in solving overdetermined sets of simultaneous equations. He noted that when the error vector $\mathbf{e} = \mathbf{Ax} - \mathbf{b}$ is transformed by a unitary matrix $\mathbf{Ue}$ the problem of minimizing the length $(\mathbf{e}^T\mathbf{U}^T\mathbf{Ue})^{1/2}$ of $\mathbf{Ue}$ by variation of $\mathbf{x}$ reduces to exactly the same problem as minimizing the length $(\mathbf{e}^T\mathbf{e})^{1/2}$ of $\mathbf{e}$ with respect to variation of $\mathbf{x}$. Thus a succession of Householder transforms could be found to reduce $\mathbf{e} = \mathbf{Ax} - \mathbf{b}$ to the form

$$
\begin{bmatrix}
\mathbf{e}_1 \\
\text{-----} \\
\mathbf{e}_2
\end{bmatrix}
=
\begin{bmatrix}
a & a & a \\
0 & a & a \\
0 & 0 & a \\
0 & 0 & 0 \\
0 & 0 & 0 \\
0 & 0 & 0
\end{bmatrix}
\begin{bmatrix}
x_1 \\
x_2 \\
x_3
\end{bmatrix}
-
\begin{bmatrix}
a \\
a \\
a \\
a \\
a \\
a
\end{bmatrix}
\qquad (6\text{-}4\text{-}4)
$$

Now for the clever observation that because of the zeros in the bottom part of the transformed $\mathbf{A}$ matrix there is no possibility of choosing any x_i values which alter $\mathbf{e}_2$ in any way. The top part of the transformed $\mathbf{A}$ matrix is an upper triangular matrix which for any value of $\mathbf{e}_1$ can be solved exactly for the x_i. The least-squares solution x_i is the one for which $\mathbf{e}_1$ has been set equal to zero.

Now we return to the task of finding the special reflection transformation, called the Householder transformation, which accomplishes (6-4-2). Observe that the left-hand operator below is a reflection transformation for any numerical choice of s.

$$
\left\{
\begin{bmatrix}
1 & & & & \\
& 1 & & & \\
& & 1 & & \\
& & & 1 & \\
& & & & 1
\end{bmatrix}
-
\frac{2}{(a_3 - s)^2 + a_4^2 + a_5^2}
\begin{bmatrix}
0 \\
0 \\
a_3 - s \\
a_4 \\
a_5
\end{bmatrix}
\begin{bmatrix} 0 & 0 & (a_3 - s) & a_4 & a_5 \end{bmatrix}
\right\}
\begin{bmatrix}
a_1 \\
a_2 \\
a_3 \\
a_4 \\
a_5
\end{bmatrix}
$$

$$
=
\begin{bmatrix}
a_1 \\
a_2 \\
a_3 \\
a_4 \\
a_5
\end{bmatrix}
-
\begin{bmatrix}
0 \\
0 \\
(a_3 - s) \\
a_4 \\
a_5
\end{bmatrix}
=
\begin{bmatrix}
a_1 \\
a_2 \\
s \\
0 \\
0
\end{bmatrix}
\qquad (6\text{-}4\text{-}5)
$$

Alternatively, if (6-4-5) is to be valid, then s must take a particular value such that

$$1 = \frac{2}{(a_3 - s)^2 + a_4{}^2 + a_5{}^2} \begin{bmatrix} 0 & 0 & (a_3 - s) & a_4 & a_5 \end{bmatrix} \begin{bmatrix} a_1 \\ a_2 \\ a_3 \\ a_4 \\ a_5 \end{bmatrix} \qquad (6\text{-}4\text{-}6)$$

or

$$1 = \frac{-2sa_3 + 2(a_3{}^2 + a_4{}^2 + a_5{}^2)}{s^2 - 2sa_3 + (a_3{}^2 + a_4{}^2 + a_5{}^2)} \qquad (6\text{-}4\text{-}7)$$

This will be true only for s given by

$$s = \pm (a_3{}^2 + a_4{}^2 + a_5{}^2)^{1/2} \qquad (6\text{-}4\text{-}8)$$

Now let us see why the left-hand operator in (6-4-5) can achieve (6-4-2). Choice of the a vector as the third column in the matrix of (6-4-2) introduces the desired zeros on the right-hand side. Finally, it is necessary also to observe that this choice of H does not destroy any of the zeros which already existed on the left-hand side in (6-4-2). A subroutine for this task is in Fig. 6-1. Householder transformations can also be used in problems with constraints. In the set

$$\begin{bmatrix} C \\ A \end{bmatrix} [x] = \begin{bmatrix} d \\ b \end{bmatrix} \qquad (6\text{-}4\text{-}9)$$

one may desire to satisfy the top block exactly and the bottom block only in the least-squares sense. Define y as a succession of Householder transforms on x; for example, $y = H_2 H_1 x$. Then substitute $x = H_1 H_2 H_2 H_1 x = H_1 H_2 y$ into (6-4-9). Householder transforms used as postmultipliers on the matrix of (6-4-9) can be chosen to introduce zeros in the top two rows of (6-4-9), for example

$$\begin{bmatrix} a & 0 & 0 & 0 \\ a & a & 0 & 0 \\ a & a & a & a \\ a & a & a & a \\ a & a & a & a \end{bmatrix} \begin{bmatrix} y_1 \\ y_2 \\ y_3 \\ y_4 \end{bmatrix} \approx \begin{bmatrix} d_1 \\ d_2 \\ b_1 \\ b_2 \\ b_3 \end{bmatrix} \qquad (6\text{-}4\text{-}10)$$

Now we could use premultiplying Householder transforms on (6-4-10) to bring it to the form

$$\begin{bmatrix} a & 0 & 0 & 0 \\ a & a & 0 & 0 \\ a & a & a & a \\ a & a & 0 & a \\ a & a & 0 & 0 \end{bmatrix} \begin{bmatrix} y_1 \\ y_2 \\ y_3 \\ y_4 \end{bmatrix} \approx \begin{bmatrix} d_1 \\ d_2 \\ b_1 \\ b_2 \\ b_3 \end{bmatrix} \qquad (6\text{-}4\text{-}11)$$

Since the top two equations of (6-4-10) or of (6-4-11) are to be satisfied exactly, then y_1 and y_2 are uniquely determined. They cannot be adjusted to help attain minimum

```
      SUBROUTINE GOLUB  (A,X,B,M,N)
C
C     A(M,N) ; B(M)  GIVEN WITH M>N   SOLVES FOR X(N) SUCH THAT
C                || B - AX || = MINIMUM
C     METHOD OF G.GOLUB, NUMERISCHE MATHEMATIK 7,206-216 (1965)
C
      IMPLICIT DOUBLE PRECISION (D)
      REAL A(M,N),X(N),B(M),U(50)
C......DIMENSION U(M)
C......PERFORM N ORTHOGONAL TRANSFORMATIONS TO A(.,.) TO
C......UPPER TRIANGULARIZE THE MATRIX
      DO 3010 K=1,N
      DSUM=0.0D0
      DO 1010 I=K,M
      DAJ=A(I,K)
 1010 DSUM=DSUM+DAJ**2
      DAI=A(K,K)
      DSIGMA=DSIGN(DSQRT(DSUM),DAI)
      DBI=DSQRT(1.0D0+DAI/DSIGMA)
      DFACT=1.0D0/(DSIGMA*DBI)
      U(K)=DBI
      FACT=DFACT
      KPLUS=K+1
      DO 1020 I=KPLUS,M
 1020 U(I)=FACT*A(I,K)
C.......I - UU' IS A SYMMETRIC, ORTHOGONAL MATRIX WHICH WHEN APPLIED
C....... TO A(.,.) WILL ANNIHILATE THE ELEMENTS BELOW THE DIAGONAL K
      DO 2030 J=K,N
c.......APPLY THE ORTHOGONAL TRANSFORMATION
      FACT=0.0
      DO 2010 I=K,M
 2010 FACT=FACT+U(I)*A(I,J)
      DO 2020 I=K,M
 2020 A(I,J)=A(I,J)-FACT*U(I)
 2030 CONTINUE
      FACT=0.0
      DO 2040 I=K,M
 2040 FACT=FACT+U(I)*B(I)
      DO 2050 I=K,M
 2050 B(I)=B(I)-FACT*U(I)
 3010 CONTINUE
C.......BACK SUBSTITUTE TO RECURSIVELY YIELD X(.)
      X(N)=B(N)/A(N,N)
      LIM=N-1
      DO 4020 I=1,LIM
      IROW=N-1
      SUM=0.0
      DO 4010 J=1,I
 4010 SUM=SUM+X(N-J+1)*A(IROW,N-J+1)
 4020 X(IROW)=(B(IROW)-SUM)/A(IROW,IROW)
      RETURN
      END
```

FIGURE 6-1
Subroutine for least squares fitting. Programmed by Don C. Riley. Note that this program does not do the square matrix case. It is necessary that $M > N$.

error in the bottom three equations. Likewise the top two equations place no restraint on y_3 and y_4, so they may be adjusted to produce minimum error in the bottom three equations. No amount of adjustment in y_3 and y_4 can change the amount of error in the last equation, so we can ignore the last equation in the determination of y_3 and y_4. The third and fourth equations can be satisfied with zero error by suitable choice of y_3 and y_4. This must be the minimum-squared-error answer. Given **y** we can go back and get **x** with $\mathbf{x} = \mathbf{H}_1 \mathbf{H}_2 \mathbf{y}$.

6-5 CHOICE OF A MODEL NORM

In recent years, a popular view of geophysical data modeling has been that the earth is a continuum and that we should regard the number of unknowns as infinite but the number of our observations as finite. In a computer one therefore approximates the situation by a highly underdetermined system of simultaneous equations. In order to get a unique answer, the solution should extremalize some integral. In practice, a sum of squares is often minimized in such a way as to produce a smooth solution. A typical mathematical formulation is to do a least-squares fitting of an equation set like

$$\begin{bmatrix} A \\ \diagdown \end{bmatrix} \mathbf{x} \approx \begin{bmatrix} \mathbf{d} \\ \mathbf{0} \end{bmatrix} \tag{6-5-1}$$

where the top block **A** denotes the underdetermined constraint equations with the data vector **d**; the unknowns are in the **x** vector; and the bottom block is a band matrix (all the nonzero elements cluster about the diagonal) which says that some filtered version of **x** should vanish. The filter is often a roughing filter like the first-difference operator. In the absence of data, the first-difference operator leads to a constant solution which is sensible. What is not sensible is that it forces the result to be smoothed even though realistic earth models often contain step discontinuities (where two homogeneous media lie in contact).

The choice of a filter is a rather subjective matter, the choice often being made on the basis of the solution it will produce. Unfortunately, the solution is often a rather sensitive function of the subjectively chosen weights and filters; this fact makes the whole business an art, a matter of experience and judgment. General-purpose theories of inversion exist, but they do not prepare the geophysicist to exploit the peculiarities of any particular stituation or data set. Inversion theories, like mathematical statistics, should be used like a lamp post—to light the way, not to lean upon.

One useful concept in inversion theory is the idea of coordinate-system invariance. The idea is that one should get the same answer in an electrical conductivity problem whether one parameterizes the earth by an unknown conductivity at every point on a sufficiently dense mesh, or one parameterizes with resistivity on the mesh, or one parameterizes by coefficients of some expansion in a complete set of basis functions. Clearly, the idea of fitting low-order coefficients in some expansion setting the high-order coefficients equal to zero is not a coordinate-invariant approach. A different origin for polynomial expansions can change everything. A different set of basis functions would change everything. Of course, it is not essential to use a coordinate-system invariant technique in data inversion. But if one does not, one should beware of the sensitivity of one's solution to changes in the coordinate system.

Let us consider some inversion procedures which are coordinate-system invariant. We will restrict ourselves to physical problems in which we can identify a positive density function p as energy or dissipation per unit volume. Let us denote

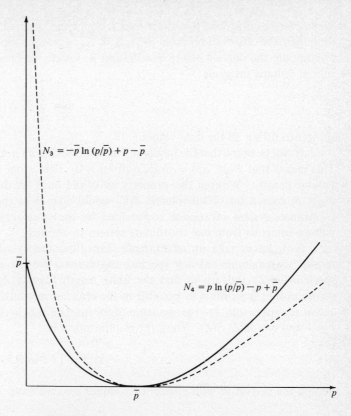

FIGURE 6-2
Minimizing either of these two functions will drive p toward $\bar{p}$.

by $\bar{p}$ the value of this power as a function of space in the default model of the earth. The default model is the one we want to find when we have no measurements. It will often be one in which the material properties are constant functions of space. Now we will need some functions which we will call *model norms*. They have the properties of being positive for all (positive) p and $\bar{p}$ and being minimized at $p = \bar{p}$. Some examples plotted in Fig. 6-2 are

$$N_1(p, \bar{p}) = |p - \bar{p}|$$

$$N_2(p, \bar{p}) = \frac{(p - \bar{p})^2}{\bar{p}}$$

$$N_3(p, \bar{p}) = -\bar{p} \ln\left(\frac{p}{\bar{p}}\right) + p - \bar{p}$$

$$N_4(p, \bar{p}) = p \ln\left(\frac{p}{\bar{p}}\right) - p + \bar{p}$$

Now let the adjustable earth properties be denoted by x, a function of space. We can choose x to minimize some volume integral of one of the model norms subject to the constraint that the model produce the required observations. Sometimes we have observations from $j = 1, 2, 3, \ldots, n_s$ source locations. We then need to compute the default power distribution $\bar{p}_j$ for each. Then we can minimize a sum of volume integrals

$$\min_x \sum_{j=1}^{n_s} \int N(p_j, \bar{p}_j) \, dV$$

subject to fitting all the data values.

It will be noted that the model-norm functions are all homogeneous of order 1. This means that $N(ap, a\bar{p}) = aN(p, \bar{p})$ for $a > 0$. This is our assurance that N is a volume density. Without this property we would have the difficulty that a sum of $N_k(p, \bar{p})$ over a set of subvolumes ΔV_k would change as the mesh were refined. Coordinate-system invariance is provided by the usual rules for conversion of volume integrals from one coordinate system to another.

Now let us take up an example from filter theory which turns out to be related to maximum-entropy spectral estimation. We are given a known input spectrum $R(Z)$ and are to find the finite length filter $X(Z) = x_0 + x_1 Z + x_2 Z^2$ whose output is as white as possible in the sense of minimizing the integral of N_3 across the spectrum. Let the spectrum of the filter be $S(Z) = \overline{X}(1/Z) X(Z)$. We have $\bar{p} = 1$ and $p = R(Z) S(Z)$. Thus, the minimization is

$$\min = \int (-\ln RS + RS) \, d\omega$$

Setting the derivative with respect to $\bar{x}_k$ equal zero we have

$$0 = \int \left(-\frac{1}{S} + R \right) \frac{\partial S}{\partial \bar{x}_k} \, d\omega$$

$$= \int \left(-\frac{1}{S} + R \right) Z^{-k} X(Z) \, d\omega$$

$$= \int Z^{-k} \left[-\frac{1}{\overline{X}(1/Z)} + R(Z)X(Z) \right] d\omega$$

Since we know that minimum-phase functions can represent any spectrum, we take $\overline{X}(1/Z)^{-1}$ to be expandable as $(\bar{b}_0 + \bar{b}_1/Z + \bar{b}_2/Z^2 + \cdots)$

$$X\left(\frac{1}{Z}\right)^{-1} = \int Z^{-k} \left[-\left(b_0 + \frac{b_1}{Z} + \frac{b_2}{Z^2} + \cdots \right) + RX \right] d\omega$$

We recall that this integral selects the coefficient of Z^0 of the argument. If we suppose that the filter is constrained to have $x_k = 0$ for $k \geq 3$, we get the familiar Toeplitz system

$$\begin{bmatrix} r_0 & r_{-1} & r_{-2} \\ r_1 & r_0 & r_{-1} \\ r_2 & r_1 & r_0 \end{bmatrix} \begin{bmatrix} x_0 \\ x_1 \\ x_2 \end{bmatrix} = \begin{bmatrix} b_0 \\ 0 \\ 0 \end{bmatrix}$$

6-6   ROBUST MODELING

The median and the mean are two kinds of statistical average. In a normal situation they behave in about the same way. At the present time physical scientists almost always use the mean and hence tend to be unaware of the dramatic ability of the median to cast off the effect of blunders in the data. As an example, consider an expensive, all-day-long experiment which yields only one number for a result. On the first day the result is 2.17, on the second day it is 2.14, and on the third and final day it is 1638.03. The mean of these results is 547.78 but the median (middle value) is 2.17. If one suspects a blunder on the third day, one will obviously prefer the median. Statisticians call this the "robust" property of the median. The objective of this section is to show how many kinds of geophysical data fitting can be made to be robust. In particular, all the calculations we now do which amount to solving overdetermined linear simultaneous equations by means of summed squared-error minimization can be made robust by minimizing summed absolute values of errors, instead. Computer costs are often comparable to those of least-squares methods. The algorithms turn out to solve a slightly broader class of problem than minimizing the summed absolute errors. Positive errors may be penalized with a different weight factor than negative errors. Such an arrangement is called an asymmetric linear norm. A special case of an asymmetric norm is an inequality. Not surprisingly, it turns out that all linear programming problems are special cases of asymmetric linear-norm problems and the solution techniques for asymmetric linear norms are similar to linear programming.

   First, we will see why means and medians relate to squares and absolute values. Let x_i be an arbitrary number. Let us define m_2 by the minimization of the sum of squared differences (called the L_2 norm) between m_2 and x_i:

$$m_2: \quad \min \sum_{i=1}^{N} (m_2 - x_i)^2 \qquad (6\text{-}6\text{-}1)$$

It is a straightforward task to find the minimum by setting the partial derivative of the sum with respect to m_2 equal to zero. We get

$$0 = \sum_{i=1}^{N} 2(m_2 - x_i)$$

or

$$m_2 = \frac{1}{N} \sum_{i=1}^{N} x_i \qquad (6\text{-}6\text{-}2)$$

Obviously, m_2 has turned out to be given by the usual definition of *mean*. Next, let us define m_1 by minimizing the summed absolute values (called the L_1 norm). We have

$$m_1: \quad \min \sum_{i=1}^{N} |m_1 - x_i| \qquad (6\text{-}6\text{-}3)$$

To find the minimum we may again set the partial derivative with respect to m_1 equal to zero

$$0 = \sum_{i=1}^{N} \text{sgn} \, (m_1 - x_i) \qquad (6\text{-}6\text{-}4)$$

Here the sgn function is $+1$ when the argument is positive, -1 when the argument is negative, and somewhere in between when the argument is zero. Equation (6-6-4) says that m_1 should be chosen so that m_1 exceeds x_i for $N/2$ terms, m_1 is less than x_i for $N/2$ terms, and if there is an x_i left in the middle, m_1 equals that x_i. This defines m_1 as a *median*. [For an even number N the definition (6-6-3) requires only that m_1 lie anywhere between the middle two values of the x_i.]

The computational cost for a mean is proportional to N, the number of points. The cost for completely ordering a list of numbers is $N \ln N$ [Ref. 22], but complete ordering is not required for finding the median. Hoare [Ref. 23] provided an algorithm for finding the median which requires about $3N$ operations. A computer algorithm based on Hoare's algorithm will be provided for weighted medians. Weighted medians are analogous to weighted sums. Ordinarily, 2.17 is taken to be the median of the numbers (2.14, 2.17, 1638.03) because we implicitly applied weights (1, 1, 1). If we applied weights (3, 1, 1) it would be like having the numbers 2.14, 2.14, 2.14, 2.17, 1638.03 and the median would then be 2.14. Formally, a weighted median may be defined by the minimization

$$m_1: \qquad \min \sum |w_i| \, |m_1 - x_i| \qquad (6\text{-}6\text{-}5)$$

Obviously if the weight factors are all unity, this reduces to the earlier definition whereas using a weight factor equal to 3, for example, is just like including the same term three times with a weight of 1. Figure 6-3 illustrates the definition (6-6-5) for a simple case. From Fig. 6-3 it is apparent that a median is always equal to one of the x_i even if the weights are not integers. If the weights are all unity and there is an even number of numbers, then the error norm will be flat between the two middle numbers. Then any value in between satisfies our definition of median by minimizing the sum.

Let us rearrange (6-6-5) by bringing $|w_i|$ into the other absolute-value function. We have

$$m_1: \qquad \min \sum_i |\, |w_i|m - |w_i|x_i| = \min \sum_i |w_i m - w_i x_i| \qquad (6\text{-}6\text{-}6)$$

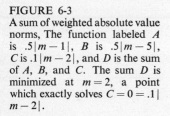

FIGURE 6-3
A sum of weighted absolute value norms. The function labeled A is $.5|m - 1|$, B is $.5|m - 5|$, C is $.1|m - 2|$, and D is the sum of A, B, and C. The sum D is minimized at $m = 2$, a point which exactly solves $C = 0 = .1| m - 2|$.

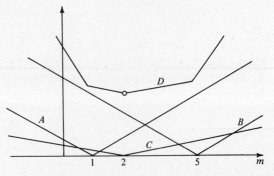

We will now relabel things from the conventions of statistics to the usual conventions of simultaneous equations and linear programming. Let

$$a_i: = w_i \qquad b_i: = w_i x_i \qquad x: = m \quad \text{(6-6-7a, b, c)}$$

With these new definitions (6-6-6) becomes

$$x: \qquad \min \sum_i |a_i x - b_i| \qquad \text{(6-6-7)}$$

The definition (6-6-7) says, in other words, to solve the rank one overdetermined equations

$$\mathbf{a}x \cong \mathbf{b} \qquad \text{(6-6-8)}$$

for x by minimizing the L_1 norm. This is, in effect, a weighted median problem. If (6-6-8) were solved by minimizing the L_2 norm (least squares) x would turn out to be the weighted average $x = (\mathbf{a} \cdot \mathbf{b})/(\mathbf{a} \cdot \mathbf{a})$.

We now consider a solution technique for the minimization (6-6-5). Essentially, it is Hoare's algorithm. On a trial basis we select a random equation from the set (6-6-8) to be exactly satisfied. This equation, called the basis equation, can be denoted $a_k x_{\text{trial}} = b_k$. Inserting x_{trial} into (6-6-8) we get equations with positive errors, negative errors, and zero errors. If we have been lucky with x_{trial}, then we find that the zero error group has enough weight to swing the balance between positive and negative weights in either direction. Otherwise, we must pick a new trial basis equation from the stronger of the positive or negative group. Fortunately, we need no longer look into the weaker group because these residuals cannot change signs as we descend into minimum. This may be seen geometrically on a figure like Fig. 6-3. We always wish to go downhill, so once it has been ascertained that a data point is uphill from the present point then it is never necessary to reinspect the uphill point. Thus, the size of the group being inspected rapidly diminishes. Figure 6-4 contains a computer program to do these operations.

The next step up the ladder of complexity is to consider two unknowns. The obvious generalization of (6-6-8) is

$$\begin{bmatrix} a_1 & c_1 \\ a_2 & c_2 \\ \vdots & \vdots \\ a_k & c_k \\ \vdots & \vdots \end{bmatrix} \begin{bmatrix} x \\ y \end{bmatrix} \cong \begin{bmatrix} b_1 \\ b_2 \\ \vdots \\ b_k \\ \vdots \end{bmatrix}, \qquad \text{(6-6-9)}$$

We will assume that the reader is familiar with the solution to (6-6-9) by the least-squares method. Solution by minimizing the sum of the absolute values of the errors begins in a similar way. We begin by defining the error

$$E = \sum_{k=1}^{N} |b_k - a_k x - c_k y| \qquad \text{(6-6-10)}$$

Then we set the x derivative of the error equal to zero and the y derivative of the error equal to zero.

$$0 = \frac{\partial E}{\partial x} = \sum_{k=1}^{N} - a_k \operatorname{sgn}(b_k - a_k x - c_k y) \quad (6\text{-}6\text{-}11a)$$

$$0 = \frac{\partial E}{\partial y} = \sum_{k=1}^{N} - c_k \operatorname{sgn}(b_k - a_k x - c_k y) \quad (6\text{-}6\text{-}11b)$$

```
      SUBROUTINE SKEWER(ND,N,W,F,GU,GD,SMALL,K,T,ML,MH)
C     SOLVE RANK 1 OVERDETERMINED EQUATIONS WITH SKEW NORM
C     INPUTS- N,W,F,GU,GD,SMALL,K.  OUTPUTS- K,T,ML,MH.
C FIND  T  TO MINIMIZE
C             N
C   LS  =    SUM   SKEWNORM(K,F(K)-W(K)*T)
C            K=1
C WHERE               ( GU(K)*(ER-SMALL) IF ER.GT.+SMALL     GU.GT.0
C       SKEWNORM(K,ER) = ( GD(K)*(ER+SMALL) IF ER.LT.-SMALL     GD.LT.0
C                       ( 0.            IF ABS(ER).LE.SMALL.GE.0.
C     GU,GD,W,AND F ARE REFERENCED INDIRECTLY AS W(K(I)),I=1,N  ETC
C     MINIMA WILL BE AT EQUATIONS K(ML),K(ML+1),...K(MH).
      DIMENSION W(ND),F(ND),K(ND),GU(ND),GD(ND)
      DIMENSION G(1000)
      LOW=1
      LARGE=N
      ML=N
      MH=1
      GN=0.
      GP=0.
      DO 50 ITRY=1,N
      L=K(LOW+MOD((LARGE-LOW)/3+ITRY,LARGE-LOW+1))
      IF(ABS (W(L)).EQ.0.) GO TO 50
      T=F(L)/(W(L))
      F(L)=W(L)*T
      DO 10 I=LOW,LARGE
      L=K(I)
      ER=F(L)-W(L)*T
      G(L)=0.
      IF(ER.GT.SMALL) G(L)=-W(L)*GU(L)
   10 IF(ER.LT.-SMALL)G(L)=-W(L)*GD(L)
      CALL SPLIT(LOW,LARGE,K,G,MLT,MHT)
      GNT=GN
      DO 20 I=LOW,MLT
   20 GNT=GNT+G(K(I))
      GPT=GP
      DO 30 I=MHT,LARGE
   30 GPT=GPT+G(K(I))
      GPLX=0.
      GMIX=0.
      DO 40 I=MLT,MHT
      L=K(I)
      IF(W(L).LT.0.)GPLX=GPLX-W(L)*GU(L)
      IF(W(L).GT.0.)GPLX=GPLX-W(L)*GD(L)
      IF(W(L).GT.0.)GMIX=GMIX-W(L)*GU(L)
   40 IF(W(L).LT.0.)GMIX=GMIX-W(L)*GD(L)
      GRAD=GNT+GPT
      IF((GRAD+GPLX)*(GRAD+GMIX).LT.0.) GO TO 60
      IF(GRAD.GE.0.)LOW=MHT+1
      IF(GRAD.LE.0.)LARGE=MLT-1
      IF(LOW.GT.LARGE) GO TO 60
      IF(GRAD.GE.0.)GN=GNT+GMIX
      IF(GRAD.LE.0.)GP=GPT+GPLX
      IF((GRAD+GPLX).EQ.0.)ML=MLT
      IF((GRAD+GMIX).EQ.0.)MH=MHT
   50 CONTINUE
```

(continues to next page)

```
      60    ML=MINO(ML,MLT)
            MH=MAXO(MH,MHT)
            RETURN
            END

            SUBROUTINE SPLIT(LOW,LARGE,K,G,ML,MH)
      C     GIVEN G(K(I)),I=LOW,LARGE
      C     THEN REARRANGE K(I),I=LOW,LARGE AND FIND ML,MH SO THAT
      C     (G(K(I)),I=LOW,(ML-1)) .LT. 0 AND
      C     (G(K(I)),I=ML,MH)=0.   AND
      C     (G(K(I)),I=(MH+1),LARGE) .GT. 0.
            DIMENSION K(LARGE),G(41)
            ML=LOW
            MH=LARGE
      10    ML=ML-1
      20    ML=ML+1
            IF(G(K(ML)))20,30,30
      30    MH=MH+1
      40    MH=MH-1
            IF(G(K(MH)))50,50,40
      50    KEEP=K(MH)
            K(MH)=K(ML)
            K(ML)=KEEP
            IF(G(K(ML)).NE.G(K(MH)))GO TO 10
            DO 60 I=ML,MH
            II=I
            IF(G(K(I)).NE.0.0) GO TO 70
      60    CONTINUE
            RETURN
      70    KEEP=K(MH)
            K(MH)=K(II)
            K(II)=KEEP
            GO TO 30
            END
```

FIGURE 6-4
A subroutine to compute weighted and skewed medians. (A "skewed median" is often called a quantile.) This subroutine is somewhat complicated because it takes special care to do the correct thing when weight factors are zero and because it provides pointers to all equations (occasionally there is more than one) which are satisfied at the final minimum.

Now we run into a snag. If the sgn function always takes the value $+1$ or -1, then (6-6-11a) implies that the a_k may be divided into two piles of equal weight. Clearly many, indeed most, collections of numbers cannot be so balanced (for example, if all the a_i except one are integers). The difficulty will be avoided if at least one of the equations of (6-6-9) is solved exactly so that sgn takes an indeterminate value for that term. Any algebraic confusion may be quickly dispelled by recollection of Fig. 6-3 and the result that even with one unknown the minimum generally occurs at a corner where the first derivative is discontinuous. The same situation must again apply to (6-6-11b). The usual situation is that for N equations and M unknowns precisely M of the N equations will be exactly satisfied in order to enable the error gradient to vanish at the minimum. Common usage in the field of linear programming is to refer to any nonsingular subset of M out of the N equations as a set of *basis equations*. The particular set of M equations which is solved when the error is minimized is called the *optimum basis*.

Although linear programming is a twentieth-century development, the basic ideas seem to have been well known before Laplace in the eighteenth century.

Indeed, in the words of Gauss' *Theoria Motus Corporum Coelestium* which appeared in 1809 [Ref. 24]:

Laplace made use of another principle for the solution of linear equations, the number of which is greater than the number of unknown quantities, which had been previously proposed by Boscovich, namely that the differences themselves, but all of them taken positively, should make up as small a sum as possible. It can be easily shown, that a system of values of unknown quantities, derived from this principle alone, must necessarily (except the special cases in which the problem remains, to some extent, indeterminate) exactly satisfy as many equations out of the number proposed, as there are unknown quantities, so that the remaining equations come into consideration only so far as they help to *determine the choice*.

Further developments and numerous geophysical applications may be found in Reference 25.

Next a simple but effective technique for descent down a multidimensional error surface will be described. The position $\mathbf{x}$ on a line through $\mathbf{x}_0$ can be indicated by a scalar parameter t. The direction of the line can be specified by an M component vector $\mathbf{g}$. Then any point $\mathbf{x}$ on the line may be represented as

$$\mathbf{x} = \mathbf{x}_0 + \mathbf{g}t \qquad (6\text{-}6\text{-}12)$$

Inserting (6-6-12) into the overdetermined set

$$\mathbf{A}\mathbf{x} \cong \mathbf{b} \qquad (6\text{-}6\text{-}13)$$

we obtain

$$\mathbf{A}(\mathbf{x}_0 + \mathbf{g}t) \cong \mathbf{b} \qquad (6\text{-}6\text{-}14a)$$

$$(\mathbf{A}\mathbf{g})t \cong \mathbf{b} - \mathbf{A}\mathbf{x}_0 \qquad (6\text{-}6\text{-}14b)$$

Defining $\mathbf{w}$ and $\mathbf{e}$ by

$$\mathbf{w} = \mathbf{A}\mathbf{g} \qquad (6\text{-}6\text{-}15a)$$

$$\mathbf{e} = \mathbf{b} - \mathbf{A}\mathbf{x}_0 \qquad (6\text{-}6\text{-}15b)$$

(6-6-14b) becomes

$$\mathbf{w}t \cong \mathbf{e} \qquad (6\text{-}6\text{-}16)$$

Solving (6-6-16) by minimizing the summed absolute errors also gives the minimum error along the line in (6-6-14a). But (6-6-16) is the weighted median problem discussed earlier. Recall that the solution t to (6-6-16) which gives minimum absolute error will exactly satisfy one of the equations in (6-6-16). Let us say $t = e_k/w_k$. For this value of t, the kth equation in (6-6-13) will also be satisfied exactly. The kth equation is now considered to be a good candidate for the basis, and we will next show how to pick the vector $\mathbf{g}$ so as to continue to satisfy the kth equation (stay on the kth hyperplane) as we adjust t in the next iteration.

Now we need a set of basis equations. This is a set of M equations which is temporarily taken to be satisfied. Then, as new equations are introduced into the basis by the weighted median solution, old equations are dropped out. The strategy of the present algorithm is merely to drop out the one which has been in longest. Let us denote our basis equations by

$$\mathbf{A'x = d'} \qquad (6\text{-}6\text{-}17)$$

$\mathbf{A'}$ is a square matrix. The inverse of the matrix $\mathbf{A'}$ will be required and will be denoted by $\mathbf{B}$. Now suppose we decide to throw out the pth equation from the basis matrix $\mathbf{A'}$. Then for $\mathbf{g}$ we select the pth column of $\mathbf{B}$. To see why this works note that since $\mathbf{A'B = I}$ the M vector $\mathbf{A'g}$ will now be the pth column from the identity matrix. Therefore, in the N vector $\mathbf{w = Ag}$ there is a component equal to $+1$, there are $M-1$ components equal to 0, and there are $N - M$ other unspecified elements. If the kth equation in (6-6-13) or (6-6-16) has been kept in the basis (6-6-17), then the kth equation in $\mathbf{Ag}t = \mathbf{d - Ax}$ now reads

$$\text{zero } t = \text{zero} \qquad (6\text{-}6\text{-}18)$$

The left-hand zero is an element from the identity matrix and the right-hand zero is from the statement that the kth equation is exactly satisfied. Clearly, we can now adjust t as much as we like to attain a new local minimum and the kth equation will still be exactly satisfied. There is also one equation of the form

$$\text{one } t = \text{zero} \qquad (6\text{-}6\text{-}19)$$

It will be satisfied only if t is zero. Geometrically, this means that if we must move to get to a minimum, then this equation is not satisfied and so we are jumping from this hyperplane. This equation is the one leaving the basis. Of course, if t turns out to be zero, then it reenters the basis. The foregoing steps are iterated until such time that for M successive iterations the equation thrown out of the basis by virtue of its age has immediately reappeared because $t = 0$. This means that the basis can no longer be improved and we have arrived at the optimum basis and the final solution.

7

WAVEFORM APPLICATIONS OF LEAST SQUARES

By the methods of calculus, one learns to find the coordinates of an extremal *point* on a curve. In the calculus of variations, one learns how to find extremal *functions*. In practice, the continuum may be approximated on a mesh and the distinction blurs. In the calculus of variations problems, however, the matrices can be immense, a disadvantage often partially offset by their orderly form. In this chapter we will take up examples in the use of least squares on waveforms and relationships between groups of waveforms. This leads to a massive full matrix called the block-Toeplitz matrix for which we have a special solution technique.

7-1 PREDICTION AND SHAPING FILTERS

A data wavelet is given by $\mathbf{b} = (b_0, b_1, \ldots, b_n)$. We plan to construct a filter $\mathbf{f} = (f_0, f_1, \ldots, f_n)$. Filtering is defined in this way: When data $\mathbf{b}$ go into a filter $\mathbf{f}$, an output wavelet $\mathbf{c}$ is produced according to the following matrix multiplication.

$$
\begin{bmatrix} c_0 \\ c_1 \\ c_2 \\ \cdot \\ \cdot \\ \cdot \\ c_{n+m} \end{bmatrix} = \begin{bmatrix} b_0 & 0 & \cdots & 0 \\ b_1 & b_0 & 0 & \\ b_2 & b_1 & b_0 & \\ \vdots & b_2 & b_1 & \ddots \\ b_n & \vdots & b_2 & \\ 0 & b_n & \vdots & \\ \vdots & & & b_{n-1} \\ 0 & 0 & \cdots & b_n \end{bmatrix} \begin{bmatrix} f_0 \\ f_1 \\ \cdot \\ \cdot \\ \cdot \\ f_m \end{bmatrix}
\tag{7-1-1}
$$

This operation is often called *complete transient convolution*. It is the same as identifying coefficients in a polynomial multiplication.

Now we introduce another wavelet **d** which will have the same number of components as **c**. We call **d** the desired output of the filter. We saw that **c** is the actual output. The actual output **c** was seen to be a function of the input **b** and the filter **f**. The problem now is to determine **f** so that **c** and **d** are very much alike. Specifically we will choose **f** so that the difference vector $c - d$ has minimum length squared (in $n + m + 1$ dimensional space). In other words, we use the method of least squares to solve the overdetermined equations

$$
\begin{bmatrix} b_0 & 0 & \cdots & 0 \\ b_1 & b_0 & 0 & \\ b_2 & b_1 & b_0 & \\ \vdots & b_2 & b_1 & \ddots \\ b_n & \vdots & b_2 & \\ 0 & b_n & \vdots & \\ \vdots & & & b_{n-1} \\ 0 & 0 & \cdots & b_n \end{bmatrix} \begin{bmatrix} f_0 \\ f_1 \\ \cdot \\ \cdot \\ \cdot \\ f_m \end{bmatrix} \approx \begin{bmatrix} d_0 \\ d_1 \\ d_2 \\ \cdot \\ \cdot \\ \vdots \\ d_{n+m} \end{bmatrix}
\tag{7-1-2}
$$

Using the "quick-and-dirty" method of the previous chapter we merely premultiply (7-1-2) by the transposed matrix. The result is a Toeplitz matrix of the form

$$
\begin{bmatrix} r_0 & r_1 & r_2 & \cdots & r_m \\ r_1 & r_0 & r_1 & & \\ r_2 & r_1 & r_0 & & \\ \vdots & & & \ddots & \\ r_m & & & & r_0 \end{bmatrix} \begin{bmatrix} f_0 \\ f_1 \\ f_2 \\ \vdots \\ f_m \end{bmatrix} = \begin{bmatrix} g_0 \\ g_1 \\ g_2 \\ \vdots \\ g_m \end{bmatrix}
\tag{7-1-3}
$$

where r_k is the autocorrelation of the input x_k and g_k is a crosscorrelation of the input x_k with the desired output d_k. For computation techniques see Chapt. 7-5.

The formulas of this section may also be used to attempt to predict a time series from its past. For example $f_1, f_2, \ldots, f_m$ is a prediction filter of x_{t+10} from $x_t, x_{t-1}, \ldots, x_{t-m+1}$ if we solve by least squares the equations

$$
\begin{bmatrix} x_{10} \\ x_{11} \\ x_{12} \\ \cdot \\ \cdot \\ \cdot \end{bmatrix} \approx \begin{bmatrix} x_0 & x_{-1} & \cdots & x_{-m+1} \\ x_1 & x_0 & & x_{-m+2} \\ x_2 & x_1 & & \cdot \\ x_3 & x_2 & & \cdot \\ \cdot & \cdot & & \cdot \\ \cdot & \cdot & & \\ \cdot & \cdot & & \end{bmatrix} \begin{bmatrix} f_1 \\ f_2 \\ \cdot \\ \cdot \end{bmatrix}
\tag{7-1-4}
$$

The matrix in (7-1-4) may be continued downward for as far as one has data. In an application, the range of t in (7-1-4) would be over past values of t. Then, after solving the equations for the filter $\mathbf{f}$ it would be hoped that the character of the time series was such that $\mathbf{f}$ could be used to predict future values of the time series which had not gone into the equation defining $\mathbf{f}$.

If the matrix of (7-1-4) is very much higher than it is wide, it may be desirable to treat the end effects differently. If one uses instead

$$
\begin{bmatrix} x_{10} \\ x_{11} \\ \cdot \\ \cdot \\ \cdot \end{bmatrix} \approx \begin{bmatrix} x_0 & 0 & 0 \\ x_1 & x_0 & 0 \\ x_2 & x_1 & \\ x_3 & x_2 & \\ \vdots & \vdots & \\ & \text{zeros} & \end{bmatrix} \begin{bmatrix} f_1 \\ \cdot \\ \cdot \\ f_m \end{bmatrix} \tag{7-1-5}
$$

one finds that the least-squares normal equation has a Toeplitz matrix whereas for (7-1-4) the matrix is not Toeplitz. As the reader is aware, the Toeplitz matrix has many advantages, both theoretical and computational.

Of special interest is the filter which is designed from the equations

$$
\begin{bmatrix} x_0 & & \text{zeros} & \\ x_1 & x_0 & & \\ x_2 & x_1 & x_0 & \\ x_3 & x_2 & \cdot & \ddots \\ \vdots & & \cdot & \\ x_n & & \cdot & \\ & x_n & & \\ & & & \ddots \\ \text{zeros} & & x_n \end{bmatrix} \begin{bmatrix} 1 \\ a_1 \\ \vdots \\ a_m \end{bmatrix} \approx \begin{bmatrix} 0 \\ 0 \\ 0 \\ \cdot \\ \cdot \\ \cdot \\ 0 \end{bmatrix} \tag{7-1-6}
$$

Such a filter is called the *prediction error filter for unit span* because the a_k operate on $(x_{t-1}, x_{t-2}, \ldots)$ attempting to cancel x_t. Thus, the a_k on the $(x_{t-1}, x_{t-2}, \ldots)$ gives the negative of a best prediction of x_t based on $(x_{t-1}, x_{t-2}, \ldots)$. The normal equations implied by (7-1-6) are the square set

$$
\begin{bmatrix} r_0 & r_1 & r_2 & \cdots \\ r_1 & r_0 & r_1 & \\ r_2 & r_1 & r_0 & \\ \cdot & & & \\ \cdot & & & \\ \cdot & & & \end{bmatrix} \begin{bmatrix} 1 \\ a_1 \\ a_2 \\ \cdot \\ \cdot \\ a_m \end{bmatrix} = \begin{bmatrix} v \\ 0 \\ 0 \\ \cdot \\ \cdot \\ 0 \end{bmatrix} \tag{7-1-7}
$$

It may be noted that the calculation of a prediction error filter depends only on the autocorrelation of the time series and not on the time series itself. As we have seen (from 3-3-3), the solutions to these equations are coefficients of a minimum-phase polynomial.

Solutions to Toeplitz equations when the right-hand side takes the more arbitrary form (7-1-3) are not generally minimum-phase, but the Levinson recursion may be generalized to make the calculation speedy. This is done in Sec. 7-5 on the multichannel Levinson recursion.

EXERCISES

1 Find a three-term zero delay inverse to the wavelet (1, 2). Compare the error to the error of (2, 1). Compare the waveform. An extensive discussion of the error in least-squares inverse filters is given in Reference 26. One conclusion is that the sum of the squared errors goes to zero as the filter length becomes infinite in two situations:

(*a*) Zero delay inverse if and only if the wavelet being inverted is minimum-phase.

(*b*) If the wavelet being inverted is not minimum-phase, the error goes to zero only if the output is delayed, that is, $d = (\ldots, 0, 0, 1, 0, 0, \ldots)$. Calculate a three-term delayed inverse to (1, 2), that is, try $d = (0, 1, 0, 0)$ or $d = (0, 0, 1, 0)$.

2 A pressure sensor in a deep well records upgoing seismic waves and, at some time t_0 later, identical downgoing waves of opposite sign. Determine delayed and non-delayed least-squares filters of length m to eliminate the double pulse. (You should be able to guess the solution to large matrices of this type. Try filters of the form $f_k = \alpha + \beta k$ where α and β are scalars.) What is the error as a function of the filter length?

3 Let $b_t = (\ldots, 1, 1, -2, 1, 1, -2, \ldots)$. Find by least squares the best one-term filter which predicts b_t, using only b_{t-1}. Find the best two-term filter using b_{t-1} and b_{t-2}. Likewise find the best three-term filter. What is the error as a function of time in each case?

7-2 BURG SPECTRAL ESTIMATION [Ref. 27]

The uncertainty principle says that if a time function contains most of its energy in the time-span Δt, then its Fourier transform contains most of its energy in a bandwidth $\Delta f \geq 1/\Delta t$. This is not the same as saying that if we have a sample of a stationary time series of length Δt, the best frequency resolution we can hope to attain will be $\Delta f = 1/\Delta t$. The difference lies in the difference between assuming a function is zero outside the interval Δt in which it is given and in assuming that it continues "in a sensible way" outside the given interval. If the data sample can be continued "in a sensible way" some distance beyond the interval in which it is given, then the frequency resolution Δf may be considerably smaller than $1/\Delta t$. A finer resolution depends upon the predictability of the data off the ends of the sample. If one has a segment of a stationary series which is short compared to the autocorrelation of the stationary series, then the spectral estimation procedure of John P. Burg will be radically better than any truncated Fourier transform method. This comes about in physical problems when one is dealing with resonances which have decay times that are long compared to the observation time or when one is looking at a function of space where each point in space represents another instrument.

If a spectrum $R(Z)$ is estimated by $\overline{X}(1/Z)X(Z)$ where $X(Z)$ is a polynomial made up from $N + 1$ known data points, then the coefficients of $R(Z)$ are computed by

$$r_k = \sum_{j=0}^{N-k} \bar{x}_{j+k} x_j \qquad (7\text{-}2\text{-}1)$$

Notice that r_0 is calculated from $N + 1$ terms, r_1 from N terms, etc. If N is not large enough, this will have an undesirable biasing effect. The biasing is removed if the r_k are computed instead by the formula

$$r_k = \frac{1}{N - k + 1} \sum_{j=0}^{N-k} \bar{x}_{j+k} x_j \qquad (7\text{-}2\text{-}2)$$

The trouble with using (7-2-2) is that data samples can easily be found for which r_k will not be a valid autocorrelation function. For example, the spectrum will not be positive at all frequencies, the solution to Toeplitz equations may blow up, etc.

Burg's approach avoids the end-effect problems of (7-2-1) and the possibility of impossible results from (7-2-2). Instead of estimating the autocorrelation r_k directly from the data he estimates a minimum-phase prediction-error filter directly from the data. The output of a prediction-error filter has a white spectrum. (If it did not, then the color could be used to improve prediction.) Since the spectrum of the output is the spectrum of the input times the spectrum of the filter, the spectrum of the input may be estimated as the inverse of the spectrum of the prediction-error filter. As we have seen, narrow spectral peaks are far more easily represented by a denominator than by a numerator.

Let the given segment of data be denoted by $x_0, x_1, \ldots, x_n$. Then a two-term prediction-error filter $(1, a)$ of the time series x_t is given by the choice of a which minimizes

$$E(a) = \sum_{t=1}^{N} |x_t + a x_{t-1}|^2 \qquad (7\text{-}2\text{-}3)$$

Unfortunately, consideration of a few examples shows that there exist time series [like $(1, 2)$] for which $|a|$ may turn out to be greater than unity. This is unacceptable because the prediction-error filter is not minimum-phase, the spectrum is not positive, etc. Recall that a prediction-error filter defined in the previous section depends only on the autocorrelation of the data and not the data per se. This means that the same filter is computed from both a time series and from the (complex-conjugate) time-reversed time series. This suggests that the error of forward prediction (7-2-3) be augmented by the error of backward prediction. That is

$$E(a) = \sum_{t=1}^{N} |x_t + a x_{t-1}|^2 + |\bar{x}_{t-1} + a \bar{x}_t|^2 \qquad (7\text{-}2\text{-}4)$$

We will later establish that the minimization of (7-2-4) always leads to an $|a|$ less than unity. The power spectral estimate associated with this value of a is

$R = 1/[(1 + \bar{a}/Z)(1 + aZ)]$. The value of Δf may be very small if a turns out very close to the unit circle.

A natural extension of (7-2-4) to filters with more terms would seem to be to minimize

$$E(a_1, a_2) = \sum_{t=2}^{N} |x_t + a_1 x_{t-1} + a_2 x_{t-2}|^2 + |\bar{x}_{t-2} + a_1 \bar{x}_{t-1} + a_2 \bar{x}_t|^2 \qquad (7\text{-}2\text{-}5)$$

Unfortunately, Burg discovered time series for which the computed filter $A(Z) = 1 + a_1 Z + a_2 Z^2$ was not minimum-phase. If $A(Z)$ is not minimum-phase, then $R = 1/[\bar{A}(1/Z)A(Z)]$ is not a satisfactory spectral estimate because $R(Z)$ is to be evaluated on the unit circle and $1/A(Z)$ would not be convergent there.

Burg noted that the Levinson recursion always gives minimum-phase filters. In the Levinson recursion a filter of order 3 is built up from one of order 2 by

$$\begin{bmatrix} 1 \\ a_1 \\ a_2 \end{bmatrix} = \begin{bmatrix} 1 \\ a \\ 0 \end{bmatrix} - c \begin{bmatrix} 0 \\ a \\ 1 \end{bmatrix}$$

Thus Burg decided that instead of using least squares to determine a_1 and a_2 as in (7-2-5), he would take a to be given from (7-2-4) and then do a least-squares problem to solve for c. This would be done in such a way as to ensure that $|c|$ comes out less than unity, which guarantees that $A(Z) = 1 + a_1 Z + a_2 Z^2$ is minimum-phase. Thus he suggested rewriting (7-2-5) as

$$E(c) = \sum_{t=2}^{N} |x_t + ax_{t-1} - c(\bar{a}x_{t-1} + x_{t-2})|^2$$
$$+ |\bar{x}_{t-2} + a\bar{x}_{t-1} - c(\bar{a}\bar{x}_{t-1} + \bar{x}_t)|^2 \qquad (7\text{-}2\text{-}6)$$

Now the error (7-2-6), which is the sum of the error of forward prediction plus the error of backward prediction, is minimized with respect to variation of c. (In a later chapter we will see fit to call c a reflection coefficient.) The quantity a remains fixed by the minimization of (7-2-4). Now let us establish that $|c|$ is less than unity. Denote by e_+ the time series $x_t + ax_{t-1}$ which is the error in forward prediction of x_t. Denote by e_- the time series $x_{t-2} + \bar{a}x_{t-1}$ of error on backward prediction. With this, (7-2-6) becomes

$$E = \sum_t |e_+ - ce_-|^2 + |\bar{e}_- - c\bar{e}_+|^2$$
$$= \sum_t \overline{(e_+ - ce_-)}(e_+ - ce_-) + \overline{(\bar{e}_- - c\bar{e}_+)}(\bar{e}_- - c\bar{e}_+) \qquad (7\text{-}2\text{-}7)$$

Setting the derivative with respect to $\bar{c}$ equal to zero

$$0 = \sum_t \bar{e}_-(e_+ - ce_-) + e_+(\bar{e}_- - c\bar{e}_+)$$

$$c = \frac{+\sum_t 2\bar{e}_- e_+}{\sum_t \bar{e}_+ e_+ + \bar{e}_- e_-} \qquad (7\text{-}2\text{-}8)$$

(One may note that $\partial E/\partial c = 0$ gives the same result.) That $|c|$ is always less than unity may be seen by noting that the length of the vector $e_+ \pm e_-$ is always positive. In particular

$$\sum_t |e_+ \pm e_-|^2 \geqslant 0$$

$$\sum_t \bar{e}_+ e_+ \pm \bar{e}_+ e_- \pm \bar{e}_- e_+ + \bar{e}_- e_- \geqslant 0$$

$$\sum_t \bar{e}_+ e_+ + \bar{e}_- e_- > 2|\bar{e}_- e_+|$$

$$|c| \leqslant 1 \qquad (7\text{-}2\text{-}9)$$

If we now redefine e_+ and e_- as

$$e_+ \leftarrow e_+ - ce_- \qquad (7\text{-}2\text{-}10a)$$

$$e_- \leftarrow e_- - \bar{c}e_+ \qquad (7\text{-}2\text{-}10b)$$

we have the forward and backward prediction errors of the three-term filter $(1, a_1', a_2') = (1, a_1 - c\bar{a}_1, -c)$. One can then return to (7-2-7) and proceed recursively. As the recursion proceeds e_+ and e_- gradually become unpredictable random numbers. We have then found a filter $A(Z)$ which filters $X(Z)$ either forward or backward and the output is white light. Since the output has a constant spectrum, the spectrum of the input must be the inverse of the spectrum of the filter.

In later chapters we will discover a wave-propagation interpretation of the Burg algorithm. In a layered medium the parameters c_k have the interpretation of reflection coefficients; the e^+ and e^- vectors have the interpretation of up- and downgoing waves; and the whole process of calculating a succession of c_k amounts to downward continuing surface seismograms into the earth, determining an earth model c_k as you go.

EXERCISE

1 Consider the time series with ten points $(1, 1, 1, -1, -1, -1, 1, 1, 1, -1)$. Compute C and A up to cubics in Z. Compare the autocorrelation r_t calculated by Burg's method with $R(Z)$ estimated from the truncated sample and with $R(Z)$ estimated by intuitively extending the data sample in time to plus and minus infinity.

2 Modify the program of Fig. 7-1 to compute and include the scale factor V which belongs in the spectrum.

7-3 ADAPTIVE FILTERS

An adaptive filter is one which changes with time to accommodate itself to changes in the time series being filtered. For example, suppose one were predicting one point ahead in a time series. One could take a lot of past data to design the filter; then one could apply the filter to present incoming data to predict future incoming

```
      SUBROUTINE BURGC(LX,X,EP,EM,LC,C,A,N2048,S)
C GIVEN A TIME SERIES X(1...LX) GET ITS LOG SPECTRUM S(1...N2048)
      DIMENSION X(LX),EP(LX),EM(LX),C(LC),A(LC),S(N2048)
      COMPLEX X,EP,EM,C,A,S,TOP,BOT,EPI,CONJG,CLOG
      DO 10 I=1,N2048
10    S(I)=0.
      A(1)=1.
      DO 20 I=1,LX
      EM(I)=X(I)
20    EP(I)=X(I)
      DO 60 J=2,LC
      TOP=0.
      BOT=0.
      DO 30 I=J,LX
      BOT=BOT+EP(I)*CONJG(EP(I))+EM(I-J+1)*CONJG(EM(I-J+1))
30    TOP=TOP+EP(I)*CONJG(EM(I-J+1))
      C(J)=2*TOP/BOT
      DO 40 I=J,LX
      EPI=EP(I)
      EP(I)=EP(I)-C(J)*EM(I-J+1)
40    EM(I-J+1)=EM(I-J+1)-CONJG(C(J))*EPI
      A(J)=0.
      DO 50 I=1,J
50    S(I)=A(I)-C(J)*CONJG(A(J-I+1))
      DO 60 I=1,J
60    A(I)=S(I)
      CALL FORK(N2048,S,+1.)
      DO 70 I=1,N2048
70    S(I)=-CLOG(S(I))*2.
      RETURN
      END
```

FIGURE 7-1
Computer program to do Burg algorithm. The program follows the notation of the text. The data X is a vector of dimension given to be LX. Choice of $LC \leq LX$ is a compromise between high resolution and high scatter. The density of points on the frequency axis, which is controlled by $N2048 \gg LX$, is chosen for plotting convenience and should be great enough to resolve narrow spectral lines.

data. As time goes on it might become desirable to recompute the filter on the basis of new data which have come in. How often should the filter be redesigned? In concept, there is no reason why it should not be recomputed very often, perhaps after each new data point arrives. In practice, this is usually prohibitively expensive. For a filter of length n it requires n multiplies and adds to apply the filter to get one new output point. To recompute the filter with Levinson recursion requires about n^2 multiply-adds. However, it is usually expected that the filter need only be changed by a very small amount when a new data point arrives. For that reason we will give the Widrow [Ref. 28] adaptive-filter algorithm which modifies the filter by means of only n arithmetic operations. Thus, a new filter is computed after each data point comes in.

For definiteness, consider a two-term prediction situation where e_t is the error in predicting a time series x_t from two of its past values

$$e_t = x_t - bx_{t-1} - cx_{t-2} \qquad (7\text{-}3\text{-}1)$$

The sum squared error in the prediction is

$$E = \sum_t e_t^2 = \sum_t (x_t - bx_{t-1} - cx_{t-2})^2 \qquad (7\text{-}3\text{-}2)$$

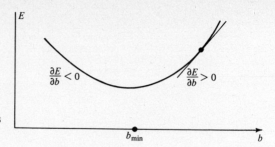

FIGURE 7-2
The sign of the partial derivative tells whether $b > b_{min}$ or $b < b_{min}$.

If the parameter b has been chosen correctly, one should find that $\partial E/\partial b = 0$. However, if the nature of the time series x_t is changing with time, $\partial E/\partial b$ may depart from zero when new data are included in the sum in (7-3-2). Since E is a positive quadratic function of b, if $\partial E/\partial b$ has become positive, then b should be reduced. If $\partial E/\partial b$ has become negative, then b should be augmented. See Fig. 7-2.

From (7-3-2) we have

$$\frac{\partial E}{\partial b} = - \sum_{i=-\infty}^{t} 2e_i x_{i-1} \qquad (7\text{-}3\text{-}3)$$

The change in $\partial E/\partial b$ from the addition of the data point x_t is just $-2e_t x_{t-1}$; thus, we are motivated to modify b and c in the following way

$$\begin{bmatrix} b \\ c \end{bmatrix} \leftarrow \begin{bmatrix} b \\ c \end{bmatrix} + ke_t \begin{bmatrix} x_{t-1} \\ x_{t-2} \end{bmatrix} \qquad (7\text{-}3\text{-}4)$$

Here the number k scales the amount of the readjustment which we are willing to make to b and c in one time step. If k is chosen very small, the adjustment will take place very slowly. If k is chosen too large, the adjustment will overshoot the minimum; however one may hope that it will bounce back, perhaps again overshooting at the next step. The choice of k is dictated in part by the nature of the time series x_t under study.

There are many variations on these same ideas. For example, we could use the L_1 norm and minimize something like

$$E(c) = \sum_t |cx_t - y_t| \qquad (7\text{-}3\text{-}5)$$

The resulting adaptation would be

$$c \leftarrow c - kx_t \, \text{sgn} \, (cx_t - y_t) \qquad (7\text{-}3\text{-}6)$$

Equation (7-3-5) is of course the weighted median. An even more robust procedure is the uniformly weighted median

$$E(c) = \sum_t \left| c - \frac{y_t}{x_t} \right| \qquad (7\text{-}3\text{-}7)$$

which leads to the adaptation

$$c \leftarrow c - k \, \text{sgn} \left(c - \frac{y_t}{x_t} \right) \qquad (7\text{-}3\text{-}8a)$$

which is identical to

$$c \leftarrow c - k \operatorname{sgn}(x_t) \operatorname{sgn}(cx_t - y_t) \quad (7\text{-}3\text{-}8b)$$

The examples (7-3-5) and (7-3-7) could be extended, in a manner like the Burg algorithm, to stationary series. Like (7-3-7) we could minimize

$$E = \sum_t \left| c - \frac{y_t}{x_t} \right| + \left| c - \frac{x_t}{y_t} \right| \quad (7\text{-}3\text{-}9)$$

This leads to a choice of c within the proper bounds because

$$-1 \leqslant \operatorname{median}\left(0, \frac{y_t}{x_t}, \frac{x_t}{y_t}\right) \leqslant +1$$

$$(\text{all } t)$$

EXERCISES

1 If x_t has physical dimensions of volts, what should be the physical dimensions for k? If x_t has an rms value of 100 V and Δt, the sampling interval, is 1 ms, what numerical value of k will allow the Widrow filter to adapt to new conditions in about a second?

2 Consider the time series $x_t = (\ldots, 1, 1, 1, 1, -4, 1, 1, 1, 1, -4, 1, 1, 1, 1, -4, \ldots)$. Consider self-prediction of the form $x_{t+1} = cx_t$. What are the results of least-squares prediction? What are the results of L_1 norm prediction of data weighted and uniformly weighted types?

7-4 DESIGN OF MULTICHANNEL FILTERS

Multichannel filters are frequently useful. For example, with a vector-prediction filter one might wish to predict a time series, using its past and the past of a group of other series. With a matrix-prediction filter one could predict a group of series, using the past of the whole group. If the series are related, the group prediction should be better than self-prediction of individual channels. For definiteness, let us take two time series x_t and y_t and suppose we are to find a vector filter which converts them into a third series d_t. If d_t is x_{t+1}, this is a unit time-span prediction filter for x_t. If d_t is a vertical seismogram and x_t and y_t are horizontals, then the two-channel filter might be called an extrapolation filter. The set of equations which we wish to solve by least squares takes the form

$$
\begin{bmatrix} d_1 \\ d_2 \\ d_3 \\ d_4 \\ \vdots \end{bmatrix}
=
\begin{bmatrix}
x_1 & y_1 & & & & \\
x_2 & y_2 & x_1 & y_1 & & \\
x_3 & y_3 & x_2 & y_2 & x_1 & y_1 \\
x_4 & y_4 & x_3 & y_3 & x_2 & y_2 & \cdot \\
\vdots & & & & & & \ddots
\end{bmatrix}
\begin{bmatrix} a_1 \\ b_1 \\ a_2 \\ b_2 \\ \vdots \\ a_m \\ b_m \end{bmatrix}
\quad (7\text{-}4\text{-}1)
$$

If this set of equations is abbreviated

$$\mathbf{d} \approx \mathbf{Bf} \qquad (7\text{-}4\text{-}2)$$

then, as we have seen in an earlier chapter, the solution is of the form

$$\mathbf{f} = (\mathbf{B}^T\mathbf{B})^{-1}\mathbf{B}^T\mathbf{d} \qquad (7\text{-}4\text{-}3)$$

We wish to inspect the matrix being inverted, call it $\mathbf{R}$. For a filter with three time lags we get

$$\mathbf{R} = \sum_t \begin{bmatrix} x_t \\ y_t \\ x_{t-1} \\ y_{t-1} \\ x_{t-2} \\ y_{t-2} \end{bmatrix} [x_t \quad y_t \quad x_{t-1} \quad y_{t-1} \quad x_{t-2} \quad y_{t-2}] \qquad (7\text{-}4\text{-}4)$$

If we define

$$r_{xx}(i) = \sum_t x_t x_{t+i}$$

$$r_{xy}(i) = \sum_t x_t y_{t+i}$$

and likewise for $r_{yx}(i)$ and $r_{yy}(i)$ the matrix (7-4-4) becomes

$$\mathbf{R} = \begin{bmatrix} r_{xx}(0) & r_{xy}(0) & r_{xx}(-1) & r_{xy}(-1) & r_{xx}(-2) & r_{xy}(-2) \\ r_{yx}(0) & r_{yy}(0) & r_{yx}(-1) & r_{yy}(-1) & r_{yx}(-2) & r_{yy}(-2) \\ \hline r_{xx}(1) & r_{xy}(1) & r_{xx}(0) & r_{xy}(0) & r_{xx}(-1) & r_{xy}(-1) \\ r_{yx}(1) & r_{yy}(1) & r_{yx}(0) & r_{yy}(0) & r_{yx}(-1) & r_{yy}(-1) \\ \hline r_{xx}(2) & r_{xy}(2) & r_{xx}(1) & r_{xy}(1) & r_{xx}(0) & r_{xy}(0) \\ r_{yx}(2) & r_{yy}(2) & r_{yx}(1) & r_{yy}(1) & r_{yx}(0) & r_{yy}(0) \end{bmatrix} \qquad (7\text{-}4\text{-}5)$$

We may take the 6×6 matrix of (7-4-5) and partition it into a 3×3 matrix of 2×2 submatrices. If we define the submatrix blocks as

$$\mathbf{R}(\tau) = \begin{bmatrix} r_{xx}(\tau) & r_{xy}(\tau) \\ r_{yx}(\tau) & r_{yy}(\tau) \end{bmatrix} = \mathbf{R}^T(-\tau) \qquad (7\text{-}4\text{-}6)$$

then (7-4-5) in terms of the blocks defined in (7-4-6) is

$$\mathbf{R} = \begin{bmatrix} R(0) & R(-1) & R(-2) \\ R(1) & R(0) & R(-1) \\ R(2) & R(1) & R(0) \end{bmatrix} \qquad (7\text{-}4\text{-}7)$$

The matrix in (7-4-7) is called *block Toeplitz* or *multichannel Toeplitz*. As with the ordinary Toeplitz matrix there is a trick method of solution. It will be taken up in the next section.

The reader should note that the matrix $\mathbf{R}$ does not depend on the desired output $\mathbf{d}$. This results in a computational saving when there is more than one possible output. An example would be when it is desired to predict several different series or distances into the future on a given series.

EXERCISE

1 In the exercises of Chap. 2, we determined $B(Z)$ and $A(Z)$ such that some given power series $C(Z)$ was expressed as $C(Z) = B(Z)/A(Z)$. Write normal equations (do not solve them) for doing this in an approximate way by minimizing

$$\min (A, B) = \sum_t (B_t - \sum_\tau C_{t-\tau} A_\tau)^2$$

where

$$A = (A_0, A_1, A_2) \qquad B = (B_0, B_1, B_2)$$

subject to the constraint $A_0 = 1$. (It can be proved that $A(Z)$ comes out minimum-phase by examining the Levinson recursion.)

7-5 LEVINSON RECURSION

The Levinson recursion is a simplified method for solving normal equations. It may be shown to be equivalent to a recurrence relation in orthogonal polynomial theory. The simplification in Levinson's method is possible because the matrix $\mathscr{R}$ has actually only N different elements when a general matrix could have N^2 different elements.

Levinson developed his recursion with single time series in mind (the basic idea was presented in Sec. 3-3). It is very little extra trouble to do the recursion for multiple time series. Let us begin with the prediction-error normal equation. With multiple time series, unlike single time series, the prediction problem is changed if time is reversed. We may write both the forward and the backward prediction-error normal equations as one equation in the form of (7-5-1).

Since end effects play an important role, we will show how, when given the solution for 3-term filters, $\mathscr{A}$ and $\mathscr{B}$

$$\begin{bmatrix} R_0 & R_{-1} & R_{-2} \\ R_1 & R_0 & R_{-1} \\ R_2 & R_0 & R_0 \end{bmatrix} \begin{bmatrix} I & B_2 \\ A_1 & B_1 \\ A_2 & I \end{bmatrix} = \begin{bmatrix} V_A & 0 \\ 0 & 0 \\ 0 & V_B \end{bmatrix} \qquad (7\text{-}5\text{-}1)$$

to find the solution $\mathscr{A}'$ and $\mathscr{B}'$ four-term filters to

$$\begin{bmatrix} R_0 & R_{-1} & R_{-2} & R_{-3} \\ R_1 & R_0 & R_{-1} & R_{-2} \\ R_2 & R_1 & R_0 & R_{-1} \\ R_3 & R_2 & R_1 & R_0 \end{bmatrix} \begin{bmatrix} I & B_3 \\ A_1 & B_2 \\ A_2 & B_1 \\ A_3 & I \end{bmatrix}' = \begin{bmatrix} V_A & 0 \\ 0 & 0 \\ 0 & 0 \\ 0 & V_B \end{bmatrix}' \qquad (7\text{-}5\text{-}2)$$

by forming a linear combination of $\mathscr{A}$ and $\mathscr{B}$. This can be done by choosing constant matrices α and β in

$$\begin{bmatrix} R_0 & R_1 & R_{-2} & R_{-3} \\ R_1 & R_0 & R_{-1} & R_{-2} \\ R_2 & R_1 & R_0 & R_1 \\ R_3 & R_2 & R_1 & R_0 \end{bmatrix} \left(\begin{bmatrix} I \\ A_1 \\ A_2 \\ 0 \end{bmatrix} \alpha + \begin{bmatrix} 0 \\ B_2 \\ B_1 \\ I \end{bmatrix} \beta \right) = \begin{bmatrix} V_A \\ 0 \\ 0 \\ E_A \end{bmatrix} \alpha + \begin{bmatrix} E_B \\ 0 \\ 0 \\ V_B \end{bmatrix} \beta \qquad (7\text{-}5\text{-}3)$$

Make $\mathscr{A}$ by choosing $\boldsymbol{\alpha}$ and $\boldsymbol{\beta}$ so that the bottom element on the right-hand side of (7-5-3) vanishes. That is, $\boldsymbol{\alpha} = \mathbf{I}, \boldsymbol{\beta} = -\mathbf{V}_B^{-1}\mathbf{E}_A$. Make $\mathscr{B}$ by choosing $\boldsymbol{\alpha}$ and $\boldsymbol{\beta}$ so that the top element on the right-hand side vanishes. That is, $\boldsymbol{\beta} = \mathbf{I}, \boldsymbol{\alpha} = -\mathbf{V}_A^{-1}\mathbf{E}_B$.

Of course, one will want to solve more than just the prediction-error problem. We will also want to go from 3×3 to 4×4 in the solution of the filter problem with arbitrary right-hand side $\mathscr{G}$. This is accomplished by choosing γ in the following construction (7-5-4) so that $\mathbf{E}_f + \mathbf{V}_B\gamma = \mathbf{G}_3$

$$
\begin{bmatrix} & \mathscr{R} & \end{bmatrix} \left\{ \begin{bmatrix} \mathbf{F}_2 \\ \mathbf{F}_1 \\ \mathbf{F}_2 \\ \mathbf{0} \end{bmatrix} + \begin{bmatrix} \mathbf{B}_3 \\ \mathbf{B}_2 \\ \mathbf{B}_1 \\ \mathbf{I} \end{bmatrix} \gamma \right\} = \left\{ \begin{bmatrix} \mathbf{G}_0 \\ \mathbf{G}_1 \\ \mathbf{G}_2 \\ \mathbf{E}_f \end{bmatrix} + \begin{bmatrix} 0 \\ 0 \\ 0 \\ \mathbf{V} \end{bmatrix} \gamma \right\} \qquad (7\text{-}5\text{-}4)
$$

7-6 CONSTRAINED FILTERS

A common geophysical situation is a plane wave (signal) incident on a group of receivers. One expects to see the same waveform at each receiver. However, there is corrupting noise present at each receiver, and the noise may or may not be coherent from one receiver to the next. In fact, we may suppose there is so much noise on each receiver that the signal might not be detectable at all if there were only one receiver. This was the situation facing M. J. Levin [Ref. 29] when he was trying to detect weak underground nuclear explosions with an array of seismometers. He suggested a multichannel filter with constraints. First suppose that either all the signals arrive at the same time or that, if the times differ, at least they are known so that the data channels may be shifted into alignment. Now the problem is to filter each channel and then add up the channels; the noise should be rejected but the signal shape should be maintained. Let $f_i(j)$ represent the filter weight on the ith channel at the jth lag. For illustration, consider two channels and three time lags. Then Levin's constraints which prevent signal distortion are

$$
\begin{aligned}
1 &= f_1(0) + f_2(0) \\
0 &= f_1(1) + f_2(1) \qquad (7\text{-}6\text{-}1) \\
0 &= f_1(2) + f_2(2)
\end{aligned}
$$

That this does not cause signal distortion follows, since if the same signal $s(Z)$ comes into each channel, the output is merely $s(Z)[f_1(Z) + f_2(Z)]$. But $f_1 + f_2$ is just $(1, 0, 0)$ in this case or a delta function in general. We call the equation set (7-6-1) constraint equations because there are fewer equations than unknowns. The constraint equations may be written in usual form as

$$
\begin{bmatrix} -1 & 1 & 1 & 0 & 0 & 0 & 0 \\ 0 & 0 & 0 & 1 & 1 & 0 & 0 \\ 0 & 0 & 0 & 0 & 0 & 1 & 1 \end{bmatrix} \begin{bmatrix} 1 \\ f_1(0) \\ f_2(0) \\ f_1(1) \\ f_2(1) \\ f_1(2) \\ f_2(2) \end{bmatrix} = \begin{bmatrix} 0 \\ 0 \\ 0 \end{bmatrix} \qquad (7\text{-}6\text{-}2)
$$

which we may abbreviate as $\mathbf{Gf} = 0$. If we use the method of least squares to minimize the total energy in the filter output, we will be attempting to suppress both signal and noise. But the constraint equations prevent the suppression of signal; hence only the noise is attenuated. If we let $\mathbf{R}$ denote the spectral matrix of the input data, then the filter $\mathbf{f}$ is determined by solving equations like

$$\begin{bmatrix} \mathbf{R} & \mathbf{G}^T \\ \mathbf{G} & 0 \end{bmatrix} \begin{bmatrix} \mathbf{f} \\ \lambda \end{bmatrix} = \begin{bmatrix} V \\ 0 \\ \vdots \end{bmatrix} \qquad (7\text{-}6\text{-}3)$$

We have solved equations of this type in preceding sections.

EXERCISES

1 In one application, where the channel amplifications were not well controlled, the lead terms of the filter were $f_1(0) = 100$ and $f_2(0) = -99$. Although this filter satisfied all that it was designed for, it was deemed inappropriate because the assumption of identical signals on each channel was a reasonable approximation but not exactly true. Can you suggest a more suitable constraint matrix?

2 Consider three seismometers in a row on the surface of the earth. The constraints considered so far have implied that all signals arrive at the same time, i.e., vertically incident waves. Define a constraint matrix to pass both the vertically incident wave and the wave which causes $x_1(t) = x_2(t+1) = x_3(t+2)$. What is the shortest filter which can both satisfy the constraints and still have some possibility of rejecting noise?

3 Consider a Levin filter on m channels with filters containing k lags. What is the size of the matrix in (7-6-3)?

8

LAYERS REVEALED BY SCATTERED WAVE FILTERING

Waves occur in almost all branches of physics. We are going to study waves, but here we will not assume knowledge of physics and differential equations. We will use only assumptions about the general principles of delay, continuity, and energy conservation. The results will be directly applicable to sound waves, water waves, light in thin films, normal incident elastic waves of both pressure and shear type, electromagnetic waves, transmission lines, electrical ladder networks, and other such things. The methods can also be applied to diffusion problems. Our first main objective is to solve the problem of calculating wave fields given reflection coefficients. Our second main objective is to gain the ability to calculate the reflection coefficients given the observed waves.

8-1 REFLECTION AND TRANSMISSION COEFFICIENTS

Consider two halfspaces (the sky above, the earth below). If a wave of unit amplitude is incident onto the boundary, there will be a transmitted wave of amplitude t and a reflected wave of amplitude c as depicted in Fig. 8-1.

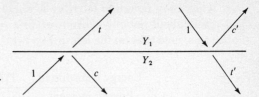

FIGURE 8-1
Waves incident, reflected c, and transmitted t at an interface.

A very simple relationship exists between t and c. The wave amplitudes have a physical meaning of something like pressure, material displacement, or tangential electric or magnetic fields; and these physical variables must be the same on either side of the boundary. Thus, we must have

$$t = 1 + c \qquad \text{(8-1-1)}$$
$$t' = 1 + c' \qquad \text{(8-1-2)}$$

It may be surprising that t may be greater than unity. However, this phenomenon may easily be seen at the ocean, where waves get larger as they approach the shore (until they break). Energy is not determined by wave height alone. Energy is equal to the squared wave amplitude multiplied by a proportionality factor Y depending upon the medium in which the wave is measured. If we denote the factor of the top medium by Y_1 and the bottom by Y_2, then the statement that the energy before incidence equals the energy after incidence is

$$Y_2 1^2 = Y_2 c^2 + Y_1 t^2 \qquad \text{(8-1-3)}$$

solving for c we get

$$0 = -Y_2 + Y_2 c^2 + Y_1(1 + c)^2$$
$$0 = (c - 1)Y_2 + (c + 1)Y_1$$
$$0 = c(Y_2 + Y_1) + (Y_1 - Y_2) \qquad \text{(8-1-4)}$$
$$c = \frac{Y_2 - Y_1}{Y_2 + Y_1}$$

In acoustics the up- and downgoing wave variables may be normalized to either pressure or velocity. When they measure velocity, the scale factor multiplying velocity squared is called the impedance I. When they measure pressure, the scale factor is called the admittance Y.

The wave c' which reflects when energy is incident from the other side is obtained from (8-1-4) if Y_1 and Y_2 are interchanged. Thus

$$c' = -c \qquad \text{(8-1-5)}$$

A perfectly reflecting interface is one which does not allow energy through. This comes about not only when $t = 0$ or $c = -1$, but also when $t = 2$ or $c = +1$. To see this, note that on the left in Fig. 8-1

$$\frac{\text{Energy transmitted}}{\text{Energy incident}} = \frac{Y_1 t^2}{Y_2 1^2} = \frac{Y_1}{Y_2}\left(1 + \frac{Y_2 - Y_1}{Y_1 + Y_2}\right)^2$$

$$= \frac{Y_1}{Y_2}\left(\frac{2Y_2}{Y_1 + Y_2}\right)^2 = \frac{4Y_1 Y_2}{(Y_1 + Y_2)^2} \qquad \text{(8-1-6)}$$

Equation (8-1-6) says that 100 percent of the incident energy is transmitted when $Y_1 = Y_2$, but the percentage of transmission is very small when Y_1 and Y_2 are very different.

A word of caution: Occasionally special applications are described by authors who do not define reflection and transmission coefficients in terms of some variable which is continuous at a boundary. This is usually an oversight which unfortunately obscures the relationship of the special application to wave theory in general and this chapter in particular. It is almost never an essential feature of the special application that $t \neq 1 + c$ but just a result of an unwise choice of variables in the description. For example, material density is an unwise variable in acoustics because it suffers a discontinuity at a material boundary. Pressure or normal velocity are better descriptors of wave strength.

Ordinarily there are two kinds of variables used to describe waves, and both of these can be continuous at a material discontinuity. One is a scalar like pressure, tension, voltage, potential, stress, or temperature. The other is a vector of which we use the vertical component. Examples of the latter are velocity, stretch, electric current, displacement, and heat flow. Occasionally a wave variable will be a tensor. When a boundary condition is the vanishing of one of the motion components, then the boundary is often said to be rigid. When it is the pressure or potential which vanishes, then the boundary is often said to be free. Rigid and free boundaries reflect waves with unit magnitude reflection coefficients.

The purpose of this chapter is to establish fundamental mathematical properties of waves in layers and to avoid specialization to any particular physical type of waves. That will be done in the next chapter. However, so as not to disguise the physical aspect of the mathematics, a precise definition of upgoing wave U and downgoing wave D will now be given in terms of classical acoustics. In acoustics one deals with pressure P and vertical component of parcel velocity W (not to be confused with wave velocity v). One possible definition for U and D (which will be developed in Chap. 9, Sec. 3) is

$$D = \frac{P + W/Y}{2} \qquad \text{(8-1-7a)}$$

$$U = \frac{P - W/Y}{2} \qquad \text{(8-1-7b)}$$

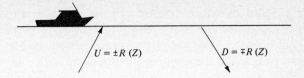

FIGURE 8-2
A waveform $R(Z)$ reflecting at the surface of the sea. Pressure equal to $U + D$ vanishes at the surface. The vertical velocity of the surface is proportional to $D - U$. Theoretically, waves are observed by measuring W at the surface; however, as a practical matter P is often observed a fraction of a wavelength below the surface.

with the inverse relations

$$P = D + U \qquad (8\text{-}1\text{-}8a)$$

$$W = (D - U)Y \qquad (8\text{-}1\text{-}8b)$$

Other definitions with different scale factors and signs are possible. With this definition, the relation $t = 1 + c$ is readily seen to be associated with (8-1-8a) and continuity of pressure at an interface. The minus signs in (8-1-7) and (8-1-8) are associated with the direction of the z axis. Reversal of the z axis changes W to $-W$ and switches the roles of U and D.

We notice that a downgoing wave D all by itself with U vanishing provides a moving disturbance of both pressure P and velocity W, and the vanishing of U assures us that the ratio between the two $W/P = Y$ is the characteristic admittance Y of the material. The energy, we have said, is proportional to either YP^2 or IW^2 from which the ratio $W/P = Y$ allows us to deduce that the impedance of a material is the inverse of its admittance $I = 1/Y$.

For sound waves in the ocean the sea surface is a nearly perfect reflector because of the great contrast between air and water. If this interface is idealized to a perfect reflector, then it is a free surface. Since the pressure vanishes on a free surface, we have that $D = -U$ at the surface so the reflection coefficient is -1. If a wave is to be seen at the surface, it is necessary to measure not pressure but something proportional to velocity. In geophysical exploration practice, pressure-sensing hydrophones are used. They must be kept at a suitable distance below the sea surface. The situation can be depicted as in Fig. 8-2. The pressure normally

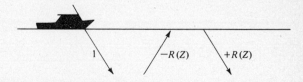

FIGURE 8-3
An initial downgoing disturbance 1 results in a later upgoing reflected wave $-R(Z)$ which reflects back down as $R(Z)$. The pressure at the surface is $D + U = 1 + R - R = 1$.

vanishes at the sea surface, but if we wish to initiate an impulsive disturbance, the pressure may momentarily take on some other value, say 1. This is depicted in Fig. 8-3. The total vertical component of velocity of the sea surface due to the source and to the resulting acoustic wave is $D - U = 1 + 2R(Z)$.

EXERCISES

1 Compute t in terms of Y_1 and Y_2.
2 In a certain application continuity is expressed by saying that $D - U$ is the same on either side of the interface. This implies that $t = 1 - c$. Derive an equation like (8-1-4) for the reflection coefficient in terms of the admittance Y.
3 What are reflection and transmission coefficients in terms of the impedance I? (Clear fractions from your result.)
4 From the principle of energy conservation we showed that $c' = -c$. It may also be deduced from time reversal. To do this, copy Fig. 8-1 with arrows reversed. Scale and linearly superpose various figures in an attempt to create a situation where a figure like the right-hand side of Fig. 8-1 has $-c'$ for the reflected wave. (HINT: Draw arrows at normal incidence.)

8-2 ENERGY FLUX IN LAYERED MEDIA

First consider wave resonance in a layer. Let the travel time through the layer and back again be given by the delay operator Z. The situation is shown in Fig. 8-4. The wave seen above the layer has the form

$$t_2 t_1 [1 + c_1 c_2' Z + (c_1 c_2')^2 Z^2 + \cdots] = \frac{t_2 t_1}{1 - c_1 c_2' Z}$$

It is no accident that the infinite series may be summed. We will soon see that for n layers the waves, which are of infinite duration, may be expressed as simple polynomials of degree n. We will consider many layers and the general problem

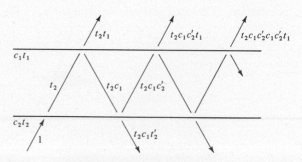

FIGURE 8-4
Some rays corresponding to resonance in a layer.

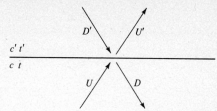

FIGURE 8-5
Waves incident and reflected from an interface.

of determining waves given reflection coefficients and determining reflection coefficients given waves.

The reflection and transmission coefficients show one how to calculate the waves resulting from a wave impinging on a layer. Equation (8-2-1) relates to Fig. 8-5 and shows how from the waves U and D' one extrapolates into the future to get U' and D.

$$U' = tU + c'D'$$
$$D = cU + t'D' \qquad (8\text{-}2\text{-}1)$$

Let us rearrange (8-2-1) to get U' and D' on the right and U and D on the left. Then we will have an equation which extrapolates from the primed medium to the unprimed medium. We get

$$-tU \qquad = -U' + c'D'$$
$$-cU + D = \qquad t'D'$$

which may be arranged in the matrix form

$$\begin{bmatrix} -t & 0 \\ -c & 1 \end{bmatrix} \begin{bmatrix} U \\ D \end{bmatrix} = \begin{bmatrix} -1 & c' \\ 0 & t' \end{bmatrix} \begin{bmatrix} U \\ D \end{bmatrix}'$$

Now premultiplying by the inverse of the left-hand matrix

$$\begin{bmatrix} U \\ D \end{bmatrix} = \frac{1}{-t} \begin{bmatrix} 1 & 0 \\ c & -t \end{bmatrix} \begin{bmatrix} -1 & c' \\ 0 & t' \end{bmatrix} \begin{bmatrix} U \\ D \end{bmatrix}'$$

$$= \frac{1}{-t} \begin{bmatrix} -1 & c' \\ -c & cc' - tt' \end{bmatrix} \begin{bmatrix} U \\ D \end{bmatrix}'$$

finally getting the result, an equation to extrapolate from the primed medium to the unprimed medium.

$$\begin{bmatrix} U \\ D \end{bmatrix} = \frac{1}{t} \begin{bmatrix} 1 & c \\ c & 1 \end{bmatrix} \begin{bmatrix} U \\ D \end{bmatrix}' \qquad (8\text{-}2\text{-}2)$$

Now let us consider the Goupillaud type [Ref. 30] layered medium shown in Fig. 8-6. For this arrangement of layers, (8-2-2) may be written

$$\begin{bmatrix} U \\ D \end{bmatrix}_{k+1} = \frac{1}{t_k} \begin{bmatrix} 1 & c_k \\ c_k & 1 \end{bmatrix} \begin{bmatrix} U \\ D \end{bmatrix}_k$$

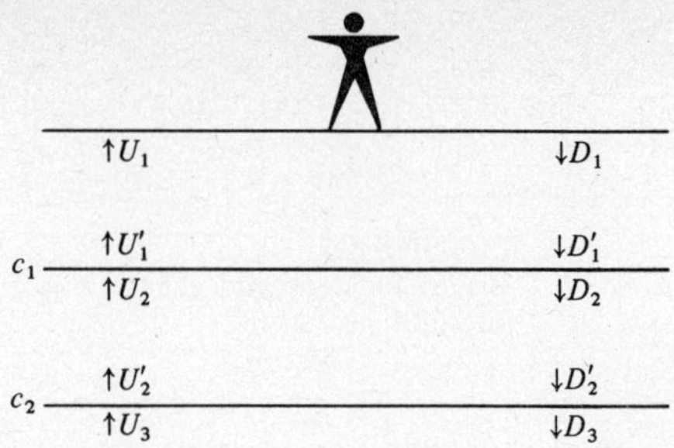

FIGURE 8-6
Goupillaud-type layered medium (layers have equal travel time).

Let $Z = e^{i\omega T}$ where T, the two-way travel time, equals the data sampling interval. Clearly, multiplication by $\sqrt{Z}$ is equivalent to delaying a function by $T/2$, the travel time across a layer. This gives in the kth layer a relation between primed and unprimed waves.

$$\begin{bmatrix} U \\ D \end{bmatrix}'_k = \begin{bmatrix} 1/\sqrt{Z} & 0 \\ 0 & \sqrt{Z} \end{bmatrix} \begin{bmatrix} U \\ D \end{bmatrix}_k \qquad (8\text{-}2\text{-}3)$$

Inserting (8-2-3) into (8-2-2) we get a *layer matrix*

$$\begin{aligned} \begin{bmatrix} U \\ D \end{bmatrix}_{k+1} &= \frac{1}{t_k} \begin{bmatrix} 1 & c_k \\ c_k & 1 \end{bmatrix} \begin{bmatrix} 1/\sqrt{Z} & 0 \\ 0 & \sqrt{Z} \end{bmatrix} \begin{bmatrix} U \\ D \end{bmatrix}_k \\ &= \frac{1}{t_k} \begin{bmatrix} 1/\sqrt{Z} & c_k\sqrt{Z} \\ c_k/\sqrt{Z} & \sqrt{Z} \end{bmatrix} \begin{bmatrix} U \\ D \end{bmatrix}_k \\ &= \frac{1}{\sqrt{Z}t_k} \begin{bmatrix} 1 & c_k Z \\ c_k & Z \end{bmatrix} \begin{bmatrix} U \\ D \end{bmatrix}_k \end{aligned} \qquad (8\text{-}2\text{-}4)$$

If there is energy flowing through a stack of layers, there must be the same total flow through the kth layer as through the $(k+1)$st layer. Otherwise, there is an energy sink or source at the layer boundary. The net upward flow of energy (energy flux) at any frequency ω in the kth layer is given by

$$\text{flux}(\omega) = Y_k \left(U(Z)\bar{U}\left(\frac{1}{Z}\right) - D(Z)\bar{D}\left(\frac{1}{Z}\right) \right)_k \qquad (8\text{-}2\text{-}5)$$

To establish that this is indeed independent of k, we take the Hermitian conjugate (transpose and conjugate with respect to real ω) of (8-2-4).

$$\left[U\left(\frac{1}{Z}\right) D\left(\frac{1}{Z}\right) \right]_{k+1} = \frac{1}{t_k} \left[U\left(\frac{1}{Z}\right) D\left(\frac{1}{Z}\right) \right]_k \begin{bmatrix} \sqrt{Z} & c_k\sqrt{Z} \\ c_k/\sqrt{Z} & 1/\sqrt{Z} \end{bmatrix} \qquad (8\text{-}2\text{-}6)$$

Now combine (8-2-4) with (8-2-6) in the form

$$\left[U\left(\frac{1}{Z}\right)D\left(\frac{1}{Z}\right)\right]_{k+1}\begin{bmatrix}1 & 0 \\ 0 & -1\end{bmatrix}\begin{bmatrix}U(Z) \\ D(Z)\end{bmatrix}_{k+1}$$

$$=\frac{1}{t_k^2}\begin{bmatrix}\bar{U} & \bar{D}\end{bmatrix}_k\begin{bmatrix}\sqrt{Z} & \sqrt{Z}c_k \\ c_k/\sqrt{Z} & 1/\sqrt{Z}\end{bmatrix}\begin{bmatrix}1 & 0 \\ 0 & -1\end{bmatrix}\begin{bmatrix}1/\sqrt{Z} & \sqrt{Z}c_k \\ c_k/\sqrt{Z} & \sqrt{Z}\end{bmatrix}\begin{bmatrix}U \\ D\end{bmatrix}_k \qquad (8\text{-}2\text{-}7)$$

$$=\frac{1}{t_k^2}\begin{bmatrix}\bar{U} & \bar{D}\end{bmatrix}_k\begin{bmatrix}1-c_k^2 & 0 \\ 0 & c_k^2-1\end{bmatrix}\begin{bmatrix}U \\ D\end{bmatrix}_k$$

Since $(1-c_k^2)/t_k^2 = t_k'/t_k = Y_k/Y_{k+1}$ this may be rewritten as the desired result, namely

$$Y_{k+1}\left[U\left(\frac{1}{Z}\right)U(Z) - D\left(\frac{1}{Z}\right)D(Z)\right]_{k+1} = Y_k\left[U\left(\frac{1}{Z}\right)U(Z) - D\left(\frac{1}{Z}\right)D(Z)\right]_k \qquad (8\text{-}2\text{-}8)$$

Equation (8-2-8) says that at each frequency ω the energy flowing through the kth layer equals the energy flowing through the $(k + 1)$st layer.

This energy flux theorem leads quickly to some sweeping statements about the waveforms scattered from layered structures. Figure 8-7 shows the basic geometry of reflection seismology. Applying the energy flux theorem to this geometry we may say that the energy flux in the top layer equals that in the lower halfspace so

$$Y_1\left\{R\left(\frac{1}{Z}\right)R(Z) - \left[1 + R\left(\frac{1}{Z}\right)\right][1 + R(Z)]\right\} = -Y_k E\left(\frac{1}{Z}\right)E(Z)$$

or

$$1 + R\left(\frac{1}{Z}\right) + R(Z) = \frac{Y_k}{Y_1}E\left(\frac{1}{Z}\right)E(Z) \qquad (8\text{-}2\text{-}9)$$

This very remarkable result says that if we were to observe the escaping wave $E(Z)$, we could by autocorrelation construct the waveform seen at the surface. We will later see that $E(Z)$ is minimum-phase so that E could be constructed from R by spectral factorization.

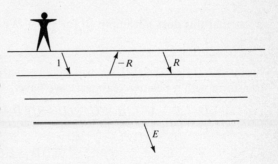

FIGURE 8-7
Basic reflection seismology geometry. The man initiates an impulse going downward. The earth sends back $-R(Z)$ to the surface. Since the surface is perfectly reflective, the surface sends $R(Z)$ back into the earth. Escaping from the bottom of the layers is a wave $E(Z)$ which is heading toward the other side of the earth.

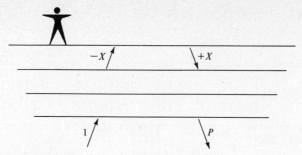

FIGURE 8-8
Earthquake seismology geometry. An impluse 1 is incident from below. The waveform $X(Z)$ is incident upon the free surface and is reflected back down. The waveform $P(Z)$ scatters back into the earth.

Now let us turn our attention to the earthquake seismology geometry depicted in Fig. 8-8. Applying the energy flux theorem to this geometry we obtain

$$Y_1\left[X\left(\frac{1}{Z}\right)X(Z) - X\left(\frac{1}{Z}\right)X(Z)\right] = Y_k\left[1 - P(Z)P\left(\frac{1}{Z}\right)\right]$$

or

$$1 = P(Z)P\left(\frac{1}{Z}\right) \qquad (8\text{-}2\text{-}10)$$

The interpretation of the result is that the backscattered waveform $P(Z)$ has the form of an all-pass filter. This result may have been anticipated on physical grounds since all the energy which is incident is ultimately reflected without attenuation; thus the only thing which can happen is that there will be frequency-dependent delay.

Finally, we will derive a theorem which relates energy flux to impedance and admittance functions (these functions have Fourier transforms with a positive real part). Suppose that a downgoing wave $D(Z)$ is stronger than an upgoing wave $U(Z)$ at all frequencies, i.e.

$$D\left(\frac{1}{Z}\right)D(Z) - U\left(\frac{1}{Z}\right)U(Z) > 0 \qquad \text{on } |Z| = 1 \qquad (8\text{-}2\text{-}11)$$

(Note that this does not imply $|d_t| > |u_t|$.) We will abbreviate (8-2-11) by

$$\bar{D}D - \bar{U}U > 0 \qquad (8\text{-}2\text{-}12)$$

From (8-2-11) or (8-2-12) we will deduce that $(D - U)/(D + U)$ has a Fourier transform with a positive real part. We have

$$2\,\text{Re}\,\frac{D - U}{D + U} = \frac{D - U}{D + U} + \frac{\bar{D} - \bar{U}}{\bar{D} + \bar{U}} = \frac{(D - U)(\bar{D} + \bar{U}) + (\bar{D} - \bar{U})(D + U)}{(D + U)(\bar{D} + \bar{U})}$$

$$= \frac{2(\bar{D}D - \bar{U}U)}{(D + U)(\bar{D} + \bar{U})} \qquad (8\text{-}2\text{-}13)$$

The numerator of (8-2-13) is positive by hypothesis (8-2-12) and the denominator of (8-2-13) is positive, since it is the spectrum of the time function $d_t + u_t$ and any spectrum is always positive. Thus $(D - U)/(D + U)$ is called "positive real." The acoustical interpretation of $(D - U)/(D + U)$ is that $(D - U)$ represents the vertical component of material velocity and $(D + U)$ represents the material pressure.

8-3 GETTING THE WAVES FROM THE REFLECTION COEFFICIENTS

A layered material may be specified by giving the reflection coefficient at each interface. Alternate descriptions are to give any one of the scattered waves $R(Z)$, $E(Z)$, $X(Z)$, or $P(Z)$. Our ultimate objective is to get such a good grip on the algebra of this kind of problem that we will be able to start with any descriptor of the layers and from it deduce all the other descriptors.

An important result of the last section was the development of a "layer matrix" (8-2-4) that is, a matrix which can be used to extrapolate waves observed in one layer to the waves observed in the next layer. This process may be continued indefinitely. To see how to extrapolate from layer 1 to layer 3 substitute (8-2-4) with $k = 1$ into (8-2-4) with $k = 2$, obtaining

$$\begin{bmatrix} U \\ D \end{bmatrix}_3 = \frac{1}{\sqrt{Zt_2}} \begin{bmatrix} 1 & Zc_2 \\ c_2 & Z \end{bmatrix} \frac{1}{\sqrt{Zt_1}} \begin{bmatrix} 1 & Zc_1 \\ c_1 & Z \end{bmatrix} \begin{bmatrix} U \\ D \end{bmatrix}_1$$

$$= \frac{1}{\sqrt{Z^2 t_1 t_2}} \begin{bmatrix} 1 + Zc_1c_2 & Zc_1 + Z^2c_2 \\ c_2 + Zc_1 & Zc_1c_2 + Z^2 \end{bmatrix} \begin{bmatrix} U \\ D \end{bmatrix}_1 \quad (8\text{-}3\text{-}1)$$

Inspection of this example suggests the general form for a product of k layer matrices

$$\frac{1}{\sqrt{Z^k} \prod_{j=1}^{k} t_j} \begin{bmatrix} F(Z) & Z^k G\left(\frac{1}{Z}\right) \\ G(Z) & Z^k F\left(\frac{1}{Z}\right) \end{bmatrix} \quad (8\text{-}3\text{-}2)$$

Now let us verify that (8-3-2) is indeed the general form. We assume (8-3-2) is correct for $k - 1$; then we multiply (8-3-2) by another layer matrix and see if the product retains the same form with $k - 1$ increased to k. The product is

$$\frac{1}{\sqrt{Zt_k}} \begin{bmatrix} 1 & c_k Z \\ c_k & Z \end{bmatrix} \frac{1}{\sqrt{Z^{k-1}} \prod_{j=1}^{k-1} t_j} \begin{bmatrix} F(Z) & Z^{k-1} G\left(\frac{1}{Z}\right) \\ G(Z) & Z^{k-1} F\left(\frac{1}{Z}\right) \end{bmatrix}$$

$$= \frac{1}{\sqrt{Z^k} \prod_1^k t_j} \begin{bmatrix} F(Z) + c_k Z G(Z) & Z^{k-1} G\left(\frac{1}{Z}\right) + c_k Z^k F\left(\frac{1}{Z}\right) \\ c_k F(Z) + Z G(Z) & c_k Z^{k-1} G\left(\frac{1}{Z}\right) + Z^k F\left(\frac{1}{Z}\right) \end{bmatrix} \quad (8\text{-}3\text{-}3)$$

By inspecting the product we see that the scaling factor is of the same form with $k - 1$ changed to k. Also the 22 matrix element can be obtained from the 11 element by replacing Z with $1/Z$ and multiplying by Z^k. Likewise, the 21 element is obtained from the 12 element; thus (8-3-2) does indeed represent a general form. The polynomials $F(Z)$ and $G(Z)$ of order k are built up in the following way [from the first column of the right-hand side of (8-3-3)]:

$$F_k(Z) = F_{k-1}(Z) + c_k Z G_{k-1}(Z) \qquad (8\text{-}3\text{-}4a)$$

$$G_k(Z) = c_k F_{k-1}(Z) + Z G_{k-1}(Z) \qquad (8\text{-}3\text{-}4b)$$

By inspecting (8-3-4) we can see some of the details of F and G. From (8-3-4a) we see that the lead coefficient f_0 of $F(Z)$ does not change with k. It is always $(f_0)_k = 1$. Knowing this from (8-3-4b) we see that $(g_0)_k = c_k$. Also with knowledge that $F(Z)$ and $G(Z)$ are of the same degree in Z, we see that (8-3-4b) implies that the highest coefficient of $G(Z)$, say $(g_k)_k$ does not change with k and therefore it equals the starting value of c_1. Finally, with this knowledge and (8-3-4a) we deduce that the highest coefficient in $F(Z)$ will always be $c_1 c_k$. Thus, in summary

$$F(Z) = 1 + f_1 Z + f_2 Z^2 + \cdots + c_1 c_k Z^{k-1} \qquad (8\text{-}3\text{-}5a)$$

$$G(Z) = c_k + g_1 Z + g_2 Z^2 + \cdots + c_1 Z^{k-1} \qquad (8\text{-}3\text{-}5b)$$

It may be noted in (8-3-5) and proved from the recurrence relations (8-3-4) that the coefficients of F contain even powers of c and that G contains odd powers of c. This means that if all c change sign, G will change sign but F is unchanged.

The polynomials $F(Z)$ and $G(Z)$ are not independent and a surprising energy-flux-like relationship exists between them. By substitution from (8-3-4) one may directly verify that

$$\left[F(Z)F\left(\frac{1}{Z}\right) - G(Z)G\left(\frac{1}{Z}\right) \right]_k = (1 - c_k^2) \left[F(Z)F\left(\frac{1}{Z}\right) - G(Z)G\left(\frac{1}{Z}\right) \right]_{k-1} \qquad (8\text{-}3\text{-}6)$$

Since $F_1(Z) = 1$ and $G_1(Z) = c_1$ we have by iterative application of (8-3-6) that

$$\left[F(Z)F\left(\frac{1}{Z}\right) - G(Z)G\left(\frac{1}{Z}\right) \right]_k = \prod_1^k (1 - c_k^2) = \prod_1^k t't \qquad (8\text{-}3\text{-}7)$$

Equation (8-3-7) is a surprising equation because on the left-hand side we have two spectra, the spectrum of f_t and the spectrum of g_t, but the right-hand side is a positive, frequency-independent constant. Since the spectrum of f_t is thus greater than the spectrum of g_t, we may apply the theorem of adding garbage to a minimum-phase wavelet to deduce from (8-3-4a) and from knowledge that $|c_k| < 1$ that $F_k(Z)$ is minimum-phase if $F_{k-1}(Z)$ is minimum-phase. Since $F_1(Z) = 1$ is minimum-phase, we see that all $F_k(Z)$ are minimum-phase. Since $F(Z)$ is minimum-phase, then $F(Z)$ may be calculated from its spectrum $F(Z)F(1/Z)$ or the spectrum of g_t (along with the single number $\pi t't$). However, we cannot get G from F. Before continuing our algebraic discussion we take up an example.

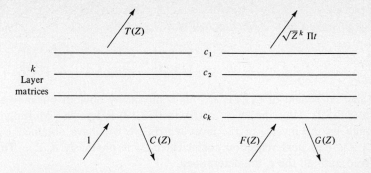

FIGURE 8-9
Waves incident, reflected, and transmitted from a stack of layers between two half-spaces.

Let a stack of layers be sandwiched in between two halfspaces (Fig. 8-9). An impulse is incident from below. The backscattered wave is called $C(Z)$ and the transmitted wave is called $T(Z)$.

Mathematically, we describe the situation with the equations

$$\begin{bmatrix} 1 \\ C(Z) \end{bmatrix} = \frac{1}{\sqrt{Z^k \Pi t}} \begin{bmatrix} F(Z) & Z^k G\left(\dfrac{1}{Z}\right) \\ G(Z) & Z^k F\left(\dfrac{1}{Z}\right) \end{bmatrix} \begin{bmatrix} T(Z) \\ 0 \end{bmatrix} \qquad (8\text{-}3\text{-}8)$$

We may solve the first of (8-3-8) for the transmitted wave $T(Z)$

$$T(Z) = \frac{\sqrt{Z^k \Pi t}}{F(Z)} \qquad (8\text{-}3\text{-}9)$$

and introduce the result back into the second of (8-3-8) to obtain the backscattered wave

$$C(Z) = \frac{G(Z) T(Z)}{\sqrt{Z^k \Pi t}} = \frac{G(Z)}{F(Z)} \qquad (8\text{-}3\text{-}10)$$

The mathematical fact that $F(Z)$ is minimum-phase corresponds to the physical fact that the $C(Z)$ and $T(Z)$ have finite energy; therefore the denominators of (8-3-9) and (8-3-10) cannot have zeros inside the unit circle. Since we know that the backscattered wave $C(Z)$ contains less energy than the incident wave by reference to (8-2-13) we know that a positive real function is given by

$$\frac{1 - C(Z)}{1 + C(Z)} = \frac{1 - G(Z)/F(Z)}{1 + G(Z)/F(Z)} \qquad (8\text{-}3\text{-}11)$$

Now let us see how to reconstruct the reflection coefficients c_j from the observed scattered wave $C(Z)$. Referring to Fig. 8-9 we have

$$\begin{bmatrix} 1 \\ C(Z) \end{bmatrix} = \frac{1}{t_k \sqrt{Z}} \begin{bmatrix} 1 & c_k Z \\ c_k & Z \end{bmatrix} \begin{bmatrix} U \\ D \end{bmatrix}_{k-1} \quad (8\text{-}3\text{-}12)$$

The first coefficient of $C(Z)$ is c_k [this is physically obvious but may also be seen from (8-3-5)]. Thus the layer matrix in (8-3-12) is known. Multiplying (8-3-12) through by the inverse of the layer matrix we will have obtained $U_{k-1}(Z)$ and $D_{k-1}(Z)$. The next reflection coefficient c_{k-1} is obviously d_0/u_0. Thus we may proceed until all the c_k are determined.

Next let us reconsider the reflection seismology geometry. We have

$$\begin{bmatrix} 0 \\ E(Z) \end{bmatrix} = \frac{1}{\sqrt{Z^k \Pi t}} \begin{bmatrix} F(Z) & Z^k G\left(\dfrac{1}{Z}\right) \\ G(Z) & Z^k F\left(\dfrac{1}{Z}\right) \end{bmatrix} \begin{bmatrix} -R \\ 1+R \end{bmatrix} \quad (8\text{-}3\text{-}13)$$

From the first equation we may solve for $R(Z)$

$$R(Z) = +\frac{Z^k G(1/Z)}{F(Z) - Z^k G(1/Z)} \quad (8\text{-}3\text{-}14)$$

The denominator occurs so often that we give it the name $A(Z)$

$$A(Z) = F(Z) - Z^k G\left(\frac{1}{Z}\right) \quad (8\text{-}3\text{-}15)$$

$A(Z)$, like $F(Z)$, is minimum-phase. The second of (8-3-13) gives the escaping wave as

$$E(Z) = \frac{Z^k F(1/Z) + [-G(Z) + Z^k F(1/Z)]R}{\sqrt{Z^k \Pi t}}$$

$$= \frac{Z^k F(1/Z)[F(Z) - Z^k G(1/Z)] + [-G(Z) + Z^k F(1/Z)]Z^k G(1/Z)}{A(Z)\sqrt{Z^k \Pi t}}$$

simplifying with (8-3-7) we get

$$E(Z) = \frac{\sqrt{Z^k \Pi t'}}{A(Z)} \quad (8\text{-}3\text{-}16)$$

The positive real function is

$$\frac{D-U}{D+U} = \frac{1+R-(-R)}{1+R-R} = 1 + 2R(Z) \quad (8\text{-}3\text{-}17)$$

$$= \frac{\text{Vertical velocity} = 1 + 2R.}{\text{Pressure} = 1}$$

As mentioned earlier, if the equations are interpreted in terms of acoustics, then $Y(D - U)/(D + U)$ is interpreted as vertical velocity divided by pressure. It is called the *admittance* which is the inverse of the impedance.

We have now completed the task of solving for the waves given the reflection coefficients. In the subsequent section we attack the inverse problems of getting the reflection coefficients from knowledge of various waves.

EXERCISES

1 In Fig. 8-9 let $c_1 = \frac{1}{2}$, $c_2 = -\frac{1}{3}$, and $c_3 = \frac{1}{4}$. What are the polynomial ratios $T(Z)$ and $C(Z)$?

2 For a simple interface, we had the simple relations $t = 1 + c$, $t' = 1 + c'$, and $c = -c'$. What sort of analogous relations can you find for the generalized interface of Fig. 8-9? [For example, show $1 - T(Z)T'(1/Z) = C(Z)C(1/Z)$ which is analogous to $1 - tt' = c^2$.]

3 Show that $T(Z)$ and $T'(Z)$ are the same waveforms within a scale factor. Deduce that many different stacks of layers may have the same $T(Z)$.

4 Let an impulse be incident on a stack of layers and let a wave $C(Z)$ be reflected. What is the reflection coefficient at the first layer encountered? What would be the reflected wave as a function of C for a situation which differs from the above by the removal of the first reflector?

5 Consider the earth to be modeled by layers over a halfspace. Let an impulse be incident from below (Fig. E8-3-5). Given $F(Z)$ and $G(Z)$, elements of the product of the layer

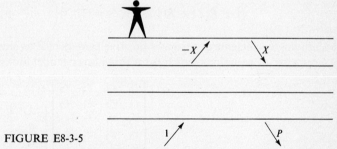

FIGURE E8-3-5

matrices, solve for X and for P. Check your answer by showing that $P(Z)\bar{P}(1/Z) = 1$. How is X related to E? This relation illustrates the principle of reciprocity which says source and receiver may be interchanged.

6 Show that $1 + R(1/Z) + R(Z) = $ (scale factor) $X(Z)\ X(1/Z)$, which shows that one may autocorrelate the transmission seismogram to get the reflection seismogram.

7 Refer to Fig. E8-3-7. Calculate R' from R.

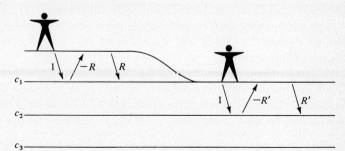

FIGURE E8-3-7

8-4 GETTING THE REFLECTION COEFFICIENTS FROM THE WAVES

The best starting point for inverse problems is the Kunetz equation [Ref. 31] (8-2-9).

$$1 + R\left(\frac{1}{Z}\right) + R(Z) = \frac{Y_k}{Y_1} E\left(\frac{1}{Z}\right) E(Z) \qquad (8\text{-}4\text{-}1)$$

We need also the expression for the escaping wave (8-3-16)

$$E(Z) = \frac{\sqrt{Z^k}\Pi t'}{A(Z)} \qquad (8\text{-}4\text{-}2)$$

We also need to recall that $Y_k/Y_1 = \pi t/t'$. With this (8-4-1) becomes

$$1 + R(Z) + R\left(\frac{1}{Z}\right) = \frac{\Pi t' t}{A(Z)A(1/Z)} \qquad (8\text{-}4\text{-}3)$$

Multiplying through by $A(Z)$ we get

$$\left[1 + R(Z) + R\left(\frac{1}{Z}\right)\right]A(Z) = \frac{\Pi t' t}{A(1/Z)} \qquad (8\text{-}4\text{-}4)$$

Since $A(Z)$ is minimum-phase, $A(Z)$ may be written as $1/B(Z)$ or $A(1/Z) = 1/B(1/Z)$. Thus (8-4-4) becomes

$$[1 + R(Z) + R(1/Z)]A(Z) = (\Pi t' t)\left(1 + \frac{b_1}{Z} + \frac{b_2}{Z^2} + \cdots\right) \qquad (8\text{-}4\text{-}5)$$

Identifying coefficients of zero and positive powers of Z as simultaneous equations, we get a set of equations which for a three-layer model looks like ($r_0 = 1$).

$$\begin{bmatrix} r_0 & r_1 & r_2 & r_3 \\ r_1 & r_0 & r_1 & r_2 \\ r_2 & r_1 & r_0 & r_1 \\ r_3 & r_2 & r_1 & r_0 \\ r_4 & r_3 & r_2 & r_1 \\ \cdot & \cdot & \cdot & \cdot \\ \cdot & \cdot & \cdot & \cdot \end{bmatrix} \begin{bmatrix} 1 \\ a_1 \\ a_2 \\ -c_3 \end{bmatrix} = \begin{bmatrix} \Pi t' t \\ 0 \\ 0 \\ 0 \\ 0 \\ \cdot \end{bmatrix} \qquad (8\text{-}4\text{-}6)$$

In (8-4-6) we see our old friend the Toeplitz matrix. It used to work for factoring spectra and predicting time series. Notice that $-c_3$ has been inserted in (8-4-6) as the highest coefficient of $A(Z)$. This is justified by reference back to the definition of $A(Z)$ in terms of $F(Z)$ and $G(Z)$ which were in turn defined from the c_k. It is by reexamining the Teoplitz simultaneous equations (8-4-6) and the Levinson method of solution (3-3-8) that we will learn how to compute the reflection coefficients from the waves.

The first four equations in (8-4-6) would normally be thought of as follows: Given the first three reflected pulses r_1, r_2, and r_3 we may solve the equations for A, incidentally getting the reflection coefficient c_3. Knowing A, the 5th equation in (8-4-6) may be used to compute r_4. If the model were truly a three-layer model, it would come out right; if not, the discrepancy would be indicative of another reflector c_4 which could be found by expanding equation (8-4-6) from 4th order to 5th order. In summary, given the reflected pulses r_k, the Levinson recursion successively turns out the reflection coefficients c_k.

Now suppose we begin by observation of the escaping wave $E(Z)$. One way to determine the reflection coefficients would be to form $1 + R(Z) + R(1/Z)$ by the autocorrelation of $E(Z)$; then, the Levinson recursion could be used to solve for the reflection coefficients. The only disadvantage of this method is that $E(Z)$ contains an infinite number of coefficients so that in practice some truncation must be done. The truncation is avoided by an alternative method. Given $E(Z)$ polynomial division will find $A(Z)$. The heart of the Levinson recursion is the building up of $A(Z)$ by $A_{k+1}(Z) = A_k(Z) - cZ^k A_k(1/Z)$. In particular, from (3-3-12) we have

$$\begin{bmatrix} 1 \\ a_1 \\ a_2 \\ a_3 \end{bmatrix}_3 = \begin{bmatrix} 1 \\ a_1 \\ a_2 \\ 0 \end{bmatrix}_2 - c_3 \begin{bmatrix} 0 \\ a_2 \\ a_1 \\ 1 \end{bmatrix}_2 \qquad (8\text{-}4\text{-}7)$$

which shows how to get $A_3(Z)$ from $A_2(Z)$ and c_3. To do it backwards, we see first that c_3 is $-a_3$. Then write (8-4-7) upside-down

$$\begin{bmatrix} a_3 \\ a_2 \\ a_1 \\ 1 \end{bmatrix}_3 = \begin{bmatrix} 0 \\ a_2 \\ a_1 \\ 1 \end{bmatrix}_2 - c_3 \begin{bmatrix} 1 \\ a_1 \\ a_2 \\ 0 \end{bmatrix}_2 \qquad (8\text{-}4\text{-}8)$$

Next multiply (8-4-7) by $1/(1 - c_3{}^2)$ and add the product to (8-4-8) multiplied by $c_3/(1 - c_3{}^2)$. Notice that the upside-down vectors on the right-hand side cancel, leaving

$$\frac{1}{1 - c_3{}^2} \begin{bmatrix} 1 \\ a_1 \\ a_2 \\ a_3 \end{bmatrix}_3 + \frac{c_3}{1 - c_3{}^2} \begin{bmatrix} a_3 \\ a_2 \\ a_1 \\ 1 \end{bmatrix}_3 = \begin{bmatrix} 1 \\ a_1 \\ a_2 \\ 0 \end{bmatrix}_2 \qquad (8\text{-}4\text{-}9)$$

Equation (8-4-9) is the desired result which shows how to reduce $A_{k+1}(Z)$ to $A_k(Z)$ while learning c_{k+1}. A program to continue this process is given in Fig. 8-10. An inverse program to get R and A from c is in Fig. 8-11.

FIGURE 8-10
A program to compute reflection coefficients c_k from the prediction-error filter $A(Z)$. The complex arithmetic is optional.

```
      COMPLEX A,C,AL,BE,TOP,CONJG
      C(1)=-1.; R(1)=1.; A(1)=1.; V(1)=1.
300   DO 310 I=1,N
310   C(I)=A(I)
      DO 330 K=1,N
      J=N-K+2
      AL=1./(1.-C(J)*CONJG(C(J)))
      BE=C(J)*AL
      JH=(J+1)/2
      DO 320 I=1,JH
      TOP=AL*C(I)-BE*CONJG(C(J-I+1))
      C(J-I+1)=AL*C(J-I+1)-BE*CONJG(C(I))
320   C(I)=TOP
330   C(J)=-BE/AL
```

```
         COMPLEX C,R,A,BOT,CONJG
         C(1)=-1.; R(1)=1.; A(1)=1.; V(1)=1.
100      DO 120 J=2,N
         A(J)=0.
         R(J)=C(J)*V(J-1)
         V(J)=V(J-1)*(1.-C(J)*CONJG(C(J)))
         DO 110 I=2,J
110      R(J)=R(J)-A(I)*R(J-I+1)
         JH=(J+1)/2
         DO 120 I=1,JH
         BOT=A(J-I+1)-C(J)*CONJG(A(I))
         A(I)=A(I)-C(J)*CONJG(A(J-I+1))
120      A(J-I+1)=BOT
```

FIGURE 8-11
A program inverse to the program of
Fig. 8-10. It computes both R and A
from c.

Finally, let us see how to do a problem where there are random sources. Figure 8-12 shows the "earthquake geometry." However, in order to introduce a statistical element, the pulse incident from below has been convolved with a white-light series w_t of random numbers. Consequently, all the waves internal to Fig. 8-12 are given by the convolution of w_t with the corresponding wave in the impulse-incident model. Now suppose we are given the top-layer waves $D = -U = XW$ and wish to consider downward continuation. We have the layer matrix

$$\begin{bmatrix} U \\ D \end{bmatrix}_{k+1} = \frac{1}{\sqrt{Z}t_k} \begin{bmatrix} 1 & cZ \\ c & Z \end{bmatrix} \begin{bmatrix} U \\ D \end{bmatrix}_k \quad (8\text{-}4\text{-}10)$$

which can be re-written as

$$\begin{bmatrix} -U \\ D \end{bmatrix}_{k+1} = \frac{1}{\sqrt{Z}t_k} \begin{bmatrix} 1 & -cZ \\ -c & Z \end{bmatrix} \begin{bmatrix} -U \\ D \end{bmatrix}_k \quad (8\text{-}4\text{-}11)$$

The Burg prediction-error scheme can be written in the form

$$\begin{bmatrix} e^+ \\ e^- \end{bmatrix}_{k+1} = \begin{bmatrix} 1 & -c \\ -c & 1 \end{bmatrix} \begin{bmatrix} e^+ \\ e^- \end{bmatrix}_k \quad (8\text{-}4\text{-}12)$$

which makes it equivalent within a scale factor to downward continuing surface waveforms. The remaining question is whether Burg's estimate of the reflection coefficient, namely,

$$\hat{c}_k = \frac{2 \sum_t e_k^+ e_k^-}{\sum_t e_k^+ e_k^+ + e_k^- e_k^-} \quad (8\text{-}4\text{-}13)$$

FIGURE 8-12
Earthquake seismogram geometry with white light incident from below. In the top layer, the sum of the waves vanishes representing zero pressure at the free surface. The difference of up- and down-going waves is the observed vertical component of velocity.

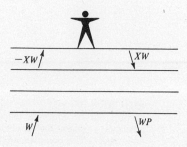

turns out to estimate the reflection coefficient c_k in the physical model. To see how Burg's $\hat{c}_k$ is related to the c_k arising in the Levinson recursion, we define $\mathbf{f}^+$ and $\mathbf{f}^-$ for $k = 2$ as

$$
\begin{bmatrix} \\ \mathbf{f}^+ \\ \\ \end{bmatrix} = \begin{bmatrix} x_0 & & \\ x_1 & x_0 & \\ x_2 & x_1 & x_0 \\ & x_2 & x_1 \\ & & x_2 \end{bmatrix} \begin{bmatrix} 1 \\ a_1 \\ 0 \end{bmatrix} \quad \text{and} \quad \begin{bmatrix} \\ \mathbf{f}^- \\ \\ \end{bmatrix} = \begin{bmatrix} x_0 & & \\ x_1 & x_0 & \\ x_2 & x_1 & x_0 \\ & x_2 & x_1 \\ & & x_2 \end{bmatrix} \begin{bmatrix} 0 \\ a_1 \\ 1 \end{bmatrix} \quad (8\text{-}4\text{-}14)
$$

Next form the dot product

$$
(\mathbf{f}^-)^T \mathbf{f}^+ = \begin{bmatrix} 0 & a_1 & 1 \end{bmatrix} \begin{bmatrix} x_0 & x_1 & x_2 & & \\ & x_0 & x_1 & x_2 & \\ & & x_0 & x_1 & x_2 \end{bmatrix} \begin{bmatrix} x_0 & & \\ x_1 & x_0 & \\ x_2 & x_1 & x_0 \\ & x_2 & x_1 \\ & & x_2 \end{bmatrix} \begin{bmatrix} 1 \\ a_1 \\ 0 \end{bmatrix}
$$

$$
= \begin{bmatrix} 0 & a_1 & 1 \end{bmatrix} \begin{bmatrix} r_0 & r_1 & r_2 \\ r_1 & r_0 & r_1 \\ r_2 & r_1 & r_0 \end{bmatrix} \begin{bmatrix} 1 \\ a_1 \\ 0 \end{bmatrix} \quad (8\text{-}4\text{-}15)
$$

Now utilize the fact that $(1, a_1)$ satisfies the 2×2 system. Following the Levinson recursion (8-4-15) can be written as

$$
(\mathbf{f}^-)^T \mathbf{f}^+ = \begin{bmatrix} 0 & a_1 & 1 \end{bmatrix} \begin{bmatrix} v \\ 0 \\ e \end{bmatrix} = e \quad (8\text{-}4\text{-}16)
$$

Likewise we can deduce that $\mathbf{f}^+ \cdot \mathbf{f}^+ = \mathbf{f}^- \cdot \mathbf{f}^- = v$. Thus, the Levinson calculation of the reflection coefficient can be written as

$$
c = \frac{2(\mathbf{f}^+ \cdot \mathbf{f}^-)}{(\mathbf{f}^+ \cdot \mathbf{f}^+) + (\mathbf{f}^- \cdot \mathbf{f}^-)} = \frac{2e}{2v} \quad (8\text{-}4\text{-}17)
$$

The Burg treatment differs from the Levinson treatment in that Burg omits end-effect terms on (8-4-14). Instead of (8-4-14) he has

$$
\begin{bmatrix} \mathbf{e}^+ \end{bmatrix} = \begin{bmatrix} x_1 & x_0 \\ x_2 & x_1 \end{bmatrix} \begin{bmatrix} 1 \\ a_1 \end{bmatrix} \quad \text{and} \quad \begin{bmatrix} \mathbf{e}^- \end{bmatrix} = \begin{bmatrix} x_1 & x_0 \\ x_2 & x_1 \end{bmatrix} \begin{bmatrix} a_1 \\ 1 \end{bmatrix} \quad (8\text{-}4\text{-}18)
$$

For a sufficiently long data sequence the Burg method and the Levinson technique thus become indistinguishable. For a data sample of finite duration we must make a choice. The Levinson technique with (8-4-14) is equivalent to assuming the data sample vanishes off the ends of the interval in which it is observed. In most applications this is untrue, and so the Burg technique is usually preferable.

EXERCISES

1 An impulse and the first part of a reflection seismogram, that is, $1 + 2R(Z)$ is $1 + 2(Z/4 + Z^2/16 + Z^3/4 + \cdots)$. What are the first three reflection coefficients? Assuming there are no more reflectors what is the next point in the reflection seismogram?

2 A seismogram $X(Z) = 1/(1 - .1Z + .9Z^2)$ is observed at the surface of some layers over a halfspace. Sketch the time function and indicate its resonance frequency and decay time. Find the reflection coefficients if $X(Z)$ is due to an impulsive source of unknown magnitude in the halfspace below the layers.

3 A source $b_0 + b_1 Z$ deep in the halfspace produces a seismogram $B(Z)X(Z) = 1 - Z + Z^2/2 - Z^3/2 + Z^4/4 - Z^5/4 + Z^6/8 - Z^7/8 + \cdots$. What are the layered structure and the source time function?

9

MATHEMATICAL PHYSICS IN STRATIFIED MEDIA

In stratified media there are many common mathematical aspects in phenomena so physically diverse as acoustics, electromagnetic waves, magnetostatics, gravitational-elastic spherical resonance, heat flow, gas diffusion, electric current in a resistive material, seismic waves, water waves, and atmospheric gravity waves, among many others. We will present the general theory and work out some of the details for the case of simple acoustics.

By a stratified medium we mean one in which material properties, compressibility, conductivity, density, etc., are functions of one spatial coordinate only. The usual situation is cartesian coordinates, but when geophysics is done on a global scale spherical coordinates may be used.

9-1 FROM PHYSICS TO MATHEMATICS

First step:
The first step is to write down all the basic partial differential equations of classical physics which relate to the problem of interest. Do not write down equations containing second space derivatives which are derived from first-derivative equations.

Write down the first-derivative equations. Write each component of vector or matrix equations.

In acoustics we have that the gradient pressure p gives rise to an acceleration of mass density ρ. For convenience we restrict motion to the x, z plane. Letting u and w represent x and z components of velocity we have

$$\rho \frac{\partial u}{\partial t} = -\frac{\partial p}{\partial x} \qquad (9\text{-}1\text{-}1)$$

$$\rho \frac{\partial w}{\partial t} = -\frac{\partial p}{\partial z} \qquad (9\text{-}1\text{-}2)$$

Another equation which is important in acoustics is the one that states that the divergence of velocity multiplied by the incompressibility K yields the rate of pressure decrease.

$$\frac{\partial p}{\partial t} = -K\left(\frac{\partial u}{\partial x} + \frac{\partial w}{\partial z}\right) + s(x, z, t) \qquad (9\text{-}1\text{-}3)$$

In (9-1-3) we included a pressure source s. This is something to be externally prescribed. The quantity s may be a source of chemical energy such as an explosion; thus it may vanish everywhere except at a point. Distributed sources are also often of interest; for example, radioactive rocks in a heat-flow model of the earth. To be more general, we could also have put momentum sources into (9-1-1) or (9-1-2), but the basic principles will be adequately exemplified with a source only in (9-1-3).

Second step:
The wave disturbance variables are taken to be unknown and the material properties known. Count equations and unknowns. We have three equations; u, w, and p are the three unknowns. We take K, ρ, and s to be known. Notice that the equations are linear in the unknowns. Now we make the stratification assumption; that is, we assume K and ρ are functions of depth z only and that they are constant in x. Since our linear equations now have constant coefficients with respect to x and t, we may always expect sinusoidal solutions in x and t. We do not know what to expect of our solutions in the z coordinate because of the arbitrary z-dependence of the coefficients K and ρ. This leads us to step three.

Third step:
Fourier transform time and the space coordinates with constant coefficients. In other words, we make the following substitution into (9-1-1), (9-1-2), and (9-1-3)

$$\begin{bmatrix} u(x, z, t) \\ w(x, z, t) \\ p(x, z, t) \\ s(x, z, t) \end{bmatrix} = \begin{bmatrix} U(k, z, \omega) \\ W(k, z, \omega) \\ P(k, z, \omega) \\ S(k, z, \omega) \end{bmatrix} e^{-i\omega t + ik_x x} \qquad (9\text{-}1\text{-}4)$$

After substitution, cancel the exponential and obtain

$$-i\omega\rho(z) = -ikP \tag{9-1-5a}$$

$$-i\omega\rho(z) = -\frac{\partial P}{\partial z} \tag{9-1-5b}$$

$$-i\omega P = -K(z)\left(ik_x U + \frac{\partial W}{\partial z}\right) + S \tag{9-1-5c}$$

Fourth step:
Eliminate algebraically the algebraic unknowns. In other words, when you examine (9-1-5) you see terms in $\partial P/\partial z$ and $\partial W/\partial z$ but you do not see $\partial U/\partial z$. This means that U is an algebraic variable which can be eliminated by purely algebraic means. We do this by substituting (9-1-5a) into (9-1-5c).

Fifth step:
Bring $\partial/\partial z$ terms to the left, bring all others to the right, and arrange terms into a neat matrix form. We have

$$\frac{\partial P}{\partial z} = i\omega\rho W$$

$$\frac{\partial W}{\partial z} = i\left(\frac{\omega}{K} - \frac{k_x^2}{\omega\rho}\right)P + \frac{S}{K}$$

and then

$$\frac{\partial}{\partial z}\begin{bmatrix} P \\ W \end{bmatrix} = i\begin{bmatrix} 0 & \omega\rho \\ \dfrac{\omega}{K} - \dfrac{k_x^2}{\omega\rho} & 0 \end{bmatrix}\begin{bmatrix} P \\ W \end{bmatrix} + \begin{bmatrix} 0 \\ \dfrac{S}{K} \end{bmatrix} \tag{9-1-6}$$

Sixth step:
Recognize that, no matter the physical problem with which you started, you should have a matrix first-order differential equation of the form

$$\frac{\partial}{\partial z}\mathbf{x} = \mathbf{A}\mathbf{x} + \mathbf{s} \tag{9-1-7}$$

where $\mathbf{x}$ is a vector containing the field variables of interest, $\mathbf{A}$ is a matrix depending on temporal and spatial frequency and on material properties, and $\mathbf{s}$ is a (possibly absent) vector function of the sources.

Before we look into techniques of solving (9-1-7) we can immediately deduce that in a source-free region the field variables $\mathbf{x}$ are smoother functions than the material properties. To see this, consider two homogeneous layers in contact. At the contact the $\mathbf{A}$ matrix has step-function discontinuities. Now let us see whether the wave fields in $\mathbf{x}$ can have step-function discontinuities. Obviously they cannot, since a step discontinuity in $\mathbf{x}$ would imply $d\mathbf{x}/dz = \infty$, whereas (9-1-7) in a source-free region states that $d\mathbf{x}/dz = \mathbf{A}\mathbf{x}$ and both $\mathbf{A}$ and $\mathbf{x}$ are supposed finite.

This does not mean that all field variables are always smooth. The algebraic variables eliminated in the *fourth step* can and often will be discontinuous at layer boundaries.

EXERCISES

1 What form does (9-1-7) take for the heat-flow equations? Include radioactive sources. [HINT: See equations (10-1-1) and (10-1-2).]

2 Using Maxwell's equations, $\nabla \times \mathbf{E} = -\mu \mathbf{H}$, $\nabla \times \mathbf{H} = \mathbf{J} + \varepsilon \mathbf{E}$, and Ohm's law, $\mathbf{J} = \sigma \mathbf{E}$ where σ is conductivity, set $\partial/\partial y = 0$ and derive (9-1-7).

3 In electrostatics the electric field in the ionosphere may be derived from a potential $\nabla \phi = -\mathbf{E}$, the divergence of electrical current vanishes $\nabla \cdot \mathbf{J} = 0$ and Ohm's law must have an extra term due to wind (a current source due to differential drag on ions and electrons across the earth's magnetic field) $\mathbf{J} = \sigma \mathbf{E} + \tau \mathbf{V}$. Assume you know $\mathbf{V}$. What form does (9-1-7) take assuming σ and τ to be scalars? Indicate how the calculation proceeds if σ and τ are matrices (assume you have the inverse of any matrix you wish).

4 In magnetostatics **curl H** = **J** and div **B** = 0, and **B** = μ**H**. Taking **J** as given, what is the form of (9-1-7)?

5 This exercise illustrates the linearization of nonlinear problems. For acoustic waves in a stratified windy atmosphere we use the trial slutions

$$\begin{bmatrix} P \\ U \\ W \end{bmatrix} = \begin{bmatrix} \bar{P} \\ \bar{U}(z) \\ 0 \end{bmatrix} + \begin{bmatrix} \tilde{P}(z) \\ \tilde{U}(z) \\ \tilde{W}(z) \end{bmatrix} e^{-i\omega t + ik_x x}$$

Reduce the partial differential equations to a matrix ordinary differential equation. HINT: The horizontal acceleration term is

$$\frac{du}{dt} = \frac{\partial u}{\partial t} + \frac{\partial u}{\partial x}\frac{\partial x}{\partial t} + \frac{\partial u}{\partial z}\frac{\partial z}{\partial t}$$

$$= \frac{\partial u}{\partial t} + \frac{\partial u}{\partial x}u + \frac{\partial u}{\partial z}w$$

with a like term for vertical acceleration. Drop second-order terms in $\tilde{P}$, $\tilde{U}$, and $\tilde{W}$.

6 Two equations come from heat flow: (H_x, H_z) equals the conductivity σ multiplied by the negative of the temperature gradient $(\partial_x, \partial_z)T$. The time derivative of temperature multiplied by the heat capacity c equals the negative of the heat-flow divergence $\partial_x H_x + \partial_z H_z$ gives another equation. Insert the trial solutions

$$\begin{bmatrix} T \\ H_x \\ H_z \end{bmatrix} = \begin{bmatrix} \bar{T}(z) \\ 0 \\ \bar{H}_z(z) \end{bmatrix} = \begin{bmatrix} \tilde{T}(z) \\ \tilde{H}_x(z) \\ \tilde{H}_z(z) \end{bmatrix} e^{-i\omega t + ik_x x}$$

(*a*) First derive steady-state equations for $\bar{T}$ and $\bar{H}$ assuming $\tilde{T}$ and $\tilde{H}$ vanish.

(*b*) Assuming $\bar{T}$ and $\bar{H}$ satisfy part (*a*), find equations for $\tilde{T}$ and $\tilde{H}$.

(*c*) Repeat (*a*) and (*b*) assuming linear temperature dependence of heat capacity and conductivity, i.e.,

$$\sigma = \sigma_0(z) + \sigma_1(z)T$$

$$c = c_0(z) + c_1(z)T$$

You will have to drop squared terms in $\tilde{T}$ and $\tilde{H}$.

7 Consider a compressible liquid sphere pulsating radially under its own gravitational attraction. What is the form of (9-1-6)?

HINTS:

$$\rho\ddot{\mathbf{v}} = \nabla p - \rho\mathbf{g} \qquad \text{momentum}$$

$$\nabla \cdot (\rho\mathbf{v}) = 0 \qquad \text{mass}$$

$$\dot{p} + K\nabla \cdot \mathbf{v} = 0 \qquad \text{state}$$

$$\nabla \cdot \mathbf{g} = 4\pi\gamma\rho \qquad \text{gravity}$$

9-2 NUMERICAL MATRIZANTS

A differential equation relates field variables at a point to field variables at neighboring points. A matrizant relates field variables at one depth in a stratified material to variables at some other depth. A matrizant may also be regarded as the integral of the matrix differential equation (9-1-7). First we will show how to get the matrizant of (9-1-7) by numerical means. That is, we will solve the problem for arbitrary depth variations in density and in compressibility. Then we will come back and develop analytical solutions for the special case of constant material properties. We have

$$\frac{\partial \mathbf{X}}{\partial z} \approx \frac{\mathbf{X}(z + \Delta z) - \mathbf{X}(z)}{\Delta z} \approx \mathbf{A}(z)\mathbf{X}(z) + \mathbf{S}(z)$$

or

$$\mathbf{X}(z + \Delta z) = (\mathbf{I} + \mathbf{A}\,\Delta z)\mathbf{X}(z) + \mathbf{S}(z)\,\Delta z \qquad (9\text{-}2\text{-}1)$$

Given $\mathbf{X}$ for some particular z it is clear that (9-2-1) may be used recursively to get $\mathbf{X}$ for any z. For simplicity we may take $\Delta z = 1$ and use subscripts to indicate the z coordinate. Let $[\mathbf{I} + \mathbf{A}(z)\,\Delta z]$ be denoted by $\mathbf{Q}(z)$, then (9-2-1) becomes

$$\mathbf{X}_{k+1} = \mathbf{Q}_k\mathbf{X}_k + \mathbf{S}_k$$

or

$$\mathbf{X}_1 = \mathbf{Q}_0\mathbf{X}_0 + \mathbf{S}_0$$

hence

$$\mathbf{X}_2 = \mathbf{Q}_1(\mathbf{Q}_0\mathbf{X}_0 + \mathbf{S}_0) + \mathbf{S}_1$$
$$= \mathbf{Q}_1\mathbf{Q}_0\mathbf{X}_0 + \mathbf{Q}_1\mathbf{S}_0 + \mathbf{S}_1$$

hence

$$\mathbf{X}_3 = \mathbf{Q}_2(\mathbf{Q}_1\mathbf{Q}_0\mathbf{X}_0 + \mathbf{Q}_1\mathbf{S}_0 + \mathbf{S}_1) + \mathbf{S}_2$$
$$= \mathbf{Q}_2\mathbf{Q}_1\mathbf{Q}_0\mathbf{X}_0 + \mathbf{Q}_2(\mathbf{Q}_1\mathbf{S}_0 + \mathbf{S}_1) + \mathbf{S}_2$$

likewise

$$\mathbf{X}_4 = \mathbf{M}\mathbf{X}_0 + \mathbf{T} \qquad (9\text{-}2\text{-}2)$$

So we have in general a numerically determinable matrix $\mathbf{M}$ (called the matrizant) and a vector $\mathbf{T}$ which relates the field variables at the top of the strata to those on the bottom by

$$\mathbf{X}_{z_{top}} = \mathbf{M}\mathbf{X}_{z_{bot}} + \mathbf{T} \qquad (9\text{-}2\text{-}3)$$

The matrix $\mathbf{M}$ is also called an integral matrix. Physical problems present themselves in different ways with different boundary conditions. For the acoustic problem discussed earlier $\mathbf{X}$ is a two-component vector involving pressure and vertical displacement. These are initially unknown at both the top and the bottom of the stratified medium. Thus (9-2-3) represents two equations for four unknowns. The solution to the problem comes only when two boundary conditions are introduced. If we are talking about sound waves in the ocean, (simplified) boundary conditions would be to prescribe zero pressure at the surface and zero vertical displacement at the sea floor. Then these boundary conditions with (9-2-3) would be two equations and two unknowns and consequently could be solved for surface displacement and bottom pressure. From these, pressure and displacement could be determined everywhere. Proper determination of boundary conditions is often the trickiest part of a problem; we will return to it for some other problems in a later section.

If portions of the material have constant material properties and contain no sources, then it is possible to find an analytical expression for the matrizant. A matrizant which takes one across such a layer of constant properties is called, appropriately enough, a *layer matrix*. It may be verified by substitution that

$$\mathbf{X}_z = e^{[\mathbf{A}(z - z_0)]}\mathbf{X}_{z_0} \qquad (9\text{-}2\text{-}4)$$

is the solution to $(\partial/\partial z)\mathbf{X} = \mathbf{A}\mathbf{X}$ where

$$e^{\mathbf{A}(z - z_0)} = \mathbf{I} + \mathbf{A}(z - z_0) + \frac{\mathbf{A}^2(z - z)^2}{2!} + \cdots \qquad (9\text{-}2\text{-}5)$$

in a region of space where $\mathbf{A}$ is constant with z. Thus, $e^{\mathbf{A}(z - z_0)}$ is the required matrizant. The matrix exponential could be computed numerically either by the method of (9-2-2) or the method of (9-2-5) or the method of Sylvester's theorem described in Chap. 5. In the next section we will see how Sylvester's theorem leads directly to the ideas of up- and downgoing waves.

EXERCISE

1 What is $\mathbf{Q}_k$ for the improved central difference approximation?

$$\mathbf{X}(z + \Delta z) - \mathbf{X}(z) = \frac{\mathbf{A}[\mathbf{X}(z + \Delta z) + \mathbf{X}(z)]}{2}$$

9-3 UP- AND DOWNGOING WAVES

We have seen a host of examples of how physical problems in stratified source-free media reduce to the form

$$\frac{d}{dz}\mathbf{X} = \mathbf{A}\mathbf{X} \qquad (9\text{-}3\text{-}1)$$

Where $\mathbf{X}$ is a vector of physical variables and $\mathbf{A}$ is a matrix which depends on z if material properties depend upon z. An important set of new variables in the vector $\mathbf{V}$ is defined by multiplying the vector of physical variables $\mathbf{X}$ by a square matrix $\mathbf{R}$

$$\mathbf{V} = \mathbf{R}\mathbf{X} \qquad (9\text{-}3\text{-}2)$$

where $\mathbf{R}$ is the matrix of row eigenvectors of the matrix $\mathbf{A}$. Inverse to $\mathbf{R}$ is the matrix $\mathbf{C}$ of column eigenvectors of $\mathbf{A}$. Premultiplying (9-3-2) by $\mathbf{C}$ and using $\mathbf{C}\mathbf{R} = \mathbf{I}$ we get the inverse relation to (9-3-2) which is useful to find the physical variables $\mathbf{X}$ from the new variables $\mathbf{V}$.

$$\mathbf{X} = \mathbf{C}\mathbf{V} \qquad (9\text{-}3\text{-}3)$$

Inserting (9-3-3) into (9-3-1) we obtain

$$(\mathbf{C}\mathbf{V})_z = \mathbf{A}\mathbf{C}\mathbf{V}$$
$$\mathbf{C}\mathbf{V}_z = \mathbf{A}\mathbf{C}\mathbf{V} - \mathbf{C}_z\mathbf{V}$$

Premultiplying by $\mathbf{R}$ and using $\mathbf{R}\mathbf{C} = \mathbf{I}$ we obtain

$$\mathbf{V}_z = (\mathbf{R}\mathbf{A}\mathbf{C})\mathbf{V} - \mathbf{R}\mathbf{C}_z\mathbf{V} \qquad (9\text{-}3\text{-}4)$$

Since we have supposed $\mathbf{R}$ and $\mathbf{C}$ to be row and column eigenvector matrices of $\mathbf{A}$ we can replace $\mathbf{R}\mathbf{A}\mathbf{C}$ by the diagonal matrix of eigenvalues $\mathbf{\Lambda}$, that is,

$$\mathbf{V}_z = \mathbf{\Lambda}\mathbf{V} - \mathbf{R}\mathbf{C}_z\mathbf{V} \qquad (9\text{-}3\text{-}5)$$

In any region of physical space where the material is homogeneous then $\mathbf{A}$, hence $\mathbf{C}$, will be independent of z and (9-3-5) will reduce to

$$\frac{d}{dz}\mathbf{V} = \mathbf{\Lambda}\mathbf{V} \qquad (9\text{-}3\text{-}6)$$

But the only matrix in (9-3-6) is a diagonal matrix, and so the problem for the different variables in the vector $\mathbf{V}$ decouples into a separate problem for each component. In wave problems it will be seen to be appropriate to call the components of $\mathbf{V}$ upgoing and downgoing wave variables. These variables flow up and down in homogeneous regions without interacting with each other. Let us consider an example.

In Sec. 9-1 we deduced that the matrix first-order differential equation for the acoustic problem in a region of no sources takes the form

$$\frac{d}{dz}\begin{bmatrix} P \\ W \end{bmatrix} = \begin{bmatrix} 0 & ia^2 \\ ib^2 & 0 \end{bmatrix}\begin{bmatrix} P \\ W \end{bmatrix} \qquad (9\text{-}3\text{-}7)$$

where

$$a^2 = \omega\rho \qquad (9\text{-}3\text{-}8)$$

$$b^2 = \frac{\omega}{K} - \frac{k_x{}^2}{\omega\rho} \qquad (9\text{-}3\text{-}9)$$

The matrix of column eigenvectors **C** and the matrix of row eigenvectors of the matrix of (9-3-7) are readily verified to be

$$\mathbf{C} = \begin{bmatrix} 1 & 1 \\ -\dfrac{b}{a} & \dfrac{b}{a} \end{bmatrix} \qquad (9\text{-}3\text{-}10)$$

$$\mathbf{R} = \tfrac{1}{2} \begin{bmatrix} 1 & -\dfrac{a}{b} \\ 1 & \dfrac{a}{b} \end{bmatrix} \qquad (9\text{-}3\text{-}11)$$

It is also readily verified that the vectors are normalized, namely $\mathbf{RC} = \mathbf{CR} = \mathbf{I}$ and that

$$\mathbf{\Lambda} = \mathbf{RAC} = \begin{bmatrix} -iab & 0 \\ 0 & +iab \end{bmatrix}$$

The downgoing wave variable D is associated with the iab eigenvalue and the upcoming wave variable U is associated with the $-iab$. We have definitions for up- and downgoing waves as

$$\begin{bmatrix} U \\ D \end{bmatrix} = \tfrac{1}{2} \begin{bmatrix} 1 & -\dfrac{a}{b} \\ 1 & \dfrac{a}{b} \end{bmatrix} \begin{bmatrix} P \\ W \end{bmatrix} \qquad (9\text{-}3\text{-}12a)$$

Of course a row eigenvector may contain an arbitrary multiplicative scaling factor if the scaling factor is divided from the corresponding column eigenvector. This means that the definition (9-3-12a) is not unique. As it happens, the present scale factors give the up- and downgoing waves the physical dimensions of P. The physical variables P and W are found from U and D by the inverse relation

$$\begin{bmatrix} P \\ W \end{bmatrix} = \begin{bmatrix} 1 & 1 \\ -\dfrac{b}{a} & \dfrac{b}{a} \end{bmatrix} \begin{bmatrix} U \\ D \end{bmatrix} \qquad (9\text{-}3\text{-}12b)$$

from which we see that the pressure P is the downgoing wave plus the upcoming wave and the vertical velocity is b/a times the difference. Equation (9-3-5) governing the propagation of U and D is

$$\frac{d}{dz}\begin{bmatrix} U \\ D \end{bmatrix} = \begin{bmatrix} -iab & 0 \\ 0 & iab \end{bmatrix}\begin{bmatrix} U \\ D \end{bmatrix} - \frac{1}{2}\frac{(b/a)_z}{b/a}\begin{bmatrix} 1 & -1 \\ -1 & 1 \end{bmatrix}\begin{bmatrix} U \\ D \end{bmatrix} \qquad (9\text{-}3\text{-}13)$$

In any region of space where b/a is not a function of z we are left with the simple uncoupled equations

$$\frac{d}{dz}\begin{bmatrix} U \\ D \end{bmatrix} = \begin{bmatrix} -iab & \\ & iab \end{bmatrix}\begin{bmatrix} U \\ D \end{bmatrix} \qquad (9\text{-}3\text{-}14)$$

Strictly, to justify the definitions of U and D as up- and downgoing waves we will have to be sure that the downgoing solution takes the form

$$d(z, t) = e^{-i\omega t + ik_z z} \qquad (9\text{-}3\text{-}15)$$

where ω and k_z must agree in sign so that constant phase is maintained as both z and t increase. The opposite sign must apply to U. In other words $k_z = ab$ must take the sign of ω. To see that this happens we take the square root of the product of (9-3-8) and (9-3-9).

$$k_z = ab = \omega\left(\frac{\rho}{K} - \frac{k_x^2}{\omega^2}\right)^{1/2} \qquad (9\text{-}3\text{-}16)$$

For vertically propagating waves we have $k_x = 0$ so that $k_z = ab$ specializes to $k_z = \omega(\rho/K)^{1/2}$. Substituting this value into (9-3-15), we see that the phase angle of the exponential is constant if $z/t = (K/\rho)^{1/2}$, making it clear that the material's intrinsic velocity is given by

$$v = \left(\frac{K}{\rho}\right)^{1/2} \qquad (9\text{-}3\text{-}17)$$

Reference to Fig. 9-1 shows that the angle θ between the vertical and a ray is defined by

$$\sin\theta = \frac{k_x v}{\omega} \qquad (9\text{-}3\text{-}18)$$

Inserting (9-3-17) and (9-3-18) into (9-3-16) we obtain

$$k_z = ab = \frac{\omega}{v}\cos\theta \qquad (9\text{-}3\text{-}19)$$

The time function (9-3-15) is complex. To get a real time function the expression (9-3-15) must be summed or integrated to include both positive and negative frequencies. Then, as we saw in the chapters on time series analysis, we must have $D(\omega) = \bar{D}(-\omega)$.

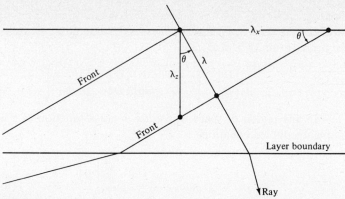

FIGURE 9–1
Rays and wavefronts in a layer. The wavelength λ_x seen on the x axis and the wavelength λ_z seen on the z axis are both greater than the wavelength λ seen along the ray. Clearly, $\lambda/\lambda_x = \sin\theta$ and $\lambda/\lambda_z = \cos\theta$ so the spatial frequencies $k_x = 2\pi/\lambda_x$ and $k_z = 2\pi/\lambda_z$ satisfy $k_x^2 + k_z^2 = (2\pi/\lambda)^2 = \omega^2/v^2$, which, besides being the pythagorean theorem (since $\sin\theta = k_x v/\omega$), is the Fourier transform of the wave equation. Snell's law that $(\sin\theta)/v$ is the same from layer to layer is thus equivalent to saying that k_x/ω is the same in each layer. That the spatial frequency k_x is the same constant in each layer is essential to the satisfaction of continuity conditions at the layer interfaces.

The quantity b/a will turn out to be the material's characteristic admittance Y. Taking the square root of the ratio of (9-3-9) over (9-3-8) we have

$$Y = \frac{b}{a} = \frac{(1 - v^2 k_x^2/\omega^2)^{1/2}}{\rho v} \tag{9-3-20}$$

$$Y = \frac{b}{a} = \frac{\cos\theta}{\rho v} \tag{9-3-21}$$

$$I = \frac{a}{b} = \frac{\rho v}{\cos\theta} \tag{9-3-22}$$

We shall now verify that this definition of impedance is the same as the one in the previous chapter. To do this we take a careful look at the matrizant to cross a layer $\exp[A_1(z_2 - z_1)] = \exp(A\,\Delta z)$. By Sylvester's theorem we have for the matrizant

$$\exp(A\,\Delta z) = C\begin{bmatrix} e^{-ik_z\,\Delta z} & 0 \\ 0 & e^{+ik_z\,\Delta z} \end{bmatrix} R \tag{9-3-23}$$

The matrizant relates the wave variables at the top z_1 of a layer to those at the bottom z_2. Thus (9-3-23) enables us to write

$$\begin{bmatrix} P \\ W \end{bmatrix}_2 = C_1 \exp(\Lambda_1\,\Delta z) R_1 \begin{bmatrix} P \\ W \end{bmatrix}_1 \tag{9-3-24}$$

Equation (9-3-24) which seems to have jumped at us from the mysteries of Sylvester's theorem actually has a simple interpretation. Starting on the right, we interpret the multiplication of $\mathbf{R}_j$ into the P and W variables as a conversion to up- and downgoing variables. Then the multiplication by $\exp(\mathbf{\Lambda}_1 \, \Delta z)$ carries these across the layer and the multiplication by $\mathbf{C}_1$ converts back to P and W variables which are continuous crossing an interface. Multiplying (9-3-24) through by R_2 and noting (9-3-11) and (9-3-12) we have

$$\begin{bmatrix} U \\ D \end{bmatrix}_2 = \mathbf{R}_2 \, \mathbf{C}_1 \, \exp(\mathbf{\Lambda} \, \Delta z) \begin{bmatrix} U \\ D \end{bmatrix}_1 \qquad (9\text{-}3\text{-}25)$$

In (9-3-25) we have now defined the up- and downgoing waves just beneath the interface as we did in the previous chapter. We should now be able to recognize the matrix as having the same form. It is

$$\frac{1}{2} \begin{bmatrix} 1 & -\dfrac{1}{Y_2} \\ 1 & \dfrac{1}{Y_2} \end{bmatrix} \begin{bmatrix} 1 & 1 \\ -Y_1 & Y_1 \end{bmatrix} \begin{bmatrix} e^{-iab\,\Delta z} & 0 \\ 0 & e^{iab\,\Delta z} \end{bmatrix} \qquad (9\text{-}3\text{-}26)$$

Defining the Z transform variable by

$$Z = \exp\left(\frac{2i\omega \, \Delta z}{v \cos \theta}\right)$$

Now we recognize that the travel time across the layer is $\Delta t = \Delta z / v \cos \theta$. The layer matrix (9-3-26) is

$$\frac{Y_1 + Y_2}{2 \, Y_2} \begin{bmatrix} 1 & \dfrac{Y_2 - Y_1}{Y_2 + Y_1} \\ \dfrac{Y_2 - Y_1}{Y_2 + Y_1} & 1 \end{bmatrix} \begin{bmatrix} Z^{-1/2} & \\ & Z^{1/2} \end{bmatrix} \qquad (9\text{-}3\text{-}27)$$

which may be compared to the matrix of (8-2-4) namely,

$$\frac{1}{t} \begin{bmatrix} 1 & c \\ c & 1 \end{bmatrix} \begin{bmatrix} Z^{-1/2} & 0 \\ 0 & Z^{1/2} \end{bmatrix}$$

establishing that the definition $Y = b/a$ has led to the familiar definition of reflection coefficient

$$c = \frac{Y_2 - Y_1}{Y_2 + Y_1} = \frac{I_1 - I_2}{I_1 + I_2}$$

$$= \frac{\rho_1 v_1 / \cos \theta_1 - \rho_2 v_2 / \cos \theta_2}{\rho_1 v_1 / \cos \theta_1 + \rho_2 v_2 / \cos \theta_2} \qquad (9\text{-}3\text{-}28)$$

EXERCISES

1 Redefine the eigenvectors so that $W = D + U$ and $P = (D - U)/Y$. This transformation would be useful if we wanted $t = 1 + c$ to refer to vertical velocity normalized
variables instead of pressure variables as in Chap. 8. Deduce changes to all the equations of this section.

2 Write the matrizant which crosses a layer in terms of *a*, *b*, and layer thickness *h*.

9-4 SOURCE-RECEIVER RECIPROCITY

The principle of reciprocity states that a source and receiver may (under some conditions) be interchanged and the same waveform will be observed. This principle
is often used to advantage in calculations and may also be used to simplify data
collection. It is somewhat amazing that this principle applies to the earth with its
complicated inhomogeneities. Intuitively, the main reason for validity of the
reciprocal principle is that energy propagates equally well along a given ray in
either direction. Either way, it goes at the same speed with the same attenuation.
This is true for all common types of waves.

Little more would need to be said if all waves were scalar phenomena with
scalar sources and scalar receivers as, for example, acoustic pressure waves with
explosive sources and pressure-sensitive receivers. The situation becomes more
complicated when the sources or receivers are moving diaphragms, because then
their orientations become important. The directional properties of the source and
receiver are often referred to as radiation patterns. To apply the reciprocity principle it is necessary to regard the radiation patterns as attached to the medium, not
as being attached to the source and receiver. Thus, when source and receiver are
said to be interchanged, it is only a scalar magnitude which is interchanged; the
radiation patterns stay fixed at the same place. These general ideas are made more
precise in the following derivation. It will be seen that the notion of rays actually
turns out to be irrelevant. Reciprocity also works in diffusion and potential
problems.

Theoretical treatments are often somewhat hard to read. They often begin
by specifying that the differential operator along with suitable boundary conditions
should constitute a self-adjoint problem. This means that when you reexpress the
differential equations in difference form you discover that the matrix of coefficients
is symmetric. Let us take the example of acoustic waves in one dimension. Newton's
equation says that mass density ρ times acceleration $\partial_{tt} u$ equals the negative of the
pressure gradient $-\partial_x p$ plus the external force F_x. Utilizing $e^{-i\omega t}$ time dependence
we have

$$-\rho\omega^2 u = -\partial_x p + F_x$$

which, defining $F = -F_x$, may be written

$$\rho\omega^2 u - \partial_x p = F \qquad (9\text{-}4\text{-}1)$$

The other important equation of acoustics says that the incompressibility K^{-1} multiplied by the pressure p plus the divergence of displacement $\partial_x u$ equals the external (relative) volume injection V, that is

$$K^{-1}p + \partial_x u = V \qquad (9\text{-}4\text{-}2)$$

We will now combine (9-4-1) and (9-4-2) in a finite difference form with, for convenience, $\Delta x = 1$. In practice, one might like to use many grid points to approximate the behavior of continuous functions, but for the sake of illustration we only need use a few grid points. Luckily, in this case reciprocity will be exactly true despite the small number of grid points. We have

$$
\begin{bmatrix}
\dfrac{1}{I_0} & 1 & & & & & \\
1 & \rho\omega^2 & -1 & & & & \\
& -1 & K^{-1} & 1 & & & \\
& & 1 & \rho\omega^2 & -1 & & \\
& & & -1 & K^{-1} & 1 & \\
& & & & 1 & I_n &
\end{bmatrix}
\begin{bmatrix}
p_0 \\ u_0 \\ p_1 \\ u_1 \\ p_2 \\ u_2
\end{bmatrix}
=
\begin{bmatrix}
V_0 \\ F_0 \\ V_1 \\ F_1 \\ V_2 \\ F_2
\end{bmatrix}
\qquad (9\text{-}4\text{-}3)
$$

The first and last rows of (9-4-3) require some special comment. The quantities I_0 and I_n are called impedances. If they vanish, we have zero pressure end conditions; if they are infinite, we have zero motion end conditions.

Now with all this fuss we have gone through to obtain the matrix (9-4-3), the only thing we want from it is to observe that the matrix is indeed a symmetric matrix (even if ρ and K^{-1} were functions of x). In the exercises it is shown that a symmetric matrix may also be attained in two dimensions. That the matrix is symmetric is partly a result of the physical nature of sound and partly a result of careful planning on the part of the author. To obtain the correct statement of reciprocity in other situations you may have to do some careful planning too. The essence of reciprocity is that since the matrix of (9-4-3) is symmetric then the inverse matrix will also be symmetric. Premultiplying (9-4-3) through by the inverse matrix we get the responses as a result of matrix multiplication on the external excitations.

$$
\begin{bmatrix}
p_0 \\ u_0 \\ p_1 \\ u_1 \\ p_2 \\ u_2
\end{bmatrix}
=
\begin{bmatrix}
\cdot & \cdot & A & B & \cdot & \cdot \\
\cdot & \cdot & C & D & \cdot & \cdot \\
A & C & \cdot & \cdot & \cdot & \cdot \\
B & D & \cdot & \cdot & \cdot & \cdot \\
\cdot & \cdot & \cdot & \cdot & \cdot & \cdot \\
\cdot & \cdot & \cdot & \cdot & \cdot & \cdot
\end{bmatrix}
\begin{bmatrix}
V_0 \\ F_0 \\ V_1 \\ F_1 \\ V_2 \\ F_2
\end{bmatrix}
\qquad (9\text{-}4\text{-}4)
$$

The letters A, B, C, and D indicate the symmetry of the matrix of (9-4-4). Now if all external sources vanish except on one end where there is a unit strength volume source $V_0 = 1$, then according to (9-4-4) the pressure in the middle p_1 will equal A. If in a second experiment all the external sources vanish except the middle volume source $V_1 = 1$, then according to (9-4-4) the pressure response p_0 at the end will also equal A. This is the reciprocal principle. Note that with the letter D in (9-4-4) a like statement applies to the forces and the displacements. A mixed statement applies with the letters C and B.

In a realistic experiment it may not be possible to have a pure volume source or a pure external force. In other words, the external source may have some finite, nonzero impedance. Then the first experiment we would perform would be with the excitation at the middle, getting for the end response:

$$\begin{bmatrix} p_0 \\ u_0 \end{bmatrix} = \begin{bmatrix} A & B \\ C & D \end{bmatrix} \begin{bmatrix} V_1 \\ F_1 \end{bmatrix} \qquad (9\text{-}4\text{-}5)$$

Interchanging source and receiver locations, we have

$$\begin{bmatrix} p_1 \\ u_1 \end{bmatrix} = \begin{bmatrix} A & C \\ B & D \end{bmatrix} \begin{bmatrix} V_0 \\ F_0 \end{bmatrix} \qquad (9\text{-}4\text{-}6)$$

The notable feature of (9-4-5) and (9-4-6) is that the matrices are transposes of one another. This feature would not be lost if we were to consider a more elaborate experiment where the vectors in (9-4-5) and (9-4-6) contained more elements. For example, a vector in (9-4-5) or (9-4-6) could contain elements of an array of physically separated volume sources or pressure sensors. In fact, if the reader is able to frame elastic, electromagnetic, diffusion, or potential problems as symmetric algebraic equations like (9-4-3), then the matrices like (9-4-5) and (9-4-6) will still be transposes of one another. The setting up of symmetric equations like (9-4-3) is often not difficult, although it may get somewhat complicated in multidimensional noncartesian geometry.

In such a more general case we may denote the right-hand vectors in (9-4-5) or (9-4-6) by $\mathbf{E}$ to denote excitation and the left-hand vectors by $\mathbf{R}$ to denote response. Using $\mathbf{M}$ for the matrix of (9-4-5) and $\mathbf{M}^T$ for the transposed matrix, (9-4-5) and (9-4-6) would be

$$\mathbf{R}_0 = \mathbf{M}\mathbf{E}_1 \qquad (9\text{-}4\text{-}7)$$

$$\mathbf{R}_1 = \mathbf{M}^T\mathbf{E}_0 \qquad (9\text{-}4\text{-}8)$$

Now let us deduce a physical statement from (9-4-7) and (9-4-8). First take the inner product of (9-4-7) with $\mathbf{E}_0^T$

$$\mathbf{E}_0^T \mathbf{R}_0 = \mathbf{E}_0^T \mathbf{M}\mathbf{E}_1$$

The right-hand side, which is a scalar, may be transposed

$$\mathbf{E}_0^T \mathbf{R}_0 = (\mathbf{E}_0^T \mathbf{M}\mathbf{E}_1)^T = \mathbf{E}_1^T \mathbf{M}^T \mathbf{E}_0$$

substituting from (9-4-8) we have

$$\mathbf{E}_0^T \mathbf{R}_0 = \mathbf{E}_1^T \mathbf{R}_1 \qquad (9\text{-}4\text{-}9)$$

Equation (9-4-9) is the basic statement of reciprocity; the inner product of the excitation vector and the response vector at place 0 equals their inner product at

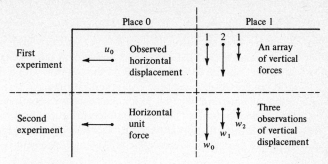

	Place 0		Place 1

| First experiment | u_0 ← • | Observed horizontal displacement | An array of vertical forces |
| Second experiment | ← • | Horizontal unit force | Three observations of vertical displacement |

FIGURE 9-2
A reciprocity example. Reciprocity says that $u_0 = w_0 + 2w_1 + w_2$.

place 1. Notice that the inner products are between vectors which occur in *different experiments.*

An example of an elastic system with vector-directed displacement and force vectors is depicted in Fig. 9-2. A laboratory example by J. E. White [Ref. 32] which combines electromagnetic, solid, liquid, and gaseous media is shown in Fig. 9-3. A geophone is a spring pendulum coupled to an induction coil. The first geophone is mounted on a pipe which rests on the bottom of a glass desiccator. The second geophone is attached to the glass with a chunk of modeling clay, below the water level. The top pair of traces shows the (source) current into the first geophone and the (open circuit) voltage at the second; the bottom traces show the current in the second geophone and the voltage at the first.

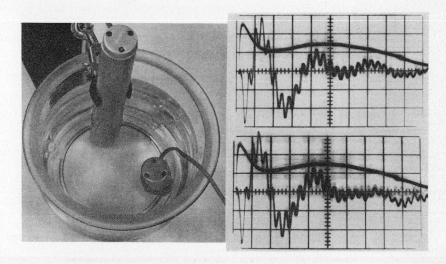

FIGURE 9-3
An example of the reciprocal principle in a combined electromagnetic, solid, liquid, and gaseous system [J. E. White, *Geophysics*, Ref. 32].

EXERCISES

1 Consider Poisson's equation $\partial_{xx} R = -E$ on five grid points where the boundary conditions are that the end points are zero. A unit excitation at the third grid point gives the solution $(0, \frac{1}{2}, 1, \frac{1}{2}, 0)$. Find the solution with a unit excitation in the second grid point. Observe reciprocity if you do it right.

2 Write an equation like (9-4-3) for the heat-flow equation. How will the introduction of imaginary numbers change the statement of the reciprocal principle?

3 Write the three first-order partial differential equations of acoustics in two-dimensional cartesian geometry. Observe the gridding arrangement below.

$$
\begin{array}{c}
\begin{array}{cc}
& \quad z \\
\uparrow & \longrightarrow \\
\downarrow \\
x
\end{array}
\end{array}
$$

$$W = 0$$

$U = 0$
$$
\begin{array}{|cccccc|}
\hline
P_1 & U_1 & P_2 & U_2 & P_3 & U_3 \\
W_1 & & W_2 & & W_3 & \\
P_4 & U_4 & P_5 & U_5 & P_6 & U_6 \\
W_4 & & W_5 & & W_6 & \\
P_7 & U_7 & P_8 & U_8 & P_9 & U_9 \\
W_7 & & W_8 & & W_9 & \\
\hline
\end{array}
$$
$$P = 0$$

FIGURE E9-4-3 $\qquad\qquad P = 0$

Write a set of 27×27 equations for the vector $(U_1, P_1, W_1, U_2, P_2, W_2, \ldots, U_9, P_9, W_9)$. Make it come out symmetric and in an obviously orderly form.

4 In Sec. 8-3, Exercises 4 and 5 taken together illustrate the reciprocity theorem which states, "If source and receiver are interchanged, the same waveform will be observed." Solve the problem of a surface source with a receiver in the middle of the layers and solve the same problem with interchanged source and receiver to test the reciprocity theorem.

9-5 CONSERVATION PRINCIPLES AND MODE ORTHOGONALITY

We showed earlier how problems in stratified media reduce to a first-order matrix differential equation of the form

$$\frac{\partial}{\partial z} \mathbf{y} = \mathbf{A}\mathbf{y} + \mathbf{s} \qquad (9\text{-}5\text{-}1)$$

It turns out that many problems in the form of (9-5-1) can be reformulated into what we will call the *Atkinson* form. It is

$$\mathbf{J} \frac{\partial}{\partial z} \mathbf{y} = [\mathbf{G}(z) + \lambda \mathbf{H}(z)]\mathbf{y} \qquad (9\text{-}5\text{-}2)$$

where $\mathbf{J}$ is a skew-Hermitian matrix ($\mathbf{J}^* = -\mathbf{J}$) independent of z, $\mathbf{G}(z)$ and $\mathbf{H}(z)$ are Hermitian matrices ($\mathbf{H}^* = \mathbf{H}$), and λ is a scalar which will come to play the role of an eigenvalue. For example, in acoustics we have

$$\frac{\partial}{\partial z}\begin{bmatrix} P \\ W \end{bmatrix} = i \begin{bmatrix} 0 & \omega\rho \\ \left(+\dfrac{\omega}{K} - \dfrac{k_x{}^2}{\omega\rho}\right) & 0 \end{bmatrix}\begin{bmatrix} P \\ W \end{bmatrix} + i\begin{bmatrix} s_p \\ s_w \end{bmatrix} \qquad (9\text{-}5\text{-}3)$$

which can be premultiplied by a skew-Hermitian matrix to give

$$\begin{bmatrix} 0 & -i \\ -i & 0 \end{bmatrix}\frac{\partial}{\partial z}\begin{bmatrix} P \\ W \end{bmatrix} = \begin{bmatrix} \left(+\dfrac{\omega}{K} - \dfrac{k_x{}^2}{\omega\rho}\right) & 0 \\ 0 & \omega\rho \end{bmatrix}\begin{bmatrix} P \\ W \end{bmatrix} + \begin{bmatrix} s_w \\ s_p \end{bmatrix} \qquad (9\text{-}5\text{-}4)$$

The significant thing about (9-5-4) is that the operators are self-adjoint, meaning that the right-hand matrix is Hermitian and so is the left-hand operator. To understand why $\mathbf{J}(\partial/\partial z)$ is Hermitian, write it out as a difference approximation

$$\mathbf{J}\frac{\partial}{\partial z} = \frac{i}{\Delta z}\begin{bmatrix} 0 & \delta_z \\ \delta_z & 0 \end{bmatrix}$$

$$= \frac{1}{\Delta z}\left[\begin{array}{ccc|ccc} & & & i & -i & \\ & & & & i & -i \\ & & & & & i & -i \\ \hline i & -i & & & & \\ & i & -i & & & \\ & & i & -i & & \end{array}\right] \qquad (9\text{-}5\text{-}5)$$

Inspecting (9-5-5) we see that it is two rows short of being square. Choosing two boundary conditions will be like obtaining two more rows. Clearly (9-5-5) is so close to being Hermitian that two more rows can be chosen to make it Hermitian. For example, the two rows

$$\begin{bmatrix} & & & & & i \\ -i & & & & & \end{bmatrix}$$

could be squeezed between the top and bottom halves of (9-5-5). Since the operator (9-5-5) can be made Hermitian by choice of suitable boundary conditions and since the other operators in (9-5-4) are already Hermitian, it seems that the Atkinson form applies to physical problems in which the reciprocity principle is applicable. Reciprocity does apply to most geophysical prospecting problems. A simple physical situation in which reciprocity does *not* apply is sound waves in a windy atmosphere. Physically it is because waves go more slowly upwind than downwind, and mathematically it is because no $\mathbf{J}$ matrix can be found to convert (9-5-1) into the form (9-5-2). Only in a source-free region can we convert (9-5-1) to (9-5-2). If we choose to let ω play the role of the eigenvalue, then taking source terms to be zero we split (9-5-4) into

$$\begin{bmatrix} 0 & -i \\ -i & 0 \end{bmatrix} \frac{\partial}{\partial z} \begin{bmatrix} P \\ W \end{bmatrix} = + \omega \begin{bmatrix} \left(\dfrac{1}{K} - \dfrac{k_x^2}{\omega^2 \rho} \right) & 0 \\ 0 & \rho \end{bmatrix} \begin{bmatrix} P \\ W \end{bmatrix} \qquad (9\text{-}5\text{-}6)$$

Here $G(z)$ has turned out to vanish and k_x^2/ω^2, which is proportional to the sine of the incident angle, is to be regarded as a constant for variable values of the eigenvalue ω. Alternatively, we could choose $-k_x^2$ to be the eigenvalue, and then (9-5-4) would become

$$\begin{bmatrix} 0 & -i \\ -i & 0 \end{bmatrix} \frac{\partial}{\partial z} \begin{bmatrix} P \\ W \end{bmatrix} = \begin{bmatrix} \dfrac{\omega}{K} & 0 \\ 0 & \omega \rho \end{bmatrix} \begin{bmatrix} P \\ W \end{bmatrix} - k^2 \begin{bmatrix} \dfrac{1}{\omega \rho} & 0 \\ 0 & 0 \end{bmatrix} \begin{bmatrix} P \\ W \end{bmatrix} \qquad (9\text{-}5\text{-}7)$$

Obviously, still another possibility is to let the angle variable $-k_x^2/\omega^2$ be the eigenvalue for fixed ω.

The Atkinson form (9-5-2) leads directly to various conservation principles. Let us compute the vertical derivative of the quadratic form $\mathbf{y^*Jy}$.

$$\frac{\partial}{\partial z} \mathbf{y^*Jy} = \mathbf{y_z^*Jy} + \mathbf{y^*Jy_z}$$

$$= -\mathbf{y_z^*J^*y} + \mathbf{y^*Jy_z}$$

$$= -(\mathbf{Jy_z})^*\mathbf{y} + \mathbf{y^*(Jy_z)}$$

$$= -(\mathbf{Gy} + \lambda \mathbf{Hy})^*\mathbf{y} + \mathbf{y^*(Gy} + \lambda \mathbf{Hy})$$

$$= (\lambda - \lambda^*)\mathbf{y^*Hy} \qquad (9\text{-}5\text{-}8)$$

Very often we take the eigenvalues ω, $-k_x^2$, or $-k_x^2/\omega^2$ to be real, and in such a case we have $\lambda - \lambda^* = 0$ and (9-5-8) shows that $\mathbf{y^*Jy}$ is a quadratic function of the wave variables which is invariant with z. In the acoustic example, this quadratic invariant is proportional to the energy flux. Specifically

$$\mathbf{y^*Jy} = -i[P^* \quad W^*] \begin{bmatrix} 0 & 1 \\ 1 & 0 \end{bmatrix} \begin{bmatrix} P \\ W \end{bmatrix} \qquad (9\text{-}5\text{-}9)$$

$$= -i(P^*W + W^*P) = -2i \operatorname{Re}(P^*W)$$

If we wish to consider a complex frequency $\omega = \omega_r + i\omega_i$, then in the first acoustic example (9-5-6) equation (9-5-8) becomes

$$-\frac{\partial}{\partial z} \operatorname{Re}(P^*W) = \omega_i \left[\left(\frac{1}{K} - \frac{k_x^2}{\omega^2 \rho} \right) P^*P + \rho W^*W \right] \qquad (9\text{-}5\text{-}10)$$

Noting that if P and W have time dependence $\exp[-i(\omega_r + i\omega_i)t] = \exp(-i\omega_r t + \omega_i t)$, then quadratics like P^*P and W^*W have time dependence $e^{2\omega_i t}$ and we see that the multiplier $2\omega_i$ can be regarded as a time derivative. Hence (9-5-10) becomes

$$-\frac{\partial}{\partial z} \operatorname{Re}(P^*W) = +\frac{\partial}{\partial t} \left\{ \frac{1}{2}\left(\frac{1}{K} - \frac{k_x^2}{\omega^2 \rho} \right) P^*P + \rho W^*W \right\} = \frac{\partial}{\partial t} E \qquad (9\text{-}5\text{-}11)$$

Equation (9-5-11) is interpreted as saying that the time derivative of the energy density E at a point is proportional to the negative of the divergence of energy flux at that point. In other problems the quadratic forms need not always turn out to involve energy. Sometimes momentum is involved.

A well-known theorem in matrix theory is that Hermitian matrices have real eigenvalues. Why then did we consider the possibility of a complex eigenvalue in (9-5-8)? The answer is that the finite difference operator matrix need not be chosen to have boundary conditions which make the operators Hermitian. In particular, for $\partial E/\partial t$ to be nonzero, energy must leak in or out at a boundary.

Now, let us suppose boundary conditions have been chosen to make $\mathbf{J}\partial/\partial z$ symmetric so the eigenvalues become real. Let $y_n(z)$ be a solution to (9-5-2) with eigenvalue λ_n, and let $y_m(z)$ be another solution with a different eigenvalue λ_m. The reasoning which led up to (9-5-8) can be used to obtain

$$\frac{\partial}{\partial z}(\mathbf{y}_m^* \mathbf{J} \mathbf{y}_n) = (\lambda_n - \lambda_m)\mathbf{y}_m^* \mathbf{H} \mathbf{y}_n \qquad (9\text{-}5\text{-}12)$$

Integrating through z from z_a to z_b, we have

$$\mathbf{y}_m^* \mathbf{J} \mathbf{y}_n \Big|_{z_a}^{z_b} = (\lambda_n - \lambda_m) \int_a^b \mathbf{y}_m^*(z, \lambda_m) \mathbf{H}(z) \mathbf{y}(z, \lambda_n)\, dz \qquad (9\text{-}5\text{-}13)$$

If boundary conditions have been chosen so that no energy gets in or out at z_a and z_b, then the left-hand side vanishes. Since by hypothesis $\lambda_n \neq \lambda_m$ we must have the right-hand integral vanishing. This states the orthogonality of the two solutions (called the two *modes*) and the idea is the same as the orthogonality of eigenvectors of the Hermitian difference operator matrices. The orthogonality of these functions is frequently useful in theoretical and computational work. Further details, including the most general form of energy-conserving boundary conditions, may be found in Reference 14, Chap. 9.

EXERCISE

1 Show that application of (9-5-8) to (9-5-7) leads to a definition of horizontal energy flux. You may wish to take $k_x = k_r + ik_i$ and assume $|k_r| \gg |k_i|$.

9-6 ELASTIC WAVES

It is now presumed that the reader has a general knowledge of classical elasticity theory. Few textbooks, if any, develop the special subject of stratified media which is so important in seismology. Many papers on that subject may be found in the *Bulletin of the Seismological Society of America* (*BSSA*). For those readers unfamiliar with the *BSSA*, we now present the results of applying the general methods of this chapter to the equations of isotropic elasticity.

The conventions in elasticity are (u, w) displacements in x and z directions,

τ is the stress matrix, λ and μ are Lame's constants and ρ is density. Hooke's law and Newton's law with $e^{-i\omega t}$ time dependence leads to

$$\frac{\partial}{\partial z}\begin{bmatrix} U \\ \tau_{zz} \\ W \\ \tau_{zx} \end{bmatrix} = \begin{bmatrix} 0 & 0 & -\partial_x & \dfrac{1}{\mu} \\[2mm] 0 & 0 & -\rho\omega^2 & -\partial_x \\[2mm] \dfrac{-\lambda}{(\lambda+2\mu)\,\partial_x} & \dfrac{1}{\lambda+2\mu} & 0 & 0 \\[2mm] -\rho\omega^2-\gamma & \dfrac{-\partial_x\lambda}{\lambda+2\mu} & 0 & 0 \end{bmatrix}\begin{bmatrix} U \\ \tau_{zz} \\ W \\ \tau_{zx} \end{bmatrix} \qquad (9\text{-}6\text{-}1)$$

where

$$\gamma = \partial_x \frac{4\mu(\lambda+\mu)}{(\lambda+2\mu)\,\partial_x} \qquad (9\text{-}6\text{-}2)$$

Define also

$$\alpha^2 = \frac{\lambda+2\mu}{\rho}$$

$$\beta^2 = \frac{\mu}{\rho}$$

$$m^2 = \frac{-\omega^2}{\alpha^2} - \partial_{xx} \qquad (9\text{-}6\text{-}3)$$

$$n^2 = \frac{-\omega^2}{\beta^2} - \partial_{xx}$$

$$l^2 = \frac{-\omega^2}{\beta^2} - 2\partial_{xx}$$

If material properties do not vary in the x direction, we have the row eigenvector transformation $\mathbf{R}$ to up- and downgoing wave variables.

$$\begin{bmatrix} p^+ \\ s^+ \\ p^- \\ s^- \end{bmatrix} = \frac{\Lambda^{-1}}{2\omega^2\rho}\begin{bmatrix} 2\mu m\,\partial_x & m & \mu l^2 & \partial_x \\ \mu l^2 & -\partial_x & -2\mu n\,\partial_x & n \\ -2\mu m\,\partial_x & -m & \mu l^2 & \partial_x \\ -\mu l^2 & \partial_x & -2\mu n\,\partial_x & n \end{bmatrix}\begin{bmatrix} u \\ \tau_{zz} \\ w \\ \tau_{zx} \end{bmatrix} \qquad (9\text{-}6\text{-}4)$$

and the column eigenvector inverse transform $\mathbf{C}$

$$\begin{bmatrix} u \\ \tau_{zz} \\ w \\ \tau_{zx} \end{bmatrix} = \begin{bmatrix} -\partial_x & -n & -\partial_x & -n \\ -\mu l^2 & 2\mu n\,\partial_x & -\mu l^2 & 2\mu n\,\partial_x \\ -m & \partial_x & m & -\partial_x \\ -2\mu m\,\partial_x & -\mu l^2 & 2\mu m\,\partial_x & \mu l^2 \end{bmatrix}\begin{bmatrix} p^+ \\ s^+ \\ p^- \\ s^- \end{bmatrix} \qquad (9\text{-}6\text{-}5)$$

where

$$\Lambda = \begin{bmatrix} m & & & \\ & n & & \\ & & -m & \\ & & & -n \end{bmatrix}$$

The matrices partition nicely into 2×2 blocks. The reader may verify that $\mathbf{CR} = \mathbf{RC} = \mathbf{I}$ and $\mathbf{C\Lambda R} = \mathbf{A}$.

10

INITIAL-VALUE PROBLEMS IN TWO AND THREE DIMENSIONS

There are whole textbooks (for example, References 33 and 34) devoted to solving initial-value problems by difference approximations to differential equations. In this section we will briefly cover the main ideas. The overall idea in two dimensions is that one partitions a computer memory into one or a few two-dimensional grids where field variables are represented as functions of two spatial dimensions. Then you insert initial conditions, turn on the computer, and see what happens. There have been numerous extensive studies devoted to the diffusion equation, but far fewer studies have been devoted to the wave equation. The problem with modeling the wave equation is that ten points per wavelength is probably not enough, and even at that you cannot fit very many wavelengths onto a reasonable grid. The energy then propagates rapidly to the edges of the grid where it bounces back, whether you want it to or not. One way to ameliorate this kind of difficulty is to develop coordinate systems which move with the waves. These coordinate systems also facilitate projection of waves from the earth's surface, where they are observed, back down into the earth. This kind of projection forms the basis for the practical reflection seismic data processing techniques described in chapter 11.

10-1 CLASSICAL INITIAL-VALUE PROBLEMS IN TIME

It is easiest to cover fundamentals with the heat-flow equation in one dimension. The heat-flow equation is derived from two intuitively obvious equations. The first says that a flow H of heat arises from a temperature gradient and is proportional to thermal conductivity σ.

$$H = -\sigma \frac{\partial T}{\partial x} \qquad (10\text{-}1\text{-}1)$$

The second says the temperature decrease is in proportion to the divergence of heat flow H and inversely proportional to the heat capacity C of the material

$$\frac{\partial T}{\partial t} = -\frac{1}{C} \frac{\partial H}{\partial x} \qquad (10\text{-}1\text{-}2)$$

The usual procedure is to insert (10-1-1) into (10-1-2) and neglect the derivative of σ.

$$\frac{\partial T}{\partial t} = \frac{\sigma}{C} \frac{\partial^2 T}{\partial x^2} \qquad (10\text{-}1\text{-}3)$$

The usual convention in difference equation theory is that temperature $T(x, t) = T(k \, \Delta x, n \, \Delta t)$ will be written as T_k^n where the superscript denotes time. With the definition $b = \sigma \, \Delta t / 2C \, \Delta x^2$ (10-1-3) may be written

$$T_k^{n+1} - T_k^n = 2b(T_{k+1}^n - 2T_k^n + T_{k-1}^n) \qquad (10\text{-}1\text{-}4)$$

If the temperature T_k^n is known at all spatial positions k for some particular time n, then (10-1-4) may be used to calculate the temperature for all time. The reader may notice that the time derivative is centered at $T_k^{n+1/2}$ whereas the space derivative is centered at T_k^n. This can cause difficulty. The heat-flow differential equation smooths out long spatial wavelengths slowly and shorter wavelengths more rapidly. The heat-flow difference equation does the same thing, except that very short wavelengths will sense the difference in centering of time and space derivatives. The result is that the very short wavelengths will not attenuate at the proper rate and they may even amplify. In fact, as Δx is reduced more and more, thereby making it possible to contain shorter and shorter wavelengths on the grid, amplification will always occur, thereby ruining the solution. This situation, called instability, is described in more detail in all the books on the subject. One might hope that centering the time difference by approximating $\partial T / \partial t$ by $(T_k^{n+1} - T_k^{n-1})/(2 \, \Delta t)$ would avoid the instability, but it turns out even worse and creates instability for any Δx. The reason is that the heat-flow differential equation is first-order in time, but using a time difference over two steps creates a difference equation which is second-order in time. A second-order equation always has two solutions. In this case, one behaves like the heat-flow equation; the other turns out to be an oscillating increasing exponential like $(1, -2, 4, -8, \ldots)$ which rapidly overwhelms the heat-flow solution.

These problems may all be avoided with the Crank-Nicolson scheme. It will always guarantee stability for any Δx and it can also be applied to the wave equations in acoustics, electromagnetics, and elasticity. In the Crank-Nicolson scheme one centers the space difference at $T_k^{n+1/2}$ in the following way:

$$T_k^{n+1} - T_k^n = b(T_{k+1}^n - 2T_k^n + T_{k-1}^n) + b(T_{k+1}^{n+1} - 2T_k^{n+1} + T_{k-1}^{n+1}) \quad (10\text{-}1\text{-}5)$$

An apparent problem with the Crank-Nicolson scheme is that the method of getting the $n+1$ time level from the n level is no longer obvious. Bringing all the $n+1$ terms in (10-1-5) to the left and the n terms to the right, we have

$$-bT_{k+1}^{n+1} + (1 + 2b)T_k^{n+1} - bT_{k-1}^{n+1} = D_k^n \quad (10\text{-}1\text{-}6)$$

The right-hand side D_k^n is a known function of T^n. What we have here is a set of simultaneous equations for the T^{n+1}. Writing this out in full, we see why the set is called a tridiagonal set of equations

$$\begin{bmatrix} (1+2b) & -b & & \text{zeros} \\ -b & (1+2b) & -b & \\ & -b & (1+2b) & \\ & & -b & \\ \text{zeros} & & & \end{bmatrix} \begin{bmatrix} T_0 \\ T_1 \\ \\ \\ T_N \end{bmatrix}^{n+1} = \begin{bmatrix} D_0 \\ D_1 \\ \\ \\ D_N \end{bmatrix} \quad (10\text{-}1\text{-}7)$$

It turns out that the simultaneous equations in (10-1-7) may be solved extremely easily. As will be shown later there is little more effort involved than in the use of (10-1-4). The scientist who wishes to solve partial differential equations numerically without becoming a computer scientist is well advised to use the Crank-Nicolson scheme. The extra effort required to figure out how to solve (10-1-7) is well rewarded by the ability to use any Δx and Δt and to forget about stability and the biasing effects of noncentral differences.

Now let us consider heat flow in two spatial dimensions. The heat-flow equation becomes

$$\frac{\partial T}{\partial t} = \frac{\sigma}{C}\left(\frac{\partial^2 T}{\partial x^2} + \frac{\partial^2 T}{\partial y^2}\right) \quad (10\text{-}1\text{-}8)$$

A simple, effective means to solve this equation is the splitting method. One uses two different equations at alternate time steps. They are

$$\frac{\partial T}{\partial t} = \frac{2\sigma}{C}\frac{\partial^2 T}{\partial x^2} \quad \text{(all } y) \quad (10\text{-}1\text{-}9a)$$

$$\frac{\partial T}{\partial t} = \frac{2\sigma}{C}\frac{\partial^2 T}{\partial y^2} \quad \text{(all } x) \quad (10\text{-}1\text{-}9b)$$

Each of these equations (10-1-9a) and (10-1-9b) may be solved by the Crank-Nicolson method.

There are much fancier methods than the splitting method, but their truncation errors (the asymptotic difference between the difference equation and the differential equation) do not go to zero any faster than the truncation error for the splitting method.

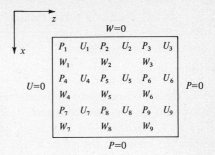

FIGURE 10-1
A grid arrangement for the acoustic equation. This arrangement avoids the necessity of taking ∂_x or ∂_z over more than one interval. It also results in (10-1-12) being a scalar equation rather than a 2×2 block matrix equation.

Now let us see how to formulate the acoustical problem in a Crank-Nicolson form. Let u and w denote velocities in the x and z directions. Let P denote pressure, ρ denote density, and K denote incompressibility. Acceleration equal to pressure gradient gives

$$\rho \frac{\partial u}{\partial t} = - \frac{\partial P}{\partial x}$$

$$\rho \frac{\partial w}{\partial t} = - \frac{\partial P}{\partial z}$$

and pressure decreasing with the divergence of velocity gives

$$\frac{\partial P}{\partial t} = - K \left(\frac{\partial u}{\partial x} + \frac{\partial w}{\partial z} \right)$$

Arranging into a matrix and letting ∂_x denote $\partial/\partial x$, etc., we have

$$\frac{\partial}{\partial t} \begin{bmatrix} P \\ U \\ W \end{bmatrix} = \begin{bmatrix} 0 & -K\partial_x & -K\partial_z \\ -\rho^{-1}\partial_x & 0 & 0 \\ -\rho^{-1}\partial_z & 0 & 0 \end{bmatrix} \begin{bmatrix} P \\ U \\ W \end{bmatrix} \quad (10\text{-}1\text{-}10)$$

The implementation of (10-1-10) by a Crank-Nicolson scheme follows in a direct analogy to the implementation of (10-1-3). The principal difference is that we have vectors and matrices in (10-1-10) but only scalars in (10-1-3). When the splitting method is applied to (10-1-10) we have

$$\frac{\partial}{\partial t} \begin{bmatrix} P \\ U \end{bmatrix} = 2 \begin{bmatrix} 0 & -K\partial_x \\ -\rho^{-1}\partial_x & 0 \end{bmatrix} \begin{bmatrix} P \\ U \end{bmatrix} \quad (10\text{-}1\text{-}11a)$$

and

$$\frac{\partial}{\partial t} \begin{bmatrix} P \\ W \end{bmatrix} = 2 \begin{bmatrix} 0 & -K\partial_z \\ -\rho^{-1}\partial_z & 0 \end{bmatrix} \begin{bmatrix} P \\ W \end{bmatrix} \quad (10\text{-}1\text{-}11b)$$

at alternate time steps. When formulating boundary conditions for (10-1-11a) it turns out to be convenient to define P and U on alternate squares of a checkerboard. See Fig. 10-1.

A final matter of great practical significance is the fast method of solution to a tridiagonal set of simultaneous equations like (10-1-6) or (10-1-7). A slightly more general set of equations is

$$A_k T_{k+1} + B_k T_k + C_k T_{k-1} = D_k \quad (10\text{-}1\text{-}12)$$

For the heat-flow equation the elements of (10-1-12) are scalars. In other physical problems we may have to regard A, B, and C as 2×2 matrices, T as a 2×1 vector for each k at the $n + 1$ time level, and D_k as a 2×1 vector function of the field variables known at the n time level. The method proceeds by writing down another equation (E_k, F_k yet unknown) with the same solution T_k as (10-1-12)

$$T_k = E_k T_{k+1} + F_k \quad (10\text{-}1\text{-}13)$$

Write (10-1-13) with shifted index

$$T_{k-1} = E_{k-1} T_k + F_{k-1} \quad (10\text{-}1\text{-}14)$$

Insert into (10-1-12)

$$A_k T_{k+1} + B_k T_k + C_k(E_{k-1}T_k + F_{k-1}) = D_k \quad (10\text{-}1\text{-}15)$$

Rearrange (10-1-15) to resemble (10-1-13)

$$T_k = -(B_k + C_k E_{k-1})^{-1} A_k T_{k+1} + (B_k + C_k E_{k-1})^{-1}(D_k - C_k F_{k-1}) \quad (10\text{-}1\text{-}16)$$

Comparing (10-1-16) to (10-1-13) we see that they are the same, so that E_k and F_k may be developed by the recursions

$$E_k = -(B_k + C_k E_{k-1})^{-1} A_k \quad (10\text{-}1\text{-}17a)$$

$$F_k = (B_k + C_k E_{k-1})^{-1}(D_k - C_k F_{k-1}) \quad (10\text{-}1\text{-}17b)$$

Naturally when doing this on a computer for any case where matrices contain zeros, as in (10-1-11), one should use this fact to simplify things.

Now we consider boundary conditions. Suppose T_0 is prescribed. Then we may satisfy (10-1-13) with $E_0 = 0$, $F_0 = T_0$. Then compute all E_k and F_k. Then if T_N is prescribed, we may use (10-1-13) to calculate successively $T_{N-1}, T_{N-2}, \ldots, T_0$. Another useful set of boundary conditions is to prescribe the ratios $r_1 = T_0/T_1$ and $r_2 = T_N/T_{N-1}$. Begin by choosing $E_0 = r_1$, $F_0 = 0$. Compute E_k and F_k. Then solve the following for T_N. From (10-1-14)

$$T_{N-1} = E_{N-1}T_N + F_{N-1}$$

$$T_N/r_2 = E_{N-1}T_N + F_{N-1}$$

$$T_N = \left(\frac{1}{r_2} - E_{N-1}\right)^{-1} F_{N-1}$$

Then compute $T_{N-1}, T_{N-2}, \ldots$ as before.

As stated earlier, there are many more details associated with numerical solutions to partial differential equations. This chapter has given only the most important tricks for initial-value problems. A program to solve tridiagonal simultaneous equations is given in Fig. 10-2.

```
SUBROUTINE TRI(A,B,C,N,T,D,E,F)
DIMENSION T(N),D(N),F(N),E(N)
N1=N-1
E(1)=1.0
F(1)=0.
DO 10 I=2,N1
DEN=B+C*E(I-1)
E(I)=-A/DEN
10   F(I)=(D(I)-C*F(I-1))/DEN
T(N)=F(N1)/(1.0-E(N1))
DO 20 J=1,N1
I=N-J
20   T(I)=E(I)*T(I+1)+F(I)
RETURN
END
```

FIGURE 10-2
A program to solve tridiagonal simultaneous equations. *A*, *B*, and *C* are assumed independent of *k* and zero-slope end conditions are used.

EXERCISES

1 Consider solving (10-1-8) by a Crank-Nicolson scheme in two dimensions on a 4×4 grid. This leads to a 16×16 set of simultaneous equations for the unknown $T_{j,k}^{n+1}$. What is the pattern of zeros in the 16×16 matrix? The difficulty in actually solving this set gives impetus to the splitting method.

2 A difference approximation to the heat-flow partial differential equation is

$$P_j{}^{n+1} - P_j{}^{n-1} = \frac{a\,\Delta t(P_{j+1}^n - 2P_j{}^n + P_{j-1}^n)}{\Delta x^2} + s_j{}^n$$

utilizing the trial solution $P_j{}^n = Q_n e^{ikj\,\Delta x}$ reduce the equation to a one-dimensional difference equation. Write the reduced equation in terms of Z transforms. Does this equation correspond to a nondivergent filter for any real values of a? for any imaginary values of a? (Use a Fourier expansion for s.)

3 Modify the computer program of Fig. 10-2 so that instead of prescribing zero-slope end conditions, (10-1-7) is solved.

4 Write a computer program to solve equation (10-1-6) with $b = .5$ and initial conditions $T(1) \cdots T(20) = 0.0$ and $T(21) \cdots T(30) = 1.0$. Use subroutine TRI.

10-2 WAVE EXTRAPOLATION IN OPTICS

In geophysics we generally have measurements along a line on the surface of the earth (x axis) from which we like to make deductions about earth properties below the surface. The first step is often to extrapolate observations at the earth's surface in a downward direction.

Before looking at numerical methods of extrapolating wave fields in space it will be valuable to review quickly the methods used in optics to extrapolate waves through microscopes and telescopes. An enjoyable, more complete account will be found in Reference 35.

We will take a wave disturbance in two-dimensional cartesian geometry $p(x, z, t)$ given at z_0 and show how it is extrapolated down the optic axis. Three common situations arise in the projection of a beam of light down an optic axis. First is the projection of a beam through an aperture or a photographic transparency. All that is required for a mathematical description is a transmittance function which ranges from 0 to 1 over the aperture or transparency. Taking the optic axis to be the z axis and restricting attention to two-dimensional geometry, the projection through an absorber $T(x)$ located at $z_0 + dz/2$ is

$$p(t, x, z_0 + dz) = T(x)p(t, x, z_0) \quad (10\text{-}2\text{-}1)$$

The second common situation is projection through a lens, often approximated as a "thin lens." Here it is necessary to define a differential delay function $\tau(x)$ which describes the time delay on propagation through the lens of a ray at x parallel to the z axis. If the lens is located at $z_0 + dz/2$, convolution of the wave field with a delayed impulse is represented as

$$p(t, x, z_0 + dz) = \int p(t - s, x, z_0)\delta[s - \tau(x)] \, ds$$

$$= p[t - \tau(x), x, z_0] \quad (10\text{-}2\text{-}2a)$$

This time shifting is simply expressed in the frequency domain where the convolution (10-2-2a) becomes a product. Then

$$P(\omega, x, z_0 + dz) = P(\omega, x, z_0)e^{i\omega\tau(x)} \quad (10\text{-}2\text{-}2b)$$

The third common situation in optics is the projection of waves across a region of empty space. Surprisingly, this is the most difficult of the three projections. First we recall the wave equation

$$\left(\partial_{xx} + \partial_{zz} - \frac{1}{v^2} \partial_{tt}\right) p(t, x, z) = 0 \quad (10\text{-}2\text{-}3)$$

Taking the velocity v to be a constant in time and space, we may use the trial solution

$$p(t, x, z) = P(\omega, k_x, z)e^{-i\omega t + ik_x x}$$

which reduces (10-2-3) to the ordinary differential equation

$$\frac{d^2}{dz^2} P = \left(-\frac{\omega^2}{v^2} + k_x^2\right) P \quad (10\text{-}2\text{-}4)$$

This equation has two solutions, $e^{ik_z z}$ and $e^{-ik_z z}$, where

$$k_z = \left(\frac{\omega^2}{v^2} - k_x^2\right)^{1/2} \quad (10\text{-}2\text{-}5)$$

One of these solutions is a wave down the z axis and the other is a wave going up the axis. Initial conditions (and the no-backscattering approximation at lenses and apertures) enable us to reject one of the solutions, leaving us with

$$P(\omega, k_x, z) = P(\omega, k_x, z_0)e^{ik_z(z - z_0)}$$

$$= P(\omega, k_x, z_0)e^{i(\omega^2/v^2 - k_x^2)^{1/2}(z - z_0)} \quad (10\text{-}2\text{-}6)$$

The right-hand side is a product of two functions of k_x. It is also the product of two functions of ω. This means that with the standard tools of Fourier analysis we

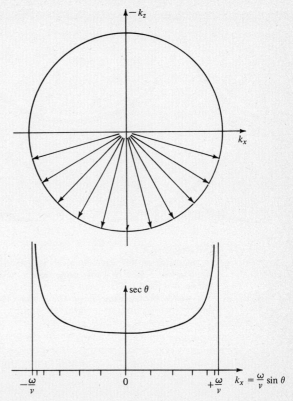

FIGURE 10-3
Power spectrum in k_x for an isotropic distribution of rays from a point source. Around $\theta = \pm 90°$ there is a clustering of rays at $k_x = \pm \omega/v$. Power as a function of k_x will be proportional to $d\theta/(dk_x/d\theta) = d\theta/(d \sin \theta/d\theta) = d\theta/\cos \theta = [1 - (vk_x/\omega)^2]^{-1/2} d\theta$. This result may be compared to the transfer function (10-2-7) which has a constant magnitude for $-\omega/v < k_x < \omega/v$.

could recast (10-2-6) to a convolution in either the time domain or the space domain x or both. Converting the "filter" transfer function

$$\exp i \left(\frac{\omega^2}{v^2} - k_x^2 \right)^{1/2} (z - z_0) \qquad (10\text{-}2\text{-}7)$$

to the space domain will give us an "impulse response" which in this case has the physical meaning of the wave field transmitted through a point aperture. A beam emerging from a point aperture behaves somewhat like a beam from a point source. To recognize the difference, note that the transfer function (10-2-7) has a unit magnitude independent of k_x but, from Fig. 10-3, the spectral magnitude of a point source is lower near $k_x = 0$ and peaks up around $k_x = \pm\omega/v$. This means that the aperture function does not radiate isotropically like the point source but

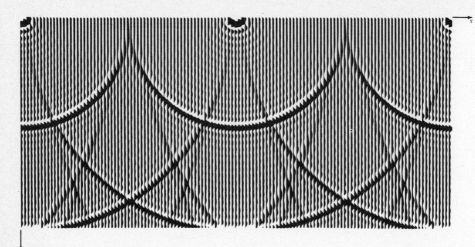

FIGURE 10-4
A snapshot of the wave-equation transfer function. A double Fourier sum of $\exp[i(\omega^2/v^2 - k_x^2)^{1/2} z]$ was done over k_x and ω. We see a display of the (x, z) plane at a fixed t. The result is semi-circular wavefronts with amplitude greatest for waves propagating along the z axis. Periodicity in x and t results from approximating Fourier integrals by sums.

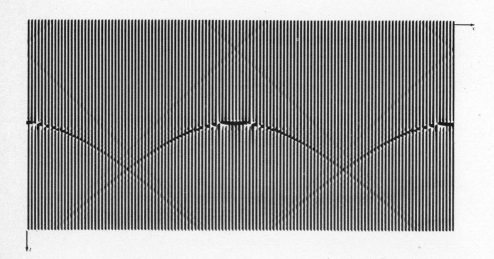

FIGURE 10-5
Seismic profile type displays of the wave-equation transfer function. A double Fourier sum of (10-2-7) was done over k_x and ω. As with a collection of seismograms, we see the (x, t) plane for a fixed z_0. The hyperbolic arrival times measure the distance from a point aperture at $(0, 0)$ to the screen (x, z_0). Ray theory easily explains the travel time, but the slow amplitude decay along the hyperbola, an obliquity function, is a diffraction phenomenon not easily computed by analytic means, especially far off axis. The obliquity function should not be confused with the hash which arises from attempted representation of a delta function on a grid.

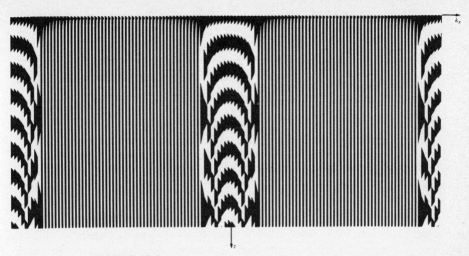

FIGURE 10-6
The real part of the exact transfer function, $\exp[i\sqrt{(\omega^2/v^2) - k_x^2}\,z]$, plotted k_x vs. z with ω taken as constant. The abrupt change in character of the function occurs at $\omega^2/v^2 = k_x^2$, the transition between propagation and evanescence.

contains more energy near $k_x = 0$, which is energy directed along the z axis. There seems to be no easy analytical procedure for the Fourier transformation of (10-2-7) into time and space domains. One of my associates, Philip Schultz, did some numerical Fourier transforms to obtain the display's real parts shown in Figs. 10-4, 10-5, and 10-6. Sample data Fourier transforms induce periodicity in all the transformed coordinates. The periodicity is quite apparent, and there has been no attempt to suppress it in the figures.

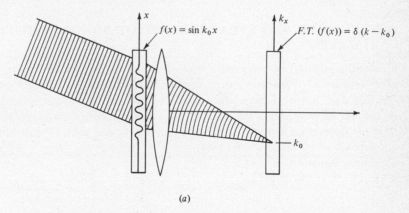

(a)

FIGURE 10-7a
Fourier transformation by a lens. A sinusoidal oscillation in the x domain results from a beam propagating through at some angle. A lens then converts the beam to a point in the k_x domain. The Fourier transform of a sinusoid is a delta function. The shift of the delta function from the optic axis is in proportion to the rate of oscillation of the sinusoid.

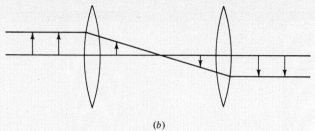

(b)

FIGURE 10-7*b*
Two lenses separated by twice their focal length can be used to invert an image.
Two Fourier transforms can be used to reverse a function.

It is well known that a lens can be used to take a Fourier transform. Actually
a Fourier transform takes place when a beam is allowed to propagate to infinity.
The lens just serves to bring infinity back into range. Suppose a monochromatic
optical disturbance $P(x, z, \omega)$ is observed at z_0. This function of x may be expanded
in a Fourier integral of components of the form $A(k_x)e^{ik_x x}$. The important thing
is to recognize that any single component represents a plane wave propagating at
an angle $\sin \theta = v k_x/\omega$ from the z axis. On propagation to infinity, all these rays
separate from one another. When they are projected on a screen the largest values
of k_x project farthest from $x = 0$. Figure 10-7*a* exhibits this idea where a lens is
used. That two lenses invert an image is the physical manifestation of the mathematical fact that a Fourier transform is not its own inverse. The inverse transform has
an opposite signed exponential. One can readily verify that transforming twice
with the same signed exponential just reverses the original waveform. The situation
is depicted in Fig. 10-7*b*.

10-3 NUMERICAL EXTRAPOLATION OF MONOCHROMATIC WAVES

The optical method of wave extrapolation is not valid in materials for which the
wave velocity $v = v(x, z)$ is space variable because then the complex exponential
function does not turn out to be a valid solution to the wave equation. For this
reason we will now seek a numerical procedure for extrapolating wave fields which
does not depend on analytic solutions or any particular velocity distribution. The
assumption of monochromatic solutions $e^{-i\omega t}$ reduces the wave equation to the
Helmholtz equation

$$P_{xx} + P_{zz} = \frac{-\omega^2}{v(x, z)^2} P \qquad (10\text{-}3\text{-}1)$$

Now let us think about using (10-3-1) to extrapolate $P(x, z_0)$ in the z direction. Say
we know P at z_0 for all x. Then we can find P_{zz} by rearrangement of (10-3-1)

$$P_{zz} = -\frac{\omega^2}{v^2} P - P_{xx} \qquad (10\text{-}3\text{-}2)$$

Given P and P_z for all x at some particular z with the help of (10-3-2) we might theoretically expect that we would be able to use a finite differencing scheme to obtain P and P_z at $z + \Delta z$. Actually, a fundamental difficulty is sneaking up on us. To understand it, let us assume v is a constant independent of x and that we have Fourier transformed the x dependence to k_x dependence. Then (10-3-2) becomes

$$P_{zz} = \left(-\frac{\omega^2}{v^2} + k_x^2 \right) P \qquad (10\text{-}3\text{-}3)$$

The behavior of (10-3-3) will be dramatically affected by the sign of the factor $-\omega^2/v^2 + k_x^2$. If it is positive, we will have growing and decaying exponential solutions. If it is negative, we will have nice, sinusoidal, wavelike solutions. Numerically the growing exponential solutions will present problems. These growing solutions can be kept from getting out of sight if we can start the growing exponential function with zero amplitude. This can be arranged by prescribing a certain ratio between P and P_z. Actually, come to think of it, geophysically we usually measure only P anyway and we do not measure P_z, so why not figure out theoretically a value of P_z from P which avoids the growing solution? Furthermore, in optics the extrapolation of $P(x, z_1)$ to $P(x, z_2)$ does not depend on knowledge of the derivative $P_z(x, z_1)$. The wave equation is second order in z and hence has two solutions (upgoing and downgoing). Thus two boundary conditions are required. In the usual boundary-value problems in physics, solutions are required in the intermediate region between z_0 and z_N and the appropriate boundary conditions are to prescribe P at z_0 and P at z_N. How does the optical method succeed in avoiding the need for either P_z at z_0 or the need for P at z_N? It succeeds because one of the two solutions was thrown away when k_z was defined by choosing only one of two possible square roots. Since one solution is left, only one boundary condition is required instead of two. Throwing away one of the solutions amounts to making an assumption about the physical situation which may or may not be valid. The validity of this assumption is always a matter of degree and depends on practical factors. Our present objective is to modify (10-3-2) to build in the common optical assumption that we are only trying to describe waves with a component along the $+z$ axis, without building in the common optical assumption of a homogeneous medium. Instead of (10-3-2), which is second order in z and describes waves which go in both plus and minus z directions, we would like to have an equation which is first order in z and describes only waves in the $+z$ direction. Since geophysically we do not observe P_z, a valuable added bonus would be that such a first-order equation would require only $P(x)$ as an initial condition, not both P and P_z. Geophysically, the "downgoing wave" assumption can often be used when we are describing the wave field emitted from active prospecting equipment, and an "upgoing wave" assumption can often be used to describe subsequent observations. Naturally, in any situation, the validity of these assumptions must be investigated. To describe a plane wave propagating in the $+z$ direction we may write

$$P(x, z) = Q_0 \, e^{i(\omega/v)z}$$

Saying that Q_0 is an unknown constant amounts to saying that the wave has unknown amplitude and phase. Next we write

$$P(x, z) = Q(x, z)e^{i(\omega/v)z}$$

Now, "$Q(x, z)$ is approximately a constant function of x and z" is a rather fuzzy statement which we will proceed to sharpen up. By restricting $Q(x, z)$ to slowly variable functions we will be restricting $P(x, z)$ to wave fields which are near to plane waves propagating in the z direction. In fact, P might represent plane waves propagating at a small angle from the z axis, or it might be a small portion of a spherical wave, or it might be the observed backscattered radiation in a seismic reflection survey, or on 90° rotation of the coordinate system it might describe surface waves.

The ratio ω/v occurs often and it is called the spatial frequency of the wave. We define

$$m = \frac{\omega}{v(x, z)} \qquad (10\text{-}3\text{-}4)$$

We also define $\bar{m}$ as a spatial average of m.

$$\bar{m} = \frac{\omega}{\bar{v}} \qquad (10\text{-}3\text{-}5)$$

In a material which is homogeneous $\bar{m}$ will equal m. With this definition we write the wave disturbance as

$$P(x, z) = Q(x, z)e^{i\bar{m}z} \qquad (10\text{-}3\text{-}6)$$

Now an additional condition to make $Q(x, z)$ slowly variable with z is that $m(x, z)$ be relatively near to $\bar{m}$. Let us compute some partial derivatives of (10-3-6)

$$P_x = Q_x e^{i\bar{m}z} \qquad (10\text{-}3\text{-}7a)$$

$$P_{xx} = Q_{xx} e^{i\bar{m}z} \qquad (10\text{-}3\text{-}7b)$$

$$P_z = (Q_z + i\bar{m}Q)e^{i\bar{m}z} \qquad (10\text{-}3\text{-}7c)$$

$$P_{zz} = (Q_{zz} + 2i\bar{m}Q_z - \bar{m}^2 Q)e^{i\bar{m}z} \qquad (10\text{-}3\text{-}7d)$$

Insert (10-3-7b) and (10-3-7d) into (10-3-1) and cancel the exponential, obtaining

$$Q_{xx} + Q_{zz} + 2i\bar{m}Q_z + (m^2 - \bar{m}^2)Q = 0 \qquad (10\text{-}3\text{-}8)$$

Now we make the very important step where we assert that for many applications Q is slowly variable and Q_{zz} may be neglected in comparison with $2i\bar{m}Q_z$. Dropping the Q_{zz} term will be called the parabolic approximation or the paraxial approximation. This gives us the desired first-order, hence initial-value, equation in z.

$$Q_{xx} + 2i\bar{m}Q_z + (m^2 - \bar{m}^2)Q = 0 \qquad (10\text{-}3\text{-}9)$$

In a homogeneous medium, (10-3-9) reduces to

$$Q_{xx} + 2i\bar{m}Q_z = 0 \qquad (10\text{-}3\text{-}10)$$

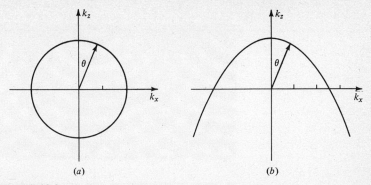

FIGURE 10-8
Graph of acceptable wave numbers to wave equation (*a*) and to one-way wave equation (*b*).

Equation (10-3-10) is really of the same form as the heat-flow equation if z is associated with time and the heat conductivity is taken to be imaginary. The equation is, in fact, known as the Schroedinger equation. It may be solved numerically by the means described for the heat-flow equation in Sec. 10-1. Ultimately (10-3-10) will be advocated for quite a number of purposes, so before we proceed let us take a look at what we have lost by dropping Q_{zz}. To facilitate comparison of (10-3-10) to the wave equation, let us convert back from the Q variable to the P variable. Rearrange (10-3-7) and form derivatives.

$$Q = Pe^{-i\bar{m}z} \qquad (10\text{-}3\text{-}11a)$$

$$Q_{xx} = P_{xx}e^{-i\bar{m}z} \qquad (10\text{-}3\text{-}11b)$$

$$Q_z = (P_z - i\bar{m}P)e^{-i\bar{m}z} \qquad (10\text{-}3\text{-}11c)$$

Insert (10-3-11b) and (10-3-11c) into (10-3-10) and cancel the exponential, getting the equation which we will call the one-way wave equation.

$$P_{xx} + 2i\bar{m}(P_z - i\bar{m}P) = 0$$

$$P_{xx} + 2i\bar{m}P_z + 2\bar{m}^2P = 0 \qquad (10\text{-}3\text{-}12)$$

One technique which may be used to solve any partial differential equation in cartesian coordinates with constant coefficients is to insert the complex exponential $e^{(ik_x x + ik_z z)}$. If k_x and k_z turn out to be real, then this trial solution may be interpreted as a plane wave propagating in the $k = (k_x, k_z)$ direction. Inserting this exponential into both the wave equation (10-3-1) and the one-way wave equation (10-3-12) and canceling the exponential, we get two algebraic equations called dispersion relations. They are

$$-k_x^2 - k_z^2 + m^2 = 0 \qquad (10\text{-}3\text{-}13)$$

$$-k_x^2 - 2\bar{m}k_z + 2\bar{m}^2 = 0 \qquad (10\text{-}3\text{-}14)$$

These two equations are graphed in Fig. 10-8, a and b.

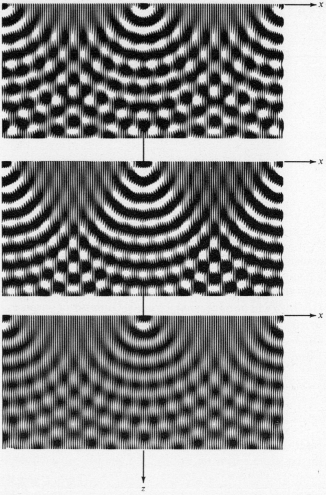

FIGURE 10-9
Snapshots of the monochromatic wave-equation transfer function. A Fourier sum over k_x, done over the exact wave-equation transfer function $\exp[i(1 - k_x^2 v^2/\omega^2)^{1/2}\,\omega z/v]$, is displayed (top) in the (x, z) plane for a fixed frequency ω_0. Middle is the same for the 15° approximate transfer function $\exp[i(1 - k_x^2 v^2/2\omega^2)\omega z/v]$. Bottom is the same for the 45° approximation

$$\exp\left[i\,\frac{\omega}{v}\,\frac{4\omega^2 - 3k_x^2\,v^2}{4\omega^2 - k_x^2\,v^2}\,z\right] \quad \text{of Exercise 2.}$$

The physical picture is of waves passing through small apertures which are periodically spaced along the x axis.

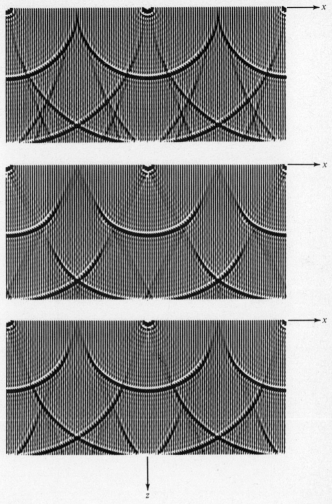

FIGURE 10-10
Snapshots of the time-dependent wave-equation transfer function and approximations. A double Fourier sum over k_x and ω of the functions of Fig. 10-9 shows the (x, z) plane at a fixed time.

The graph for the wave equation is a circle and illustrates what we already know, namely that the magnitude of the wave number in an arbitrary direction, that is, $(k_x{}^2 + k_z{}^2)^{1/2}$ is equal to the constant ω/v. Such is not the case, however, for the one-way wave equation. Here we have only the approximation $k_x{}^2 + k_z{}^2 \approx \omega^2/v^2$ for small angles θ. Figure 10-8a also illustrates geometrically that (10-3-14) is an initial-value problem in z because Fig. 10-8a gives two values for k_z corresponding to any k_x, but Fig. 10-8b gives only one value for k_z. Figures 10-9,

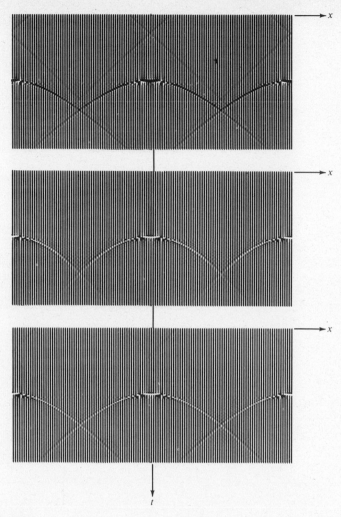

FIGURE 10-11
Seismic profile-type displays of the wave-equation transfer function and two approximations to it. Exact, 15° approximate, and 45° approximate forms of the wave-equation transfer function were Fourier summed over k_x and ω. As with a seismic profile, we see a display of the (x, t) plane for a fixed z. The exact solution (top) is a delta function along a hyperbola. The 15° approximation (middle) is a parabola. The approximations die out more rapidly with angle than the exact solution.

FIGURE 10-12
Monochromatic wave-equation transfer functions displayed in the plane of (k_x, z). The real part only is shown. Top is the exact transfer function. Note the abrupt change to evanescence at $|k_x v/\omega| = |\sin 90°| = 1$. The exponential decay for $k_x > \omega/v$ is perceptible near $z = 0$. The 15° approximation (middle) and the 45° approximation (bottom) are all-pass filters and have replaced the evanescent region by an interesting design. In order to eliminate a massive amount of short horizontal wavelength fuzz in the spatial domain on the previous two figures, this evanescent zone was removed with a step function. The implication in a data processing application is that occasionally the approximate transfer functions may well be augmented by a fan-filter. (See Ref. [36].)

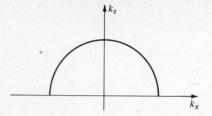

FIGURE 10-13
The dispersion relation for an ideal one-way wave equation is a semicircle.

10-10, 10-11, and 10-12 show the wave equation transformation function $e^{ik_z z}$ and approximations $e^{ik_z z}$ and Fourier transformations thereof.

What we really want is a one-way wave equation which has the semicircle of Fig. 10-13 for its dispersion relation. The equation for the perfect semicircle is given by

$$k_z = \sqrt{m^2 - k_x^2} \quad (10\text{-}3\text{-}15)$$

This of course is the basic relation used for extrapolation in optics. By the binomial expansion, (10-3-15) may be written

$$k_z = m\left(\frac{1 - k_x^2}{2m^2} - \frac{k_x^4}{8m^4} + \cdots\right) \quad (10\text{-}3\text{-}16)$$

This expression converges for all $0 < k_z < m$.

Now for the sudden flash of insight which enables us to write the partial differential equation with this semicircle as its dispersion relation, from (10-3-16) we are inspired to write

$$\partial_z P = im\left(1 + \frac{\partial_{xx}}{2m^2} - \frac{\partial_{xxxx}}{8m^4} + \cdots\right)P \quad (10\text{-}3\text{-}17)$$

Clearly, insertion of the plane wave $\exp(ik_x x + ik_z z)$ into (10-3-17) immediately gives the desired semicircular dispersion relation (10-3-16). Thus, the greater the angular accuracy desired the more terms of (10-3-17) are required in the calculation. As a shorthand we may choose to write (10-3-17) as

$$\partial_z P = i(m^2 + \partial_{xx})^{1/2} P \quad (10\text{-}3\text{-}18)$$

It will be of no help to us, but it turns out that (10-3-18) is the relativistic Schroedinger equation.

It is easy to obtain the wave equation from (10-3-18). Just differentiate with respect to z

$$\partial_{zz} P = i\,\partial_z(m^2 + \partial_{xx})^{1/2} P$$

Taking m independent of z, we may interchange the order of differentiation

$$\partial_{zz} P = i(m^2 + \partial_{xx})^{1/2}\,\partial_z P$$

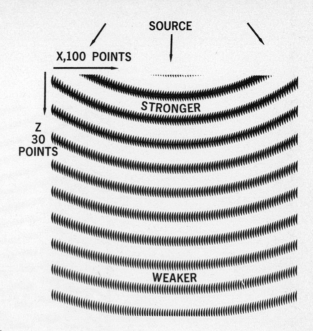

FIGURE 10-14

An expanding monochromatic cylindrical wave. The wavefronts are concentric circles of decreasing amplitude. The computation begins with an analytic solution at the top of the figure in a 100-point linear grid. Using difference equations, we stepped the grid downward, thirty steps making up the whole figure. About six complex multiplications are required per point; this amounts to about five seconds of time on our computer. The display is the (x, z) plane, although a multichannel seismogram plotter has been used. (From Ref. [3], p. 408.)

inserting (10-3-18)

$$\partial_{zz} P = -(m^2 + \partial_{xx})^{1/2}(m^2 + \partial_{xx})^{1/2} P$$
$$= -(m^2 + \partial_{xx})P$$

which is the wave equation.

Figures 10-14, 10-15, and 10-16 show finite-difference solutions to the parabolic approximated wave equation in homogeneous media.

Next, let us turn to the question of using the parabolic approximation in the presence of space variations in material velocity. The exercises go into considerable detail on this matter, but we can easily make some improvements over (10-3-9). The main idea is to approximate a circle by a parabola; the actual radius of the circle does not have anything to do with the approximation. This leads to the suggestion that (10-3-12) or (10-3-14) could be used with $\overline{m}$ replaced by m, as in (10-3-17); hence (10 3 12) would be

$$P_{xx} + 2im(x, z)P_z + 2m^2(x, z)P = 0 \quad (10\text{-}3\text{-}19)$$

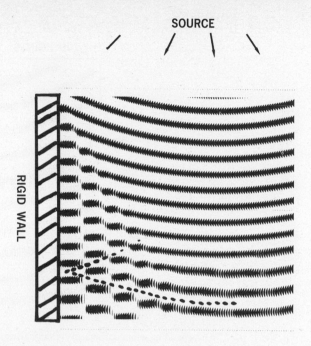

FIGURE 10-15
Like Fig. 10-14, but the left-hand boundary is a rigid wall. Waves may be seen reflecting back into the medium from the boundary. The reflected wavefront is indicated by the shorter of the two dashed lines. (From Ref. [3], p. 409.)

FIGURE 10-16
Expanding cylindrical wave. A theoretical solution was put in at the top boundary and extrapolated downward with the equation of Exercise 2. The wavefronts are not quite circular as they would be were it feasible to use (10-3-18). Notice also that the theoretical $r^{-1/2}$ amplitude decay is not exhibited for waves about 60° off the vertical. Such waves attenuate less rapidly because at 60° the phase curve is flatter than a circle. (From Ref. [5], p. 476.)

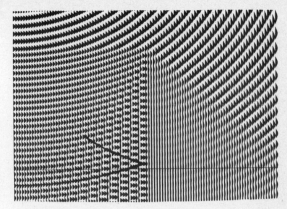

FIGURE 10-17
Waves impinging on a buried block of low-velocity material. Waves enter at the top of the block and are completely internally reflected from the side of the block. This leaves a shadow on the outside of the block. (From Ref. [5], p. 474.)

With (10-3-19) we no longer need to assume that $m \approx \overline{m}$ so we can now deal with a wide range of velocities. Actually, as the exercises will show, the validity of (10-3-19) depends also on the approximation that the logarithmic space gradients of material velocity are small compared with the logarithmic gradients of the waves. In other words, the waves change faster than the material does.

Figures 10-17, 10-18, and 10-19 illustrate the propagation of waves in inhomogeneous materials.

The approximation is evidently best at high frequencies (short wavelengths). This approximation is well known in wave theory. Although it is sometimes called a ray approximation, the reader should not fear that the theory has degenerated to geometrical optics. Actually all the phenomena of physical optics (for example: interference, diffraction, and finite size focus) are still present. In fact we need not go to the physical optics limit at all. Some of the exercises are examples that include the velocity gradients found in lower frequency terms. Whether many or none of these terms is important in practice is a question which is particular to each application.

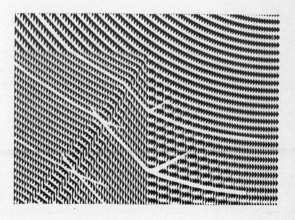

FIGURE 10-18
A low-velocity block is illuminated from the side. There is partial reflection from the side of the block and interference between waves entering the block through different faces. (From Ref. [5], p. 474.)

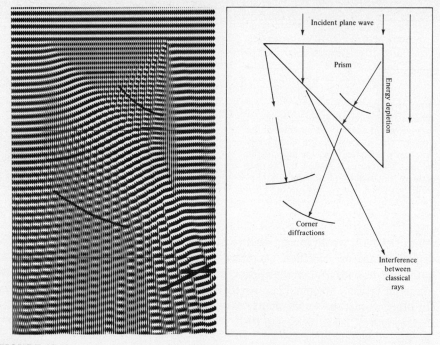

FIGURE 10-19

Plane waves propagating through a right 45° prism. Waves are incident from the top of the page. Shortened wavelength is shown inside the prism. When the waves emerge from the prism they are bent toward the right-hand side of the page. As they emerge, their amplitude increases because they are compressed into a narrower beam. At the bottom right they interfere with waves which have passed along the side of the prism, causing amplitude modulation. Curved wavefronts are the result of diffraction from corners of the prism. Especially interesting is the diffraction from the upper right-hand corner. This is best seen by viewing the figure edge-on from the right-hand edge of the page. The energy for this diffraction is removed from the wave along the right-hand vertical edge of the prism. This calculation requires ten seconds of computer time on the Stanford IBM 360-67. (From Ref. [5], p. 475.)

EXERCISES

1 The variable Q has the practical advantage over P because, being more slowly variable with the z coordinate, it may be sampled less densely, thereby conserving computational effort. Convert (10-3-19) to an equation in Q by means of (10-3-11). Compare the result to (10-3-9). Of the two equations, yours and (10-3-9), which do you believe to be more accurate? Why?

2 An excellent square root approximation is given by the rational expression

$$(1 + x)^{1/2} \approx \frac{1 + 3x/4}{1 + x/4}$$

What "one-way wave equation" is suggested by this approximation? Make a graph of the dispersion relation. For selected angles of propagation how does accuracy compare to that of (10-3-14)?

3   The algebraic equation $a + bx + cx^2 = 0$ has two roots. If b is sufficiently large, we may approximate the smallest root with the linear relation $a + bx = 0$. An improved approximation which is still linear in x may be found by substituting $x = -a/b$ back into the quadratic

$$a + bx + c\left(\frac{-a}{b}\right)x = 0$$

$$ab + (b^2 - ac)x = 0$$

Define $k'_z = m - k_z$ and substitute $k_z = m - k'_z$ into $k_x^2 + k_z^2 = m^2$. Find the smallest root for k'_z. Show that this gives the same partial differential equation as Exercise 2.

4   Let the velocity $v = v(x) \neq v(x, z)$ be a function of x and define $m = \omega/v(x)$. Define the operator

$$\mathrm{Op} = m + \frac{1}{2m}\partial_{xx} - \frac{m_x}{2m^2}\partial_x$$

Note that

$$\partial_z P = i\,\mathrm{Op}\,P$$

$$\partial_{zz} P = i\,\partial_z\,\mathrm{Op}\,P = i\,\mathrm{Op}\,\partial_z P = -\mathrm{Op}^2\,P$$

$$(\partial_{zz} + \mathrm{Op}^2)P = 0 = \text{wave equation} + \text{error}$$

Examine each error term and decide whether it is important (1) at high frequencies (collect terms proportional to nth power of wavelength) and (2) at small or large angles from the z axis.

5   Review the section on Sylvester's matrix theorem. How is the square root of a matrix analogous to the square root of an operator?

6   Deduce the "outgoing wave equation" in cylindrical coordinates.

7   Deduce the "outgoing wave equation" in spherical coordinates.

8   Exercise 3 gave a good wide-angle approximation but Exercise 4 works for $m = m(x)$. To utilize the method of Exercise 3 for $m = m(x)$ it is necessary to note that although $bx - xb = 0$, it is not true that $(m\,\partial_x - \partial_x m)P = 0$ unless $m \neq m(x)$. Salvage the method of Exercise 3 by avoiding the use of commutivity as much as possible.

9   Consider surface waves propagating on the surface of an imperfect sphere. Deduce an equation, first-order in ϕ, the longitude coordinate, second-order in θ, the latitude coordinate, for waves beamed roughly along the equator. Assume all quantities are independent of the radial coordinate axis.

10 Modify the program of Sec. 10-1 in Exercise 4 to compute the solution to (10-3-10). You will need to review the compiler conventions of complex arithmetic. Also, after computing $Q(x, z)$ multiply it by e^{imz} to give $P(x, z)$. Print only the real part of $P(x, z)$. A physical interpretation of this result is light behind an edge of an opaque screen. Waves diffracted into the shadow zone should have semicircular wavefronts if you have arranged your display to preserve $\Delta z = \Delta x$ on the output.

11 Let $Z = e^{ik_x \Delta x}$ denote a discretization of the x coordinate. Define $A(Z) = \sum a_n Z^n$ by finding a_n such that

$$a_0 + \sum_{n=1}^{\infty} a_n\left(Z^n + \frac{1}{Z^n}\right) = |k_x| \qquad \text{for } |k_x|\Delta x \leq \pi$$

Show that either solution to

$$\frac{\partial P(Z)}{\partial z} = \pm A(Z)P(Z)$$

is a solution to Laplace's *differential* equation $P_{xx} + P_{zz} = 0$. These solutions may be used for upward and downward continuation.

10-4 EXTRAPOLATION OF TIME-DEPENDENT WAVEFORMS IN SPACE

In Sec. 10-3 we learned how to extrapolate monochromatic waves in space. To extrapolate a time-dependent waveform in space, one could first Fourier transform it into monochromatic waves, then extrapolate them as in the previous section, and finally Fourier transform back into the time domain. Thus, although this section solves, in principle, the same problem as the last section, a direct time-domain method will often be preferable for practical reasons. Although a time-domain study is necessarily more complicated than one in the frequency domain (all time points must be considered together, but each frequency is isolated from the others) there is a great deal more understanding to be gained in the time domain, especially as regards causality. We will discover that wave-extrapolation procedures are like filters (in fact, they are a special kind of multidimensional all-pass filter) and that the feedback parts of these filters must be minimum-phase. There are two independent time-domain derivations.

The first derivation begins by transforming the scalar wave equation

$$0 = P_{xx} + P_{zz} - v^{-2}P_{tt} \qquad (10\text{-}4\text{-}1)$$

into a coordinate frame which translates along the z axis at the speed $\bar{v}$ which we will generally take to equal or exceed v. It does not matter which way energy is propagating in the fixed frame; when it is seen in the moving frame it will remain stationary or fall backward. The coordinate transformation

$$x' = x \qquad (10\text{-}4\text{-}2a)$$
$$z' = \bar{v}t - z \qquad (10\text{-}4\text{-}2b)$$
$$t' = t \qquad (10\text{-}4\text{-}2c)$$

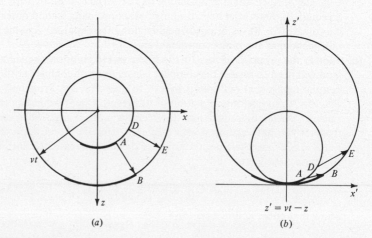

(a)   (b)

FIGURE 10-20
Expanding spherical wave in (a) fixed coordinates (left) and in (b) coordinates which translate in the z direction with the velocity of the wave (right).

is depicted in Fig. 10-20 for $\bar{v} = v$. In the primed frame all waves have a velocity component in the plus z' direction. Knowledge of P for present and past time at all x' for fixed z' should be sufficient to determine P for present and past values of time at $(x', z' + \Delta z')$ because before anything happens at $z' + \Delta z'$ something has to happen at z'. Thus, because of the restriction $\bar{v} \geq v$ we anticipate that the linear operators which we will develop to extrapolate P in the plus z' direction should be causal. Let P' denote the disturbance in the moving frame. We have

$$P(x, z, t) = P'(x', z', t') \quad (10\text{-}4\text{-}3)$$

It will be convenient to use subscripts to denote partial derivatives. Obviously,

$$P_x = P'_{x'} \quad \text{and}$$
$$P_{xx} = P'_{x'x'} \quad (10\text{-}4\text{-}4)$$

Also

$$P_z = P'_{x'} x'_z + P'_{z'} z'_z + P'_{t'} t'_z = -P'_{z'}$$

so

$$P_{zz} = P'_{z'z'} \quad (10\text{-}4\text{-}5)$$

and

$$P_t = P'_{x'} x'_t + P'_{z'} z'_t + P'_{t'} t'_t = \bar{v} P'_{z'} + P'_{t'}$$

so

$$P_{tt} = \bar{v}(\bar{v} P'_{z'z'} + P'_{t'z'}) + \bar{v} P'_{z't'} + P'_{t't'}$$
$$= \bar{v}^2 P'_{z'z'} + 2\bar{v} P'_{z't'} + P'_{t't'} \quad (10\text{-}4\text{-}6)$$

Now we may insert (10-4-4), (10-4-5), and (10-4-6) into (10-4-1) and we obtain

$$P'_{x'x'} + \left[1 - \left(\frac{\bar{v}}{v} \right)^2 \right] P'_{z'z'} - 2\frac{\bar{v}}{v^2} P'_{z't'} - \frac{1}{v^2} P'_{t't'} = 0 \quad (10\text{-}4\text{-}7)$$

We will take up the constant velocity case $v(x, z) = \bar{v}$. The case $v \neq \bar{v}$ is left for the exercises. Our main interest in (10-4-7) is with those waves which propagate with approximately the velocity of the new coordinate frame. In the moving frame such waves are doppler shifted close to zero frequency. This suggests omitting the $P'_{t't'}$ term from (10-4-7). Thus (10-4-7) becomes

$$P'_{t'z'} = \frac{v}{2} P'_{x'x'} \quad (10\text{-}4\text{-}8)$$

If we Fourier transform out the time coordinate equation (10-4-8) becomes $-i\omega P'_{z'} = (v/2)P'_{x'x'}$ which is identical to the monochromatic equation

$$Q_{xx} + 2im Q_z = 0$$

derived in the preceding chapter. Thus, dropping the $P'_{t't'}$ term is the familiar approximation of a circle by a parabola.

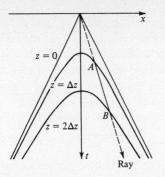
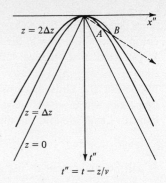

FIGURE 10-21
A point-source at $x = 0$, $z = 0$, $t = 0$. Hyperbolas at left indicate arrival times t at $z = 0$, Δz, $2\Delta z$. When time is a function of position as given by $t'' = t - z/v$ the arrival times t'' are as indicated on the right. Energy moves in the direction of $+t''$, since on a wavefront $z = vt \cos \theta$ and we have $t'' = t - z/v = t(1 - \cos \theta)$.

In solving (10-4-8) in a computer we can take either of two points of view. The first point of view is that P' is prescribed initially on a grid over x' and z' and then the equation is used for extrapolation in t'. The second point of view is that P' is prescribed initially on a grid over x' and t' and then (10-4-8) is used for extrapolation in z'.

Before developing a numerical method for the solution to (10-4-8) we will derive it by means of an entirely different coordinate transformation. Let us take the new coordinate frame fixed in space relative to the old one. However, let a different clock be used at each point in space in the new frame. The clocks all run at the same speed, but they are initialized in such a way that a plane wave traveling in the $+z$ direction will have the same arrival time measured at all clocks. (This is somewhat like a westward moving jet plane.) The transformation equations are

$$x'' = x \qquad (10\text{-}4\text{-}9a)$$

$$z'' = z \qquad (10\text{-}4\text{-}9b)$$

$$t'' = t - \frac{z}{v} \qquad (10\text{-}4\text{-}9c)$$

A disturbance initiated at $(x, z, t) = 0$ is depicted in Fig. 10-21. Referencing time with respect to the time of the earliest possible ray is a great computational convenience. It means the wave onset does not move off the finite, perhaps short, computational grid on which a wave packet has been defined. Define the disturbance in the new frame by P'' where

$$P(x, z, t) = P''(x'', z'', t'') \qquad (10\text{-}4\text{-}10)$$

Proceeding as before, we obtain

$$P_{xx} = P''_{x''x''} \qquad (10\text{-}4\text{-}11)$$

$$P_{zz} = P''_{z''z''} - 2v^{-1}P''_{t''z''} + v^{-2}P''_{t''t''} \qquad (10\text{-}4\text{-}12)$$

$$P_{tt} = P''_{t''t''} \qquad (10\text{-}4\text{-}13)$$

Inserting these into the wave equation (10-4-1) we obtain

$$P''_{t''z''} = \frac{v}{2}(P''_{x''x''} + P''_{z''z''}) \quad (10\text{-}4\text{-}14)$$

The last term of (10-4-14) is higher-order small for waves traveling at small angles from the z axis; this recalls that the solution to the wave equation for waves in the $+z$ direction is an arbitrary function $f(t - z/v) = f''(t'')$. Thus $\partial f''/\partial z''$ vanishes for a wave along the z'' axis. Neglecting $P''_{z''z''}$ we find that (10-4-14) reduces to

$$P''_{t''z''} = \frac{v}{2}P''_{x''x''} \quad (10\text{-}4\text{-}15)$$

which is the same equation as (10-4-8). Use of $e^{-i\omega t}$ time dependence in either (10-4-8) or (10-4-15) yields the equation (10-3-10) which was developed for extrapolation of monochromatic waves. Another point of view is that we could have obtained the time-dependent equations of this chapter by merely replacing $-i\omega$ in the monochromatic equations with ∂_t.

Now we develop a differencing scheme for the solution to (10-4-8) or (10-4-15). Drop primes. Let $j \Delta t$ refer to time. Let $n \Delta z$ refer to the coordinate z. Let δ denote a difference operator. Let $\mathbf{P}^n_j$ be a vector at each value of n and j. Running down the vector will be values of pressure along the x axis. By using matrix algebra we avoid writing a subscript for the x dependence. Let $\mathbf{T}$ denote a tridiagonal matrix with the negative of the second difference operator $-(1, -2, 1)$ on the diagonal. With all these definitions (10-4-8) or (10-4-15) becomes

$$\delta_z \delta_t \mathbf{P}^n_j = -\frac{v \Delta z \Delta t}{8 \Delta x^2} \mathbf{T} 4 \mathbf{P}^n_j \quad (10\text{-}4\text{-}16)$$

Let us define $a = v \Delta z \Delta t/8 \Delta x^2$. Now we must decide more precisely what first-difference approximations to use in (10-4-16). We will use the Crank-Nicolson scheme which is equivalent to the bilinear transform. First do centered time differencing

$$\delta_z(\mathbf{P}^n_{j+1} - \mathbf{P}^n_j) = -a \mathbf{T} 2(\mathbf{P}^n_{j+1} + \mathbf{P}^n_j)$$

and then do centered space differencing

$$(\mathbf{P}^{n+1}_{j+1} - \mathbf{P}^{n+1}_j) - (\mathbf{P}^n_{j+1} - \mathbf{P}^n_j) = -a\mathbf{T}(\mathbf{P}^{n+1}_{j+1} + \mathbf{P}^{n+1}_j + \mathbf{P}^n_{j+1} + \mathbf{P}^n_j) \quad (10\text{-}4\text{-}17)$$

From the point of view of computation we assume the unknown is $\mathbf{P}^{n+1}_{j+1}$ and that all else is known. Bringing the unknown to the left and the known to the right, we have

$$(\mathbf{I} + a\mathbf{T})\mathbf{P}^{n+1}_{j+1} = \mathbf{P}^{n+1}_j + \mathbf{P}^n_{j+1} - \mathbf{P}^n_j - a\mathbf{T}(\mathbf{P}^{n+1}_j + \mathbf{P}^n_{j+1} + \mathbf{P}^n_j) \quad (10\text{-}4\text{-}18)$$

For each n and j, the right side collapses to a known vector. The left side is the tridiagonal matrix $(\mathbf{I} + a\mathbf{T})$ multiplying the unknown vector $\mathbf{P}^{n+1}_{j+1}$. The solution of these equations is extremely simple and may be done as was the heat-flow equation in Sec. 10-1. Boundary conditions in x are contained on the ends of $\mathbf{T}$. For z and t boundary conditions it is sufficient to give, at all x, $\mathbf{P}^n_0$ for all n and $\mathbf{P}^0_j$ for all j. Other boundary arrangements are possible.

A very important question is the one of stability. We will now establish that the recursion (10-4-18) is stable for any positive value of a. If eigenvalues and eigenvectors of $\mathbf{T}$ were known and if all the $\mathbf{P}_j^n$ were expanded in terms of the eigenvectors of $\mathbf{T}$, then (10-4-18) would decouple into many separate equations, one for each of the eigenvalues of $\mathbf{T}$. The eigenvectors of $\mathbf{T}$ have components which are sinusoidal functions of x. If there are boundaries in x, then a discrete set of frequencies is allowed, otherwise there is a continuum. To see this observe that for the unbounded case $\mathbf{TP}$ is $(-1, 2, -1)$ convolved with $e^{ik_x m \, \Delta x}$ giving

$$(-e^{ik_x \, \Delta x} + 2 - e^{-ik_x \, \Delta x})e^{ik_x m \, \Delta x}$$

Thus the eigenvalue is $2 - 2 \cos k_x \, \Delta x = (2 \sin k_x \, \Delta x / 2)^2$. Since any eigenvalue must be between 0 and 4 it is sufficient to study (10-4-18) where the vector $\mathbf{P}_j^n$ has become a scalar P_j^n function of k_x, $\mathbf{I}$ is replaced by 1, and $\mathbf{T}$ is replaced by T, an arbitrary number between 0 and $+4$. It can be shown that for energy-conserving boundary conditions the eigenvalues are also between 0 and 4. Now, suppose P_j^n is known for all j at some particular value of n and we will investigate the stability of finding P_j^{n+1} for all j. Now, in (10-4-17) bring unknowns to the left.

$$(1 + aT)P_{j+1}^{n+1} - (1 - aT)P_j^{n+1} = (1 - aT)P_{j+1}^n - (1 + aT)P_j^n \quad (10\text{-}4\text{-}19)$$

The important thing for stability in (10-4-19) is that if we are successively increasing j, then the magnitude of the coefficient of P_{j+1}^{n+1} must exceed that of the coefficient of P_j^{n+1}. If we are decreasing j, the reverse should be true. The stability may be studied by the Z-transform methods discussed in earlier chapters. By the Z transform of (10-4-19) we mean that the coefficient of Z^j of

$$[(1 + aT) - Z(1 - aT)]P(Z)^{n+1} = [(1 - aT) - Z(1 + aT)]P(Z)^n \quad (10\text{-}4\text{-}20)$$

gives (10-4-19). The filter function for computing $P(Z)^{n+1}$ from $P(Z)^n$ is

$$\frac{(1 - aT) - Z(1 + aT)}{(1 + aT) - Z(1 - aT)} \quad (10\text{-}4\text{-}21)$$

We note that for positive a and for all T between 0 and 4, the denominator is a minimum-phase polynomial. This means that the time recurrence implied by (10-4-19) will be stable. The fact that (10-4-21) takes the form of an all-pass filter means that the depth recurrence on n will also be stable.

We have just completed a rather laborious stability proof. The reader will undoubtedly discover that his own application involves a slightly different equation, perhaps $v = v(x, z)$ or increased angular accuracy. What general advice can be given about formulating problems so that they will be stable for extrapolation? To begin with, it helps if you have a physical feeling that all of the information must be flowing one way. Then, if trouble occurs, it is most likely to be at unsuspected values of ω, k_x, k_z, or ratios thereof. Note that (10-4-15) in Fourier transform domain is

$$\omega k_z = -\frac{v}{2} k_x^2 \quad (10\text{-}4\text{-}22)$$

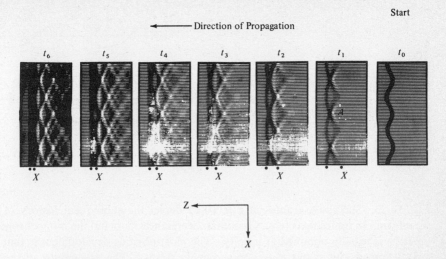

FIGURE 10-22
Disturbed plane wave propagating through a homogeneous medium. The first arrival of a disturbed plane wave heals itself during propagation. The wave coda or trail gets more and more complicated and energetic. In the trail, energy moves back away from the first arrival while phase fronts (marked by X) move forward. Beam-steer signal processing (sum over the x coordinate) enhances the first arriving signal but tries to destroy later arriving signals (the trail). Although this calculation was done beginning with frame t_0 and ending with frame t_6, the calculation could be done backwards, starting with t_6 and ending with t_0. After time realignment, beam-steer on frame t_0 could collect all signal energy.

If we are intending to extrapolate in the z'' direction we will be forming essentially $\exp(ik_z z'')$ or $\exp(-ik_x^2 z''/\omega)$. The reader should recall all the important facts about all-pass filters and spectral factorization. When wave propagation is to be modeled by all-pass filters and if the all-pass filters are supposed to be realizable or causal, then the phase derivative or group delay should be positive for all frequencies. We have in this case for the phase derivative

$$\frac{d}{d\omega}(k_z z) = -z\frac{v}{2}k_x^2\frac{d}{d\omega}\left(\frac{1}{\omega}\right) = z\frac{v}{2}\frac{k_x^2}{\omega^2} \quad (10\text{-}4\text{-}23)$$

which is, as required, positive for all ω. The fact that it is positive for all ω and all k_x is important. Merely to be positive for values of ω and k_x of practical interest is not enough. If for any value of ω or k_x the group delay were negative, then the time domain extrapolation equations would blow up.

Finally, let us consider the example depicted in Fig. 10-22. In the first frame, a planar wavefront is deformed, as if by propagation through a region of velocity which varies periodically in the x direction. In optical terminology, the first frame of Fig. 10-22 would represent an impulsive plane wave just after emergence from a phase grating. In terms of atmospheric acoustics, the disturbance might arise from passage of a plane wave through the periodic circulation cells depicted in

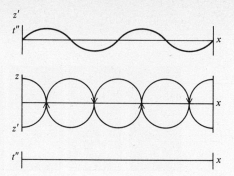

FIGURE 10-23
Possible means of producing a disturbed plane wave. Incident plane wave at bottom is altered by a material inhomogeneity. For example, circulating air cells (center), resulting in the disturbed wave at the top.

Fig. 10-23. Successive frames in Fig. 10-22 depict the subsequent history of the waveform. In optics texts (e.g., Goodman, Reference 35, p. 69) the monochromatic solution is usually obtained at infinity. The most obvious development is that the energy spreads out as one moves to successive frames. The single pulse of the top frame has become an extended oscillatory arrival by the last frame. As time goes on, less and less energy is in the first pulse and more and more is in the oscillatory tail. Another very notable feature is that after some long time the first arrivals tend to be aligned again so that disturbances in a wavefront may be said to heal themselves as time goes on. In contrast, the coda (wave tail) develops into a spatially incoherent wave. (This mimics the behavior of most geophysical wave observations.) We may note several other less apparent aspects to Fig. 10-22. Although energy moves back from the first arrival, a point of constant phase in the wave tail (indicated by X) moves forward toward the wave onset. Also the dip, or apparent direction of propagation, tends to increase going down a frame. This represents the ray interpretation that late arrivals have taken longer ray paths. Also the $\pi/2$ phase shift of a two-dimensional focus which causes doublets to form may be seen at A in the second frame.

In order to represent a disturbance of infinite extent in x on a finite computer grid, the problem was initialized with a periodic disturbance having zero slope at the side boundaries. Zero-slope boundary conditions are then equivalent to infinite periodic extension in x. A value of $v \, \Delta t \, \Delta z / \Delta x^2 = \frac{1}{2}$ was chosen to give an appropriate variation in progressive frames with each frame in Fig. 10-22 representing five computational iterations. The solution may be rescaled in several ways because of the interdependence of $v \, \Delta t$, Δx, and Δz.

It might be valuable to consider various data enhancement processes in the light of Fig. 10-22. In the process called "beam-steering," observations such as those in Fig. 10-22 would be summed over the x coordinate in an effort to enhance signal and reject noise. Clearly beam-steering will enhance the first arrival while rejecting random noise. It will also tend to cancel signal energy which resides in the oscillatory wave tails. If one is really interested in enhancing signal-to-noise ratio it would hardly seem desirable to use a processing scheme which cancels signal energy. As z' or t'' is increased the situation becomes increasingly severe, since signal energy moves from the initial pulse toward the oscillatory wave tails. What has often been regarded as "signal-generated-noise" may turn out to be signal in a potentially valuable form. One can indeed expect dramatic results if enhancement techniques are based on entire waveforms rather than only on the initial pulse.

EXERCISES

1 State all the assumptions which must be made to specialize (10-4-7) to

$$[v(z) - \bar{v}]P'_{z'} = P'_{t'}.$$

Derive the analogous equation for the double-prime coordinates.

2 Find a difference scheme for the equation of Exercise 1 which extrapolates from z' to $(z' + \Delta z)$. Show that past time is required if $\bar{v} > v$ and future time if $\bar{v} < v$.

3 Let a coordinate transformation be defined by

$$x' = x$$
$$z' = z$$
$$t' = t - \int_0^z v^{-1}(z)\,dz$$

Put the scalar wave equation into these coordinates.

4 Show that if the transformation velocity $\bar{v}$ in (10-4-9a), (10-4-9b), and (10-4-9c) takes any value less than the v in the wave equation, then stable difference equations will result.

5 Consider the difference equation $(1 + \delta_{xx}/12)\,\delta_{zt}P = b\,\delta_{xx}P$. For what value of b does it reduce to an explicit scheme? Is the time recurrence stable for that value of b?

10-5 BEAM COUPLING

Much of our information about the interior of the earth arises from interfaces within the earth which convert downgoing waves to upgoing waves. In layered media a mathematically strict decomposition of disturbances into downgoing waves $[\exp(ik_z z)]$ and upcoming waves $[\exp(-ik_z z)]$ was possible, but at present no such decomposition has been developed for two- or three-dimensional inhomogeneity. What we have is a collection of ad hoc techniques whose rigorous justification depends on the absence of horizontally propagating or evanescent energy. As a practical matter, what we are really interested in is not just the decomposition of waves into downgoing and upgoing parts. We are interested in describing the interactions between more-or-less collimated beams. In holography, these are the *incident* (or reference) beam and the *scattered* beam. In global seismology, these could be the incident compressional wave beam and the scattered shear wave beam. They need not have any particular orientation to each other or to the vertical.

The wave-extrapolation techniques described earlier can be used to describe beams collimated roughly along the z axis. Now we take up the task of describing the interaction between two such beams. For simplicity, these will initially be taken to be two more-or-less vertically propagating beams, one going down, the other up, interacting at a planar horizontal interface. The technique developed can then be applied to a great many less restrictive geometries. The accuracy of results in more general geometries is then a practical question whose answer varies from one situation to the next. Accuracy limitations come from many sources, which include

1 Angular dependence of velocity in the collimated beam which arises from Fresnel-like approximations
2 Neglect of evanescent energy
3 Possible inability of two collimated-beam equations to describe all important beams generated at a complicated interface
4 Approximation of elastic compressional waves by the scalar wave equation.

The significance of accuracy limitations must be evaluated in terms of accuracy of experimental work, required accuracy, and accuracy and cost of competitive techniques. Such evaluations are completely beyond the scope of our present efforts.

In this section we will describe only the *primary* reflected seismic energy in reflection seismic exploration. Large-amplitude waves are initiated at the earth's surface by means of dynamite or other high-energy sources. These waves penetrate into the earth where a small fraction of the energy echoes at weak reflectors and gets sent back to sensitive surface geophones. Occasional situations where a noticeable amount of energy scatters up and down several times (called multiple reflections or just multiples) will be discussed in a later section. For a plane layered medium we can use equation (9-3-13).

$$\frac{d}{dz}\begin{bmatrix} U \\ D \end{bmatrix} = \begin{bmatrix} -iab & \\ & iab \end{bmatrix}\begin{bmatrix} U \\ D \end{bmatrix} - \frac{1}{2}\frac{Y_z}{Y}\begin{bmatrix} 1 & -1 \\ -1 & 1 \end{bmatrix}\begin{bmatrix} U \\ D \end{bmatrix} \qquad (10\text{-}5\text{-}1)$$

Because the practical situation which we are trying to describe satisfies the inequality $U \ll D$, we will approximate the lower equation in (10-5-1) by

$$D_z = iab\,D - \frac{Y_z}{2\,Y}\,D \qquad (10\text{-}5\text{-}2)$$

To get a physical understanding of (10-5-2) which is applicable even when a, b, and Y are z-variable, note that the solution to (10-5-2) which can be verified by direct substitution, is

$$D = D_0\,Y^{-1/2}\exp\left(i\int_0^z ab\,dz\right) \qquad (10\text{-}5\text{-}3)$$

In other words, iab controls the phase (or velocity) of the wave and Y_z/Y controls amplitude change. Thus, we can interpret the Y_z/Y term as providing the physical effect associated with a transmission coefficient. It often happens that the velocity information in ab is approximately known, but the location of interfaces in the earth given by discontinuities in Y_z/Y are totally unknown. This means that we need not abandon our calculation of D if we are prepared to admit that its amplitude errs by the unknown transmission coefficients.

The basic thrust of Sec. 10-3 was that we can treat nonplanar waves by regarding iab as the square root of the differential operator $-(\omega^2/v^2 + \partial_{xx})$. For a beam collimated downward along the z axis a first approximation to the square root is given by $i\omega/v[1 + v^2(\partial_{xx}/2\omega^2)]$. With the beam-collimation assumption ($\partial_{zz} \approx 0$) and the unknown admittance gradient taken as zero, the downgoing wave D can be calculated with the equation

$$D_z = \frac{i\omega}{v} D + \frac{iv}{2\omega} D_{xx} \qquad (10\text{-}5\text{-}4)$$

This would more closely resemble the bulk of our earlier work if we assumed homogeneous velocity $v = \bar{v}$ and then made the transformation $D = D'e^{imz}$ where $m = \omega/v$, in which case (10-5-4) would reduce to

$$D'_z = \frac{iv}{2\omega} D'_{xx} \qquad (10\text{-}5\text{-}5)$$

To solve (10-5-4) or (10-5-5) inside the earth it is only necessary to know values for D along the surface of the earth (all x, $z = 0$). In a reflection seismic prospecting situation, D could usually be approximated by a delta function at the shot location.

Now let us turn to the calculation of the upgoing wave U. From the top row of (10-5-1) we have

$$U_z = -iab\, U - \frac{Y_z}{2Y}(U - D) \qquad (10\text{-}5\text{-}6)$$

If we care to neglect the transmission coefficient effect on U while retaining the reflection coefficient interaction of U and D, this becomes

$$U_z = -iab\, U + \frac{Y_z}{2Y} D \qquad (10\text{-}5\text{-}7)$$

Because reflection coefficient c is defined as

$$c = \frac{Y_2 - Y_1}{Y_2 + Y_1} = -c'$$

we can [for $Y(z)$ differentiable] write (10-5-7) as

$$U_z = -iab\, U - c'(z)D \qquad (10\text{-}5\text{-}8)$$

As with the downgoing waves, we can generalize from plane waves to beams with the square root approximation, obtaining

$$U_z = -\frac{i\omega}{v} U - \frac{iv}{2\omega} U_{xx} - c'(x, z)D \qquad (10\text{-}5\text{-}9)$$

A change of variables to $U = U''e^{-imz}$ and $D = D'e^{imz}$ with the homogeneous-velocity, inhomogeneous admittance assumption converts (10-5-9) to

$$U''_z = -\frac{\bar{v}/2}{-i\omega} U''_{xx} - c'(x, z)D'e^{2imz} \qquad (10\text{-}5\text{-}10)$$

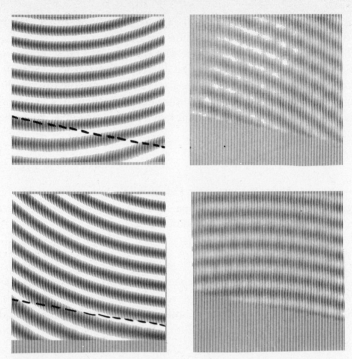

FIGURE 10–24
Two examples of down- and upgoing waves. The two left-hand frames show downgoing spherical waves from two different source locations. These waves illuminate a dipping interface. At the interface is both an impedance contrast and a velocity contrast. Waves of longer wavelength are seen below the interface. The right-hand frames show the upcoming waves. They vanish beneath the interface. Along the interface, the phase of the upcoming wave equals that of the downgoing wave.

It is important to understand how we can calculate the solution to (10-5-10). First of all, D' must have been calculated by some other equation before we start on U''. In the solution of (10-5-10) we will regard $c(x, z)D'$ as a source term for the generation of U''. Now there are two important cases. The first one is data synthesis. This is called the *forward* problem. The other case, called the *inverse* problem, is where the data sample U'' is given at the earth's surface, $z = 0$, and the problem is to deduce both $c(z)$ and $U''(z)$ as you integrate U'' downward. The inverse problem is more fully treated in the chapter on seismic data processing. Here we will stick to the forward problem. A boundary condition on U'' which will enable us to use (10-5-9) to find U'' everywhere is to prescribe that U'' vanishes over all x inside the earth at some depth z_N which is suitably great, say beneath all detectable reflectors. Then (10-5-10) is stepped up from z_N to z_{N-1}, z_{N-2}, etc. U'' remains zero until we come up to the first illuminated reflector; that is, the deepest place where both $c(x, z)$ and D' are nonvanishing. At this point, the source term in

(10-5-10) is turned on and U'' becomes nonzero from then on upward. This calculation is illustrated in Fig. 10-24.

The calculation can also be done in the time domain. We have the downgoing wave transformation

$$x' = x \qquad \text{(10-5-11}a)$$
$$z' = z \qquad \text{(10-5-11}b)$$
$$t' = t - \frac{z}{v} \quad \text{(10-5-11}c)$$

and the upgoing wave transformation

$$x'' = x \qquad \text{(10-5-12}a)$$
$$z'' = z \qquad \text{(10-5-12}b)$$
$$t'' = t + \frac{z}{v} \quad \text{(10-5-12}c)$$

And we have the possibility of expressing U and D in either frames (10-5-11) or frames (10-5-12)

$$U(x, z, t) = U'(x', z', t') = U''(x'', z'', t'') \quad \text{(10-5-13}a)$$
$$D(x, z, t) = D'(x', z', t') = D''(x'', z'', t'') \quad \text{(10-5-13}b)$$

The chain rule for differentiation gives

$$\partial_x D = \partial_{x'} D' \qquad \text{(10-5-14}a)$$
$$\partial_z D = \left(\partial_{z'} - \frac{1}{v}\partial_{t'}\right)D' \quad \text{(10-5-14}b)$$
$$\partial_t D = \partial_{t'} D' = \partial_{t''} D'' \quad \text{(10-5-14}c)$$

and

$$\partial_x U = \partial_{x'} U' \qquad \text{(10-5-15}a)$$
$$\partial_z U = \left(\partial_{z'} + \frac{1}{v}\partial_{t'}\right)U' \quad \text{(10-5-15}b)$$
$$\partial_t U = \partial_{t'} U' \qquad \text{(10-5-15}c)$$

Taking velocity-homogeneous media $v = \bar{v}$, multiplying (10-5-4) and (10-5-9) through by $-i\omega$, and then identifying $-i\omega$ with a time derivative, we obtain

$$D_{zt} = -\frac{1}{\bar{v}} D_{tt} + \frac{\bar{v}}{2} D_{xx} \qquad \text{(10-5-16}a)$$

$$U_{zt} = \frac{1}{\bar{v}} U_{tt} - \frac{\bar{v}}{2} U_{xx} - c'(x, z)D_t \quad \text{(10-5-16}b)$$

Equations (10-5-16a) and (10-5-16b) are readily converted by means of (10-5-14) and (10-5-15) to

$$D'_{z't'} = \frac{\bar{v}}{2} D'_{x'x'} \tag{10-5-17a}$$

$$U''_{z''t''} = -\frac{\bar{v}}{2} U''_{x''x''} - c'(x'', z'')D''_{t''} \tag{10-5-17b}$$

Now (10-5-17a) can be used to compute D' but (10-5-17b) calls for D''. Subtracting (10-5-12c) from (10-5-11c) we get

$$t' = t'' - \frac{2z}{\bar{v}}$$

So, using (10-5-13b) we find (10-5-17b) can be expressed in terms of D' as

$$U''_{z''t''} = -\frac{\bar{v}}{2} U''_{x''x''} - c'(x'', z'') \, \partial_{t''} D'\left(x'', z'', t'' - \frac{2z''}{\bar{v}}\right) \tag{10-5-18}$$

This time-domain result is the transform of (10-5-10).

EXERCISES

1 Show that $\frac{1}{2}[(d/dz) \ln I]$ is the reflection coefficient c' as seen from above the interface.
2 Recall from (9-3-20) that the definition of Y includes k_x. This was neglected in the derivation of (10-5-10). Improve (10-5-10) to include the implied $\partial D'/\partial x$ terms. This improvement allows reflection coefficient to be a function of angle.

10-6 NUMERICAL VISCOSITY

Positive numerical viscosity means that the short wavelength deviation of a difference equation from a differential equation is such that the short wavelengths tend to dissipate as the calculation proceeds. The numerical viscosity may also turn out to be negative, causing short wavelengths to amplify rather than attenuate. Whether or not there are good scientific reasons to study numerical viscosity, scientists often get dragged into this study for several reasons: First, even if differential equations do not violate causality there may be instability due to negative viscosity in the difference equations. Second, the realities of computer economics (especially in a multidimensional problem such as $P_{zt} = (v/2)P_{xx}$ may require that waveforms be sampled with as few points as practicable. Third, when observational data are to be processed, as when $P(x, t)$ is to be extrapolated from z_1 to z_2, then the data may be inconsistent with certain assumptions upon which the extrapolating equation is based.

For example, suppose that $P(x, t)$ has Fourier transform $P'(k_x, \omega)$. Then, since $k_x^2 + k_z^2 = \omega^2/v^2$, freely propagating waves are characterized by $|k_x| < \omega/v$

Points per wavelength, $2\pi/\omega\,\Delta t$	$\omega\,\Delta t$ or $k_x\,\Delta x$, radians	Relative error of $2\tan\omega\,\Delta t/2$	Relative error of $2\sin k_x\,\Delta x/2$	Relative error of (10-6-8)
$\pi \times 10^n$	2×10^{-n}	$10^{-2n}/3$	$10^{-2n}/6$	$0(10^{-4n})$
20.000000	0.314159	0.008272	−0.004116	−0.000021
16.000000	0.392699	0.012968	−0.006434	−0.000051
12.000000	0.523599	0.023218	−0.011449	−0.000159
10.000000	0.628318	0.033675	−0.016504	−0.000330
8.000000	0.785398	0.053325	−0.025834	−0.000812
6.000000	1.047197	0.097645	−0.046109	−0.002613
4.000000	1.570796	0.240396	−0.104913	−0.013849
3.000000	2.094395	0.492833	−0.189390	−0.046111
2.100000	2.991992	1.596763	−0.400123	−0.203548

FIGURE 10–25
The relative error at short wavelengths often associated with expressing differential equations in difference form.

so $P'(k_x, \omega)$ should vanish unless $|k_x| < \omega/v$. In the derivation of $P_{zt} = v/2\,P_{xx}$ it was further assumed that the waves have small angles of propagation; hence, the inequality becomes stronger, $|k_x| \ll \omega/v$. Since observational data will certainly not satisfy these conditions exactly we have two options. First, we can hope to ignore the illegal part of the (k_x, ω) space if the data do not have much energy there and if our difference equation does not unacceptably amplify it. Second, we can modify our difference or differential equations so that there is a controlled positive numerical viscosity in the illegal part of the transform space. This kind of operation is sometimes called fan-filtering because of the wedge-shaped region of attenuation in (ω, k_x) space.

The operator ∂_{xx} has the Fourier transform $-k_x^2$. The operator δ_{xx} amounts to a convolution on the x axis with the coefficients $(1, -2, 1)/\Delta x^2$; thus its Fourier transform is $[\exp(-ik_x\,\Delta x) - 2 + \exp(ik_x\,\Delta x)]/\Delta x^2$. We write this as

$$FT(\partial_{xx}) = -k_x^2 \tag{10-6-1a}$$

$$FT\left(\frac{\delta_{xx}}{\Delta x^2}\right) = -\hat{k}_x^2$$

$$= -\frac{2(1 - \cos k_x\,\Delta x)}{\Delta x^2}$$

$$FT\left(\frac{\delta_{xx}}{\Delta x^2}\right) = -\frac{2^2}{\Delta x^2}\sin^2\left(\frac{k_x\,\Delta x}{2}\right) \tag{10-6-1b}$$

The approximation $\hat{k}_x$ to k_x is given by

$$\hat{k}_x = \frac{2}{\Delta x}\sin k_x\frac{\Delta x}{2} \tag{10-6-2}$$

The error in the approximation $\hat{k}_x \approx k_x$ is tabulated in Fig. 10-25.

The Crank-Nicolson method amounts to another approximation. Here the operator $\partial/\partial t$ which has the Fourier transform $-i\omega$ is approximated by the bilinear transformation. The approximation $\hat{\omega}$ to ω is given by

$$-i\hat{\omega}\,\Delta t = \frac{2(1 - e^{i\omega\,\Delta t})}{1 + e^{i\omega\,\Delta t}}$$

Multiplying top and bottom on the right by $e^{-i\omega\,\Delta t/2}$ we get

$$-i\hat{\omega}\,\Delta t = 2\,\frac{e^{-i\omega\,\Delta t/2} - e^{i\omega\,\Delta t/2}}{e^{-i\omega\,\Delta t/2} + e^{i\omega\,\Delta t/2}}$$

$$= -2i \tan \frac{\omega\,\Delta t}{2} \qquad (10\text{-}6\text{-}3)$$

$$\hat{\omega} = \frac{2}{\Delta t} \tan \frac{\omega\,\Delta t}{2}$$

This approximation is also tabulated in Fig. 10-25.

To see how higher-order difference approximations may be built up, we solve (10-6-2) for ik_x getting

$$ik_x = \frac{2}{\Delta x} \operatorname{arcsinh}\left(\frac{i\hat{k}_x\,\Delta x}{2}\right) \qquad (10\text{-}6\text{-}4)$$

Recall the power series for arcsinh

$$\operatorname{arcsinh} u = u - \frac{1}{2}\frac{u^3}{3} + \frac{1\cdot 3}{2\cdot 4}\frac{u^5}{5} - \frac{1\cdot 3\cdot 5}{2\cdot 4\cdot 6}\frac{u^7}{7} + \cdots \qquad (10\text{-}6\text{-}5)$$

The inverse Fourier transform of (10-6-4) using (10-6-5) provides a power series expansion for ∂_x in terms of powers of δ_x.

At the present time, reflection seismic data often come close to being undersampled in the horizontal x coordinate. Hence, it is worthwhile to devise a more accurate approximation than δ_{xx} to ∂_{xx}. Squaring (10-6-4) and retaining only the first two terms in the arcsinh expansion gives

$$-k_x{}^2 \approx \frac{4}{\Delta x^2}\left(u^2 - \frac{u^4}{3}\right) \qquad (10\text{-}6\text{-}6)$$

where $u = i\hat{k}_x\,\Delta x/2$. Taking the inverse transform we have

$$\partial_{xx} \approx \frac{\delta_{xx}(1 - \delta_{xx}/12)}{\Delta x^2} \qquad (10\text{-}6\text{-}7)$$

It is most often convenient to use this in the rational form

$$\partial_{xx} \approx \frac{\delta_{xx}/\Delta x^2}{1 + \delta_{xx}/12} \qquad (10\text{-}6\text{-}8)$$

By means of a trick, the rational form can be used without going to higher-order difference operators. Note that (10-6-8) into a differential equation of the type $P_t = P_{xx}$ leads to

$$\left(1 + \frac{\delta_{xx}}{12}\right)\delta_t P = \frac{\Delta t}{\Delta x^2}\delta_{xx}P \qquad (10\text{-}6\text{-}9)$$

The new term δ_{xxt} fits on the old computation star and thus amounts to a just readjustment of coefficients; that is, hardly any increase in computer costs. Reference to Fig. 10-25 shows an astonishing increase in accuracy. On the basis of Fig. 10-25 and the acceptable error for some particular application, say 3 per cent, one determines a minimum acceptable number of points per wavelength, say 10 points per wavelength on z and t axes and $3\frac{1}{2}$ points per wavelength on the x axis. Then the useful bandwidth $-2\pi/10 < \omega \, \Delta t < +2\pi/10$ is markedly less than the total bandwidth available (the $2\pi/10$ is the periodicity interval for transforms of sampled data). In this case, the ratio of useful bandwidth to total bandwidth is 1/5. In order to use more of the available bandwidth it is necessary to put up with more error or to develop more elaborate difference approximations to differential operators. Figure 10-26 depicts the paltry portion of (ω, k_x) space which is usable.

For examples of the manipulation of numerical viscosity let us take the differential equation $P_{zt} = v/2P_{xx}$ and modify it to attenuate energy outside the usable bandwidth, say where $|k_x \, \Delta x| > \pi/5$. We simply add a term to the right-hand side. That is, we modify

$$\partial_z P = \frac{v}{-2i\omega}\partial_{xx}P \qquad (10\text{-}6\text{-}10)$$

by judicious choice of an additional term

$$\partial_z P = \frac{v}{-2i\omega}\partial_{xx}P + a\,\partial_{xx}P \qquad (10\text{-}6\text{-}11)$$

To see what numerical value to take for the constant a, we transform the x coordinate in (10-6-11)

$$\partial_z P = \left(\frac{vk_x^2}{2i\omega} - ak_x^2\right)P \qquad (10\text{-}6\text{-}12)$$

Equation (10-6-12) has the solution

$$P(z) = P(z_0)\exp\left[\left(\frac{vk_x^2}{2i\omega} - ak_x^2\right)(z - z_0)\right] \qquad (10\text{-}6\text{-}13)$$

The imaginary part of the exponent merely gives the phase angle, which we will ignore because we are interested only in magnitude. Let $z - z_0 = d$. Then (10-6-13) becomes

$$\left|\frac{P(z)}{P(z_0)}\right| = \exp\left(-ak_x^2 d\right) \qquad (10\text{-}6\text{-}14)$$

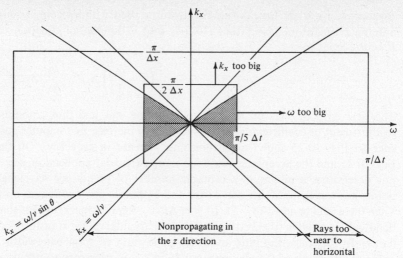

FIGURE 10-26

The (ω, k_x) plane. Field data may be expected to have some energy everywhere in the (ω, k_x) plane. Only in the speckled region will our difference equations properly simulate the wave equation. Energy with $|k_x| > |\omega/v|$ does not represent free waves; it represents either surface waves or errors in data collection (often static errors, random noise, or gain not smoothly variable from trace to trace). Such energy can mean nothing in a migration program, hence it should be rejected by filtering. This may be done by fan-filtering (as in Reference 36) or, as is done here, by means of numerical viscosity. Actually, for practical reasons one frequently may wish to reject rays outside a certain dip angle. This gives the larger fan-filter reject region $|k_x| > |\omega/v \sin(\text{dip})|$. Although information can be carried up to the folding frequency in both ω and k_x, in practice the use of operators of finite length narrows the useful bandwidth. The use of simple time-difference operators results in a practical bandwidth restriction to about a quarter of the folding frequency. This presents no problem in principle; data may be interpolated before processing, or more elaborate (i.e., longer) difference operators may be used. Finally, this figure was drawn with $\Delta x > v \, \Delta t$ because it represents the usual case in practice where extra points in time are more cheaply obtained than extra points in space.

Now we have to decide how much attenuation is wanted. Say when $k_x \, \Delta x = \pi/4$ we wish (10-6-14) to imply attenuation to e^{-1}. Thus, for the exponential of (10-6-5), we have

$$-ak_x^2 d = -1$$

$$a(k_x \, \Delta x)^2 \frac{d}{\Delta x^2} = 1$$

$$a\left(\frac{\pi}{4}\right)^2 \frac{d}{\Delta x^2} = 1$$

$$a = \frac{16 \, \Delta x^2}{d\pi^2} \qquad (10\text{-}6\text{-}15)$$

Thus, the term we added to (10-6-10) to get (10-6-11) has a coefficient which goes to 0 as the squared grid spacing Δx^2. Inclusion of this term gives the gaussian attenuation function of spatial frequency of (10-6-14). The inclusion of the viscosity term seems to add virtually no cost to a computer program.

Next, let us modify the extrapolation equation so that excessive dips [sin (dip) $= kv/\omega$] will be attenuated. This is not exactly numerical viscosity because we will alter the basic differential equation. It is like numerical viscosity in that it is an ad hoc modification intended to correct a certain deficiency. Here we modify the differential equation (10-6-10) to read

$$\partial_z P = \frac{v \, \partial_{xx}}{2(-i\omega + \omega_0)} P \quad (10\text{-}6\text{-}16)$$

To see what numerical value to pick for ω_0, we rationalize the denominator

$$\partial_z P = \frac{v}{2} \frac{i\omega + \omega_0}{(\omega^2 + \omega_0^2)} \partial_{xx} P \quad (10\text{-}6\text{-}17)$$

Now we may ignore the imaginary part of the right-hand side of (10-6-17) because it contributes only the phase of P. Fourier transforming the x coordinate, we have

$$\partial_z P = -\frac{v \, \omega_0 k_x^2}{2(\omega^2 + \omega_0^2)} P \quad (10\text{-}6\text{-}18)$$

There are two cases. We will pick ω_0 very small so that in the uninteresting case where $\omega < \omega_0$ (10-6-9) reduces to spatial frequency dissipation but in the interesting case $\omega > \omega_0$ (10-6-18) amounts to

$$\partial_z P = \frac{-v\omega_0}{2} \frac{k_x^2}{\omega^2} P \quad (10\text{-}6\text{-}19)$$

This is obviously attenuation, which is a gaussian function of dip. It is left for the exercises to find a numerical choice for ω_0.

EXERCISES

1 What value of ω_0 in (10-6-16) will attenuate waves propagating from z_1 to z_2 at a 30° angle from the z axis to e^{-1} times the original amplitude? So that ω_0 may be said to be small, it is necessary to compare it to something with physical dimensions of inverse time. Give examples of a situation where ω_0 is small and a situation where it is not.

2 Show that the parameter b in $P_z = iv/2\omega(\partial_{xx} + b \, \partial_{xxz})P$ may be used to produce a viscosity decay of approximate form $\exp[-bk_z^2(z - z_0)]$. This may be useful when Δz is taken too large.

3 Consider extrapolation one step in the z direction with the equation $P_z = -a\omega^2 P$. Insert the bilinear transformation $-i\omega = 2(1 - Z)/(1 + Z)$ and deduce that the equation cannot be used since a polynomial with a nonminimum-phase divisor results.

4 Show that the equation $P_z = a(-\omega^2 \, \Delta t^2/2 + i\omega \, \Delta t)P$, unlike the equation of Exercise 3, leads to a causal time-domain filter. (Do the extrapolation in z by the Crank-Nicolson method, i.e., the bilinear transform method.)

5 A given set of data $P(x, t)$ is believed to satisfy the equation $P_{zt} = P_{xx}$. It is observed that transformed data $Q(x, t)$, where $Q(x, t) = P(x, t)e^{\alpha t}$, fits into a reasonably small numerical range so that Q may be represented using integer arithmetic. What differential equation does Q satisfy?

11

SEISMIC DATA PROCESSING WITH THE WAVE EQUATION

The coordinate frames used by theoreticians to describe wave propagation do not include frames in common use by geophysical prospectors to describe observations. Whereas the theoretician generally considers a single source (or shot) location at a time, the experimentalist deals simultaneously with waves which have been generated separately by many shots. Our task in this section is to put the wave equation into some prospectors' coordinate frames.

11-1 DOWNWARD CONTINUATION OF GATHERS AND SECTIONS

Suboceanic prospecting is generally carried out by a ship which carries a repetitive energy source and which trails a cable that is 2 to 3 kilometers long and packed with sonic receivers. Ideally, the ship's course is a straight line which we can take to be the x axis. Ideally, all the seismic waves of interest propagate in a vertical plane through the line of the ship's course. This plane is called the plane of the seismic section. Despite the fact that it is no great problem to describe waves in three dimensions once difference techniques have been mastered for two dimensions, we will restrict the theory to two dimensions in order to keep it compatible with the bulk of present-day surveying practice and the capability of most present-day computing machines. An impulsive wave from a point source spreading out in three dimensions will decay in amplitude in inverse proportion to the travel time. (The area of a spherical wavefront increases as t^2, so the energy per unit area decreases as t^2, and so the wave amplitude is proportional to t^{-1}.) An impulsive wave from a line source (the line would be on the ocean's surface perpendicular to the ship's course) has an amplitude decay proportional to $t^{-1/2}$. Thus, the attempt to compress three-dimensional reality into a one- or two-dimensional mathematical form often begins (and almost always ends) with a $t^{1/2}$ or a t scaling factor. Of far more practical importance than this scaling is the attempt to keep all the seismic rays which emanate from and return to the ship's traverse line confined to a single plane. In other words, we hope to avoid recording side echoes. Often side echoes can be reduced or eliminated by careful choice of the ship's course. But, once the data have been recorded, you have to live with whatever side echoes are there. One way to think about these side echoes is to imagine the ship's traverse line as the axis of a cylindrical coordinate system. Instead of considering that the time delay of an echo is a measure of the depth to a reflector, one now imagines that the travel time is a measure of the radius in the cylindrical coordinate system. Interpretation is easy if the plane of the seismic section is merely somewhat tipped away from the vertical. Interpretation problems arise when the earth is so three-dimensionally complex that several wobbly planes are involved and the observed data have become a superposition of many of them. In short, where the earth gets three-dimensionally inhomogeneous you cannot get along very well with two-dimensional experimental and calculational techniques.

Figure 11-1 shows a most important relationship between two coordinate systems. The coordinates of the shot sound source s and geophone sound receiver g are taken along the ship's course, which is the x axis. Also along the x axis are the shot-to-geophone distance offset coordinate f and the midpoint y between the shot and the geophone.

We are going to describe waves of pressure $P(s, g)$ where the shot and geophone coordinates are taken to be independent variables. In reality, the shots and geophones are not distributed in a continuum along the x axis, but they are usually close enough together that it is merely a matter of interpolation to find P for any s or g. If the data cannot be interpolated, they will not be satisfactory for use in differential equations.

Another independent variable is time t. The origin point on the time axis is chosen so that time t equals 0 when the shot goes off. After the echoes from the shot at s have died out completely (ordinarily about 6 seconds), the time axis is again reset to 0 for the next shot at $s + \Delta s$. In Fig. 11-1 the t axis may be taken to be out of the plane of the paper. Both the (s, g) coordinates and (y, f) coordinate sets are orthogonal. Nonorthogonal coordinates such as (s, f) are used in marine data recording, but they are rarely used in data analysis and we will ignore them.

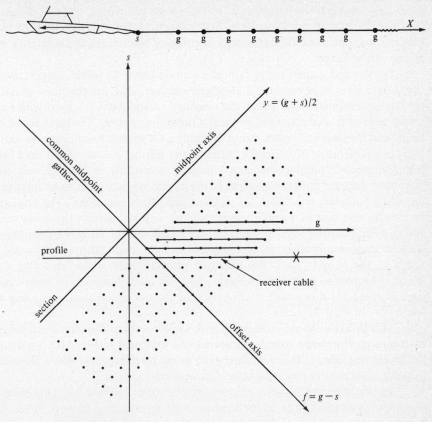

FIGURE 11-1
The relationships among sound source coordinate s, geophone sound receiver
coordinate g, offset coordinate $f = g - s$, and midpoint coordinate $y = (g + s)/2$.
Theoreticians generally use s and g as coordinates of the wave-pressure field, but
interpreters generally use f and y.

Theoreticians usually work in the (g, t) plane for a fixed s. In exploration
seismology, data in the (g, t) plane are called a *profile*. Seismic data interpreters
usually work with wave amplitude in both the (y, t) plane and the (f, t) plane.
A display in the (y, t) plane is called a *seismic section*. A display in the (f, t) plane
is called a *common midpoint gather* or a *common reflection point gather*, which,
unfortunately, in industry is often called a *common depth point gather*. This terminology originated in the days when analytical methods usually modeled the earth
as a stratified medium in which the reflection point, called the *depth point*, lay
directly beneath the midpoint. It is unfortunate because of the considerable confusion we now have with the depth z axis. In this book we will avoid the term
depth point. A data display over midpoint y at fixed offset f, that is, the (y, t)
plane, is called a *seismic section* and it is the only one of the orthogonal planes

which may continue for hundreds or thousands of kilometers. *Profiles* and *gathers* continue for only a few kilometers, first because of the limited length of the receiver cable, but more fundamentally because of the limited distance over which a shot can be heard.

Profiles and gathers today typically contain about 48 seismic traces. Because useful data often lie beyond the 3-kilometer receiver cable, another form of data recording is sometimes used. This is the *sonobuoy*. A sonobuoy is a buoy with a single sound receiver (called a *hydrophone*) and a radio transmitter. The buoy is cast overboard and the ship sails away, firing at about a six-second repetition rate until the buoy is out of range of either the radio or the seismic signals. Such data form a common receiver point gather and comprise about 200 to 2000 seismic traces. Sonobuoy data are conceptually one of the easiest kinds of data to be imagined as providing boundary conditions for wave-extrapolation equations. The principle of reciprocity says that we could imagine that the ship had carried the sonic receiver and that the buoy had carried the repetitive sound source. (It is not done this way because the sonic receiver is an inexpensive, lightweight, throwaway item, well suited to the buoy.) Thus, the principle of reciprocity which says $P(g, s, t) = P(s, g, t)$ enables us to imagine a single shot with many hundreds of receivers, just what we need for downward continuation of the upcoming and downgoing wave fields.

This brings us to the concept of how we learn about the interior of the earth by means of downward continuing waves. As indicated in Fig. 11-2, an obvious, but important, idea is that *reflectors exist in the earth at places where the onset of the downgoing wave is time coincident with an upcoming wave.*

To best illustrate this idea, monochromatic waves will be used. This enables us to compute and think in (x, z) for fixed ω rather than to have to work in the three-dimensional space of (x, z, t). A penalty we pay by going to a single frequency is that the idea of time coincidence of two time-dependent waveforms becomes for monochromatic waves the idea of both waves being coherent with a fixed phase shift (usually zero or 180 degrees) at points in space where reflection can occur. We will see that phase equivalence at a single frequency at a single point in space does not by itself provide time coincidence. The same must be true of other frequencies. Generally, the more frequencies are involved in phase coherency, the better will be the spatial resolution.

The reader will recall that a number of assumptions in Sec. 10-5 led us to the idea that up- and downgoing monochromatic waves can be calculated with equations like

$$D'_z = \frac{iv}{2\omega} D'_{xx} \tag{11-1-1}$$

$$U'_z = -\frac{iv}{2\omega} U'_{xx} - c(x, z)D'e^{2imz} \tag{11-1-2}$$

In a *forward* problem, synthetic data are calculated from a model. In an *inverse* problem, a model is calculated from the data. To do a forward problem we

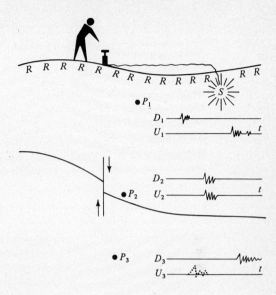

FIGURE 11-2
Illustration of the basic principle of reflector mapping. There are a near-surface source S and many surface receivers R. At a shallow depth above the reflector at the typical point P_1 the downgoing wave D_1 occurs much earlier than the upcoming wave U_1. The upgoing wave U_1 represents energy which has traveled from the source to one or more places on the reflector and then back up to the point P_1. At a point P_2, which is at or near a reflecting interface of arbitrary shape, there will be overlap in time of the down- and upgoing waves D_2 and U_2. The time overlap may be used in the construction of a map of reflector positions. Below the reflector at the point P_3, there is, in principle, no upcoming wave. However, practical schemes for estimating the upcoming waves U at various depths in large part amount to shifting the upcoming waves seen at R to earlier and earlier times corresponding to greater and greater depths. Since the practical schemes will have no knowledge of the interface, they will predict an erroneous upcoming wave U_3 at P_3, indicated by dots before the arrival of the downgoing wave. This error has no bad effect on the reflector mapping formulas which utilize time coincidence of up- and downgoing waves. These ideas are valid in situations where there are many reflectors at many depths. (From Ref. [5], p. 468.)

first assume a shot location at, say $z = 0$ and $x = 0$. This provides a delta function initial condition for the downgoing wave. Assumption of a velocity model then allows use of an equation like, say (11-1-1) to continue the downgoing wave downward to arbitrary depth. At a sufficiently great depth, the upcoming wave U is taken to be 0. It is integrated upward with (11-1-2) where the product of the reflection coefficient c and the downgoing wave D act as a source for the upgoing wave.

Now we approach the *inverse* problem where we are trying to determine the reflection coefficient $c(x, z)$, but we are given the observation of the upgoing wave U at the earth's surface ($z = 0$, all x). We calculate D as before. Since $c(x, z)$ is

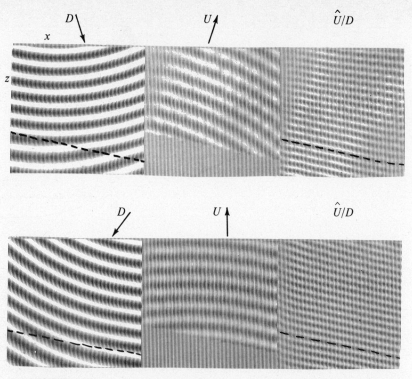

FIGURE 11-3
Location of a dipping bed with a single source emitting a single frequency. Top and bottom represent two possible source locations. The left-hand panel shows the real part of downgoing wave D. The dashed line represents a transition from low to high velocity as can be seen by the greater wavelength below. The center panel represents the upcoming reflected wave U, which originates at the velocity jump. On the right-hand panel is plotted the real part of the ratio $\hat{U}/D$. The right-hand panel tells where downgoing and upcoming waves are in phase. It gives the correct dip of the bed. With only a single frequency the depth cannot be deter-minded except to within multiples of a half-wavelength. The estimated upcoming wave $\hat{U}$, is computed from the true upcoming wave U, observed at the surface and the velocity v of the medium; Y_z/Y is not used. (From Ref. [3], p. 417.)

unknown in (11-1-2) we can compute $\hat{U}$, an approximation to U, by marching the equation

$$\hat{U}'_z = -\frac{iv}{2\omega}\,\hat{U}'_{xx} - 0 \qquad (11\text{-}1\text{-}3)$$

down from the surface $z = 0$ where $\hat{U}$ is given.

Now the question is, how will $\hat{U}$ depart from U? Between the earth's surface and the shallowest reflector (shallowest nonzero c) there will be no difference be-tween (11-1-2) and (11-1-3). In that region, both U and $\hat{U}$ will be waves of identical speeds, directions, rates of divergence, and all other properties. As (11-1-2) is projected downward (opposite to the direction of propagation), the source terms in (11-1-2) serve to "turn off" the upcoming waves until U, unlike $\hat{U}$, totally vanishes beneath the deepest reflector. Now the question is whether there is practical significance to the fact that $\hat{U}$ does not vanish beneath reflectors. Our purpose in computing $\hat{U}$ is to determine the location of reflectors by the time-coincidence idea in Fig. 11-2. For this time-coincidence idea, it does not matter that the upcoming wave is not turned off beneath an interface. Figure 11-3 illustrates these concepts for a monochromatic wave. To display the earth model, the reflection coefficient $c(x, z)$ is estimated by displaying $\hat{c}$ where

$$\hat{c}(x, z) = \frac{\hat{U}(x, z)}{D(x, z)} \qquad (11\text{-}1\text{-}4)$$

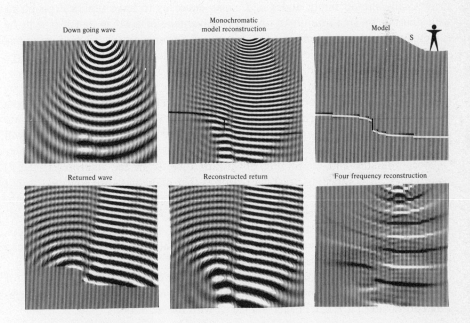

FIGURE 11-4
Synthesis of waves from a reflector which is warped and offset by a fault and the reconstruction of an image of the fault. Top left is the wave traveling away from a point source. Bottom left shows the upgoing reflected wave. It vanishes below the interface. At the bottom center, we see the reconstruction from surface observations of the upcoming wave. In doing the reconstruction, one does not assume knowledge of the reflectors, since they are what we are looking for. Hence, in the reconstructed return, one has an upgoing wave below the reflector. At the top center panel we show the product of the upgoing wave with the complex conjugate of the downgoing wave. The reflector exists along some line of zero phase, but with a single frequency, one cannot tell which line represents the reflector. The top center panel was summed over four frequencies to get the lower right panel, which gives a better indication of the reflector position. More frequencies would define it more clearly. (From Ref. [5], p. 479.)

We notice that for a monochromatic wave there will be phase coincidence of $\hat{U}$ and D not only at the reflector but also at half-wavelength intervals both above and below the reflector. (If we had U instead of $\hat{U}$, there would be 0 below.) Now the idea is that we can repeat the whole calculation for monochromatic waves at other frequencies and sum the results. The summation will always be in phase at the reflector but of variable phase away from the reflector.

Figure 11-4 shows a calculation like that in Fig. 11-3, but more terms were included in the equations (in order to better represent waves traveling at larger angles from the vertical) and more frequencies were used (in order to illustrate constructive interference at the reflector). Another difference between Figs. 11-3 and 11-4 is that the reflector estimate is $\hat{U}/D$ in (11-1-3), but in (11-1-4) it is $\hat{U}D^*$ (where D^* is the complex conjugate of D). Both $\hat{U}/D$ and $\hat{U}D^*$ have the same phase, but they have a different amplitude. The advantage of $\hat{U}/D$ is that it has the magnitude of the reflection coefficient. The disadvantage of $\hat{U}/D$ is that nodes in D, which theoretically should imply nodes in U, may cause us to have practical problems with division by small numbers. Another advantage of $\hat{U}D^*$ is that if it is to be summed not only over many frequencies, but also over many shot locations, it has the desirable characteristic that it is small where illumination is poor and large where illumination is good.

Next we attack the important practical matter of how to continue sections downward. The obvious approach is to solve a separate problem in the (g, t) plane for each shot point. Alternatively, we could use the reciprocity principle and use each receiver point as a separate problem in the (s, t) plane. For reasons we will come to recognize, there are considerable practical advantages in leaving the theory (g, s) coordinates and continuing downward directly in the interpreter's (y, f) coordinates. A compelling practical reason is that many data are collected with a single shot and a single receiver which move together across the surface of the earth. For a fixed shot point there is only one receiver point, so there are clearly insufficient data to initialize a downward continuation in the (g, t) plane. Nonetheless, when all shot points are considered, it turns out that with good accuracy we can continue the constant offset section downward.

We will develop two separate equations, one for downgoing waves and one for upcoming waves. The conversion from field coordinates (s, g, e, t) to interpretation coordinates (y, f, z, t') is accomplished with the definitions

$$y = \frac{s+g}{2} \qquad (11\text{-}1\text{-}5a)$$

$$f = g - s \qquad (11\text{-}1\text{-}5b)$$

$$z = e \qquad (11\text{-}1\text{-}5c)$$

$$t' = t - \left(\pm \frac{e}{v}\right) \qquad (11\text{-}1\text{-}5d)$$

The first two definitions (11-1-5a, 11-1-5b) are merely the transformation from shot s and geophone g coordinates to midpoint y and offset f coordinates described earlier. Equation (11-1-5c) indicates that the receiver elevation e is a point on the vertical z axis. In (11-1-5d) we have a definition of receiver-elevation-dependent time t'. Comparing to Sec. 10-4, it is clear that the plus sign is for waves propagating in the $+z$ direction (down) and the minus sign giving $t + z/\bar{v}$ is for waves going up. It is important to distinguish the constant velocity $\bar{v}$ in the coordinate transform (11-1-5) from the spatially variable velocity $\tilde{v}(x, z)$ which will be used in the wave equation. Although the coordinate transformation is based on a constant-velocity medium, the transformed wave equation can still describe waves in a variable-velocity medium.

Now we state that we are trying to describe the same disturbance in the new coordinate system as that in the old one.

$$P(g, s, t, e) = Q(y, f, t', z) \qquad (11\text{-}1\text{-}6)$$

Next, we compute the partial derivative of P with respect to its independent variables.

$$P_g = Q_y y_g + Q_f f_g + Q_{t'} t'_g + Q_z z_g \qquad (11\text{-}1\text{-}7a)$$
$$P_g = (.5\partial_y + \partial_f)Q$$

$$P_e = Q_y y_e + Q_f f_e + Q_{t'} t'_e + Q_z z_e \qquad (11\text{-}1\text{-}7b)$$
$$P_e = (t'_e \partial_{t'} + \partial_z)Q$$

$$P_t = (t'_t \partial_{t'})Q \qquad (11\text{-}1\text{-}7c)$$

Now we need the second partial derivatives of P with respect to its independent variables. Since all the coefficients of $\partial_y, \partial_f, \partial_{t'}$, and ∂_z in (11-1-7) are constants, then the second derivatives can be found by squaring the operators in parentheses in (11-1-7).

We are familiar with the wave equation in the form

$$P_{xx} + P_{zz} = \tilde{v}^{-2}P_{tt} + \text{source} \qquad (11\text{-}1\text{-}8)$$

We can imagine that the geophones used to observe P could be placed anywhere in the (x, z) space. A quantity like P_{xx} on the surface of the earth could be measured by setting out geophones at $g = x$ and measuring P_{gg}. We regard the coordinates of the geophone location $(x, z) = (g, e)$ as independent variables. Thus, we may write the wave equation as

$$P_{gg} + P_{ee} = \tilde{v}^{-2}P_{tt} + \delta(g - s, e, t) \qquad (11\text{-}1\text{-}9)$$

where we have used a source term defined as a delta function at $(x, z, t) = (s, 0, 0)$. Since we do not intend to use this equation in the vicinity of sources, we can drop the delta function at the source. Now we take the operators in (11-1-7), square them and insert them into (11-1-9). This gives

$$\tfrac{1}{4}Q_{yy} + Q_{ff} + Q_{yf} + (t'_e)^2 Q_{t't'} + 2t'_e Q_{t'z} + Q_{zz} = \frac{(t'_t)^2 Q_{t't'}}{\tilde{v}^2} \qquad (11\text{-}1\text{-}10)$$

First we will neglect Q_{zz} in a Fresnel-like approximation. (Higher accuracy can be achieved as in Sec. 10-3 if Q_{zz} is estimated.) Specializing to homogeneous media, $v = \bar{v}$ and using $t'_e = -(\pm 1/\bar{v})$ and $t'_t = 1$ our equation has reduced to

$$\pm \frac{2}{\bar{v}} Q_{t'z} = \frac{1}{4} Q_{yy} + Q_{ff} + Q_{yf} \quad (11\text{-}1\text{-}11)$$

Equation (11-1-11) is a partial differential equation in four dimensions starting from initial conditions which are data in three dimensions. Frequently a problem of this magnitude will not be computationally feasible, so we now consider how to remove the offset dimension. Let us integrate (11-1-11) over offset. We obtain

$$0 = \left(\frac{\mp 2}{\bar{v}} \partial_{zt'} + \frac{1}{4} \partial_{yy} \right) \int Q\, df + \partial_y \int \frac{\partial Q}{\partial f}\, df + \int \frac{\partial^2 Q}{\partial f^2}\, df \quad (11\text{-}1\text{-}12)$$

Now if Q and Q_f should vanish at the great offsets which we take to be the limits on the integrals, then the two terms farthest right vanish. Let us define a *vertically stacked section* by

$$S = \int Q\, df \quad (11\text{-}1\text{-}13)$$

This stacking is done without time shifting; hence, it is similar to but not precisely the same thing as the familiar common reflection-point stack. Thus, for vertically stacked sections we have

$$S_{zt'} = \pm \frac{\bar{v}}{8} S_{yy} \quad (11\text{-}1\text{-}14)$$

We could also have obtained (11-1-14) from (11-1-11) by merely asserting that for zero-offset data the offset derivatives may be neglected and that (11-1-14) applies to near-trace sections.

Of course (11-1-14) is no stranger to us. Earlier we learned how the equation $P_{zt} = .5\, v P_{xx}$ controls propagation of a wave field (like a common shot-point gather). This means that (11-1-14) should convert hyperbolas to other hyperbolas; but in fact, because of the Fresnel-like approximation (which is improvable) it converts parabolas to other parabolas. In other words, (11-1-14) does just the kind of thing which is of use in seismic migration. Further details are in succeeding sections.

11-2 WAVE EQUATION MIGRATION

The construction of a cross section of reflectivity within the earth from a seismic section is called *migration*. At many locations on the earth the subsurface consists of horizontally layered sedimentary rocks. At such locations the migration of seismic data can be extremely simple because waves propagate vertically to the reflectors and they will have a round-trip travel time which is in direct proportion to depth. Migration is, then, just applying the proportionality factor to the time axis.

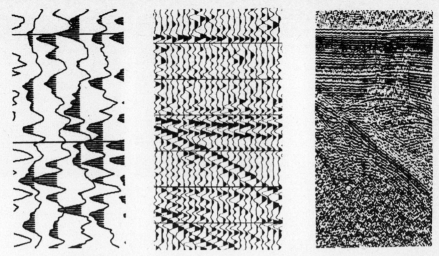

FIGURE 11-5
Display of wave fields. Side-by-side display of a collection of shaded seismograms with information density increasing from left to right becomes a picture
of a wave field. The picture usually represents acoustic pressure $P(x, t)$ as a
function of the horizontal space coordinate x and the vertical time coordinate t.
When the time axis t is taken to be the vertical travel time of echoes, then the
picture shows a cross section through the earth.

Thus, a picture of the waves, as in Fig. 11-5, which may actually show $P(x, t)$ at
$z = 0$ can be regarded as a cross section through the earth, say $P(x, z/2v)$. Ideally
the contours between light and dark delineate boundaries between different types
of sedimentary rocks. Unfortunately, the appearance of alternating strata (between
black rocks and white rocks?) is usually deceptive. It is unavoidably caused by
filtering effects in the sources, receiving equipment, or even the earth. The most
obvious, and perhaps the most important, information carried in the seismic
section is in the departure of the earth from horizontally stratified models. Seismic
sections are usually displayed with some vertical exaggeration. Such vertical
exaggerations commonly range from a factor of 1 to a factor of 20. The right-hand
frame of Fig. 11-5 happens to have a vertical exaggeration of 5 so that the obvious
strong reflector which appears to have a 45° slope actually in the earth has approximately a 9° slope. From such pictures we may attempt to deduce the present state
of the earth's subsurface and perhaps some of its history.

The purpose of migration calculations is to account for the fact that waves will
not go just straight up and down where the strata within the earth are not just horizontally layered. Such calculations can incorporate a seemingly endless list of
complicating factors, but luckily a straightforward application of our equation
$U_{tt} = U_{rr}$ can in a practical way accommodate the important departures from
horizontal layering which are found under many regions of the earth. Figure 11-6 is
an example of the migration techniques to be described in this section. It illustrates that the difference in appearance between a seismic time section and a depth

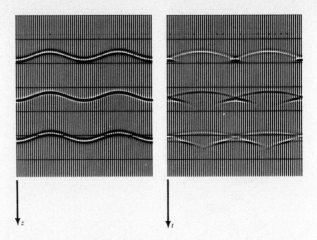

FIGURE 11-6
Three sinusoidal reflectors at increasing depths (left) and the calculated zero-offset reflection seismic section (right). There is no vertical exaggeration. Since each reflector has the same shape, it must have the same maximum dip (about 15°) as the others. The departure of the time section from the depth section obviously increases as one looks down the frames. A rule of thumb is that significant departure begins to occur when the depth becomes comparable to the smallest radius of curvature of the reflector (buried focus). The radius about equals the depth at the shallowest reflector where strong focusing is apparent. (From Ref. [8], p. 758.)

section need not arise solely from the dip of the strata; in fact, the curvature is usually the major factor. Hence, migration can be important even in relatively flat areas.

An impulse incident on an interface gives an impulsive reflection; however, there is always some filtering effect in the source, receiver, or earth which converts the supposed impulsive waveform to a little wavelet. Ideally, the cross section through the earth should display an impulse at the interface, but the migration computation carries the wavelet from the time section into the depth section. In this way, a resolving power limitation on the time axis is converted to a resolving power limitation on the depth axis. An interesting question is that of the resolving power on the horizontal x axis. Waves propagating at an angle convert the resolving power of the wavelet to an angular axis rather than the vertical axis. Clearly, the best horizontal resolving power you could get would be from waves traveling horizontally. For 45° waves, horizontal resolving distance would be $1/\sin 45°$ times as great as the vertical resolving distance. This is illustrated in Fig. 11-7 and in other figures.

The basic operation of migration can be understood in simple terms without reference to the wave equation. Figures 11-8 and 11-9 illustrate the transformations between time and depth for the two special cases where one domain or the other contains only an impulse function. Real data comprise a continuum in (x, t),

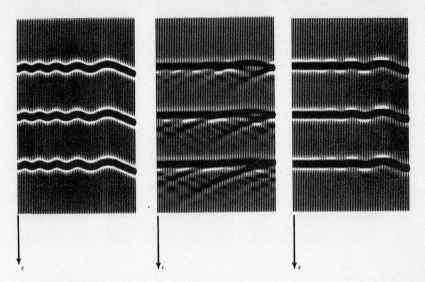

FIGURE 11-7
Reflections from oscillatory interfaces illustrating lateral resolving-power limitations. On the left is a starting model of some oscillatory interfaces. Center is the construction of a zero-offset section by the method of this section extended to the 45° approximation and modified by numerical viscosity to reject energy with dips beyond about 30°. The right-hand frame is the attempted reconstruction of starting model. It is no longer possible to resolve short-wavelength oscillations on the left-hand side of the interface. (From Ref. [8], p. 759.)

but Fig. 11-8 suggests a migration technique based on the linear superposition principle; namely, take every data point in the (x, t) domain and throw it out to a circular arc in the (z, t) domain. The migrated section is the superposition of all these arcs. A natural question one might ask is: Why bother to use the wave equation for migration when it can be done with the circular arcs? The answer to such a question depends upon a multitude of practical factors, some of which are data-dependent. One consideration favoring the wave equation approach is that velocity inhomogeneity is more easily and accurately described by wave equations than by ray tracing. Both methods are presently commercially available.

Figure 11-10 is an example which illustrates the origin of the term *migration*. It shows that a dipping interface which terminates at a point will appear in the time data to have the termination point *migrated* down-dip from its true location. In an example of this type, the departure of the depth section from the time data is really rather modest and it becomes even more slight as the dip is decreased.

This does not imply that data with slight dips need not be migrated because, as we have seen, of the dominating importance of curvature. An interesting practical example of this is shown in Fig. 11-11.

Field data examples of many of the synthetic data analyses are shown in the data of Fig. 11-12.

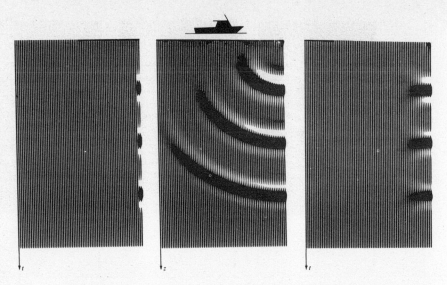

FIGURE 11-8
The depth response to time-domain impulses and reconstruction of the impulses. The fact that the left frame is mostly blank depicts a situation in which no echo is received when a source and receiver move together in the horizontal direction until they reach the right-hand edge of the frame where the three blips indicate that there are three echoes at successively increasing times. With these as observed data, the logical conclusion is that the reflection structure of the earth is three concentric circles with centers on the right-hand margin. The center frame shows the circles. (For economy, the right-hand edge of the frame is a plane of symmetry.) It will be noticed that the bottom of the circles is darker than the top. This is indicative of the 45° phase shift of bringing two-dimensional waves from a focus away from the focus. Waves with dips greater than about 45° have been filtered away by numerical viscosity. The loss of this energy plus the loss of the energy of waves which propagate at complex angles results in a reconstruction (right-hand frame) in which the impulses are somewhat spread out in the horizontal direction. (From Ref. [8], p. 750.)

We begin the explanation of how to migrate sections by recalling the basic result of Sec. 11-1 (and replacing t' by t).

$$S(y, z, t) = \int Q(y, f, z, t)\, df \quad (11\text{-}1\text{-}13)$$

$$S_{zt} = \pm \frac{v}{8} S_{yy} \quad (11\text{-}1\text{-}14)$$

Equation (11-1-13) defines a sum over offset of all the waves. Equation (11-1-14) shows how the sum S can be extrapolated down into the earth from the surface where it is known. We recall that the plus sign or the minus sign is chosen according to whether the extrapolation is done on the downgoing wave or the upgoing wave. Recall a source term $\delta(g - s, e, t)$ in (11-1-9). In midpoint-offset coordinates this source term becomes $\delta(f)$ const$(y)\delta(z)\delta(t)$. Since the source term is a constant

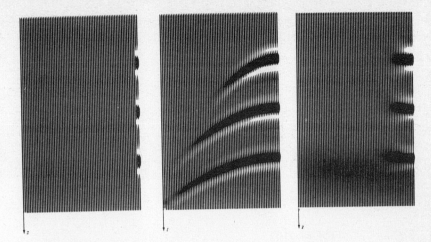

FIGURE 11-9
The time response to depth-domain impulses and reconstruction of the impulses. The left-hand frame depicts a model of the earth which consists of three point scatterers beneath one another along the right-hand edge. The second frame shows the synthetic time data created from the model. Basically one observes the hyperbolic travel-time curves to the reflecting points. The third frame represents migration of the synthetic data back to the point scatterers. As in Fig. 11-8 there is a reduced resolution because, in principle, horizontal resolution cannot be better than vertical resolution (which is controlled by the frequency content of the waves) and in practice we have included only rays up to angles of about 40°. (From Ref. [8], p. 751.)

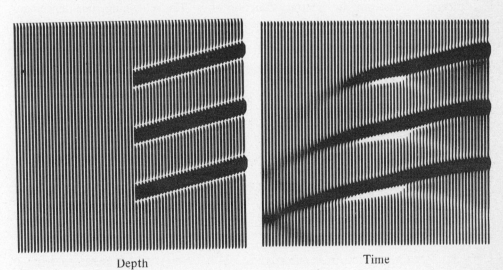

Depth   Time

FIGURE 11-10
An illustration of the origin of the term *seismic migration*. Left is a depth section consisting of three terminating interfaces. On the right is the synthetic time data. Note that besides the broadening of the termination point (by spreading it into a hyperbola) there is a general *migration* of the termination "point" in the down dip direction. (From Ref. [8], p. 756.)

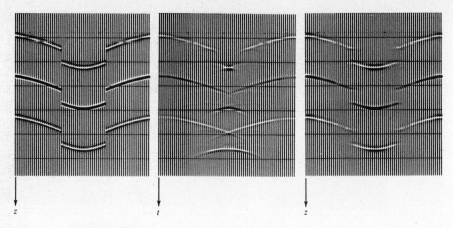

FIGURE 11-11
The classic graben model. Of practical significance is the fact that a concave structure appears convex in the time section. Thus, a geologic syncline may be confused with an anticline and the conditions for petroleum accumulation could be erroneously inferred. (From Ref. [8], p. 757.)

function of midpoint y, we can expect (11-1-14) for the downgoing wave to reduce to $S_{zt} = 0$. The general solution to $S_{zt} = 0$ is an arbitrary function of z added to an arbitrary function of t. Because we are interested in wave oscillations, not potential theory, we cast out the arbitrary function of z. The remaining time-dependent function which is independent of z must be equal to the time dependence of the

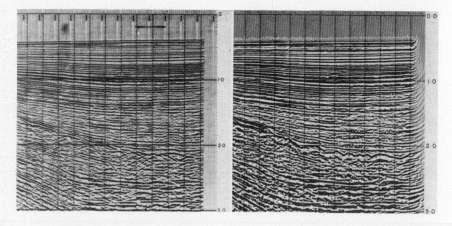

FIGURE 11-12
Example of migration of marine reflection seismic data. Left-hand frame data (courtesy of Royal Dutch Shell) were migrated with the wave equation (by Digicon, Inc.) and the result is shown in the right-hand frame. A prominent irregular reflector just below two seconds in the migrated section (right) is highly diffracted on the surface data (left).

source function; that is, a delta function of time. Since we now have an analytic function for the downgoing S at all depths, we can next turn to the numerical downward extrapolation of the upcoming S. Equation (11-1-14) with the minus sign enables us to do the downward extrapolation of the upcoming wave. Because the downgoing wave is a plane wave, the upcoming wave, if seen down at the reflectors, will take on the shape of the reflectors. Thus, the migrated section comprises the surface observations extrapolated downward to the proper depth.

As a practical matter, it turns out that investigators do not presently use the vertical stack defined by (11-1-13). They inject a time shift as a function of offset before doing the summation. The main effect of this is to make all offset traces more closely resemble the zero offset trace. The time delay of waves recorded on a zero offset section is equally divided between the downgoing path and the upgoing path. The two paths have unequal delays when the downgoing wave is a plane wave and the upcoming wave represents complicated scattering. From these considerations, we can now guess a result from (11-3-18) which is that the migration equation for a zero offset section, or an NMO stack resembles (11-1-14) but contains an extra factor of 2. Thus, we begin with

$$Q_{zt} = -\frac{\bar{v}}{4} Q_{yy} \qquad (11\text{-}2\text{-}1)$$

There may be some practical justification in allowing $\bar{v}$ to be variable in some of the three coordinates (y, z, t); however, by keeping the velocity constant in y we can simplify our first encounter with the details of migration. Fourier transforming the y variation with e^{iky} we reduce (11-2-1) to

$$0 = \left(\frac{v}{4} k_y{}^2 - \partial_{tz}\right)Q \qquad (11\text{-}2\text{-}2)$$

Now we will discretize the z and t coordinates. Then the function Q in (11-2-2) may be tabulated in the (z, t) plane and the differential operator in (11-2-2) becomes a 2×2 convolution operator in this two-dimensional plane. The operator is

$$\frac{vk_y{}^2}{16} \begin{array}{|c|c|} \hline 1 & 1 \\ \hline 1 & 1 \\ \hline \end{array} - \frac{1}{\Delta z\, \Delta t} \begin{array}{|c|c|} \hline 1 & -1 \\ \hline -1 & 1 \\ \hline \end{array} \qquad (11\text{-}2\text{-}3)$$

Defining a scale factor

$$a = \frac{\Delta z\, \Delta t\, vk_y{}^2}{16} \qquad (11\text{-}2\text{-}4)$$

the operator multiplied by $\Delta z\, \Delta t$ is

$$\begin{array}{|c|c|} \hline (a-1) & (a+1) \\ \hline (a+1) & (a-1) \\ \hline \end{array} \qquad (11\text{-}2\text{-}5)$$

When the operator (11-2-5) is laid upon an arbitrary place in the Q table and the four numbers in the operator are multiplied onto the four Q numbers beneath the operator, the meaning of (11-2-2) is that the sum of these four products should vanish. If we find a place in the Q table where only three of the four numbers of Q

are actually known, then we can calculate the fourth, unknown number. In fact, Q may be given only along a few boundaries in the (z, t) plane and we may be able to fill in the rest of the plane.

Migration and its inverse may be thought of on the following grid.

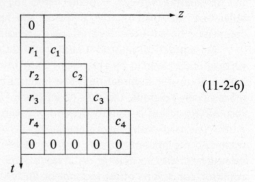

$$(11\text{-}2\text{-}6)$$

On this grid $r_0, r_1, \ldots, r_4$ represents the observed surface seismogram and c_0, $c_1, \ldots, c_4$ represents the migrated section. The zeros in the bottom row represent the idea that seismograms vanish at a sufficiently late time. Notice that in filling in the table there is a lot more work (diffraction) in going from r_4 to c_4 than in going from r_2 to c_2. This is because for fixed dip, deep events migrate farther than shallow events. Letting $k_y^2 = 0$, we have $a = 0$ and we are describing a stratified earth. Starting with surface data r_1, r_2, r_3, and r_4 and the bottom row of zeros we can use (11-2-5) to fill in the table (11-2-6) and it becomes

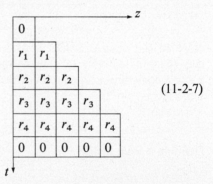

$$(11\text{-}2\text{-}7)$$

Inspecting the table (11-2-7) we see that numbers remain absolutely constant as we move in the z direction. If we had not taken $k_y^2 = 0$ but had instead taken k_y^2 to be small, we would see the numbers in the table changing gradually in the z direction. When k_y^2 is not zero some caution must be exercised in the order in which the (z, t) plane is filled up. The number a is always positive. Obviously, if the number a happens to equal $+1$, then it will be impossible to find unknown numbers in the Q table if they should lie under the multiplier $1 - a$. It turns out that, regardless of what numerical (positive) value a takes, the process of recursively seeking numbers in Q which underlie $1 - a$ will be unstable as is polynomial division by a nonminimum-phase filter. One reason is that such a process actually is polynomial division but the polynomials are two-dimensional polynomials. These complications are all the mathematical manifestations of the physical idea of causality.

From the point of view of solving hyperbolic differential equations it is most economical if you can get roughly the same number of points per wavelength (typically eight) on each coordinate axis. This means that in the development till now we have drastically over-sampled the z axis. The 15° limitation of the Fresnel approximation implies that Δz could always be taken five times or more coarser than $v \, \Delta t$. The subject of optimal selection of grid spacings is somewhat involved. Suffice it to say here that some field data have been migrated satisfactorily at a $\Delta z / v \, \Delta t$ ratio as great as 100 with a corresponding reduction in cost. Anyway, for the purpose of illustrating the point we will now redraw the section table with twice as coarse a sampling on the z axis. Numbers in the two tables below indicate two different possible orderings, both causal, of the use of (11-2-5) for the migration calculation. In either case the resulting migrated section is interpolated (perhaps very crudely) off the diagonal.

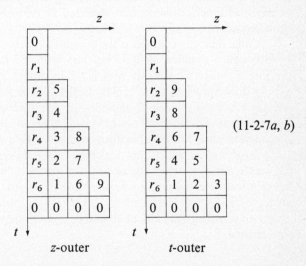

$$(11\text{-}2\text{-}7a, b)$$

z-outer   t-outer

The reader should check in each of the above tables that the number computed at any stage is based on three already known table entries and that the unknown being computed multiplies $(1 + a)$ and not $(1 - a)$. To synthesize data from a hypothetical depth section the calculation proceeds in reverse numerical order.

Study of the two optional procedures (11-2-7a) and (11-2-7b) reveals that the one labeled "t-outer" has an extraordinary advantage over the one labeled "z-outer" in the much smaller computer memory requirement.

Another practical reality is the need to be able to handle depth-variable velocities. This can be achieved by taking the migrated data $c_1, c_2, \ldots$, not from the diagonal in the (z, t) plane, but from another curve through the (z, t) plane as illustrated in Fig. 11-13. A nice feature of wave-equation migration is that velocity

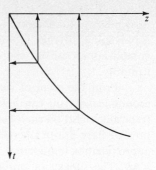

FIGURE 11-13
Plane for migration with depth-variable velocity. The curve in this plane is chosen to be the curve of two-way vertical travel time. The midpoint axis y comes out of the page. Section data along the curve may be projected to either the z axis or the t axis creating a migrated depth section or a migrated time section.

readjustment may be easily made without having to redo the calculation. Once the (z, t) plane has been filled, it is just a question of deciding which curve to display.

Dips in the (z, t) plane tend to be small. This means that changes in the velocity model, i.e., the curve in Fig. 11-13, result in more exaggerated changes to the migrated depth section than to the migrated time section. For such reasons the migrated time section is often preferred by data interpreters over the migrated depth section.

A final factor to consider is the possibility of lateral y variations in the velocity v. Although a detailed analysis has not been given, it seems clear that simply inserting a y-variable velocity into (11-2-1) should give a valid technique if the velocity does not change too rapidly in the y direction. How much change is too much is a question which is undoubtedly data-dependent and beyond the scope of our present efforts. It should be noted, however, that if a y-dependent velocity is to be used, then Fourier transformation over the y axis does not reduce (11-2-1) to (11-2-2). As a result of this, it is then necessary to regard every element in the (z, t) table not as a scalar but as a vector whose different components arise from different locations along the y axis. We then regard $1 + a$ not as a scalar divisor but as a tridiagonal matrix of the form $(I - \delta_{yy})$ which must be inverted. This creates no practical difficulty at all and is, in fact, the way the figures in this section were created.

11-3 VELOCITY ESTIMATION

Previous chapters focused on the task of delineating earth structure. Mathematically this has meant that we have taken the material velocity as known (and, for convenience, constant) but the impedance as having unknown discontinuities at interfaces of unknown shape between geologic structures. Now we seek to find the material velocity. Traditionally this has been done by assuming the earth structure consists of plane horizontal layers. Then the material velocity is deduced from the offset-dependent time shift (called the *normal moveout correction* or NMO) which best flattens the events on the common midpoint gathers. In this section, it will be shown how the assumption of flat layers may be eliminated. We will see how velocity can be estimated even in an earth consisting of random point scatterers. This can be expected to be useful in fractured zones or even perhaps in "no record" areas. An NR or no record area is one where the best-processed section shows no coherence along the midpoint y coordinate. An area may be NR not only because of poor data quality but also because the geologic structure itself has no continuity. But, as we will see, there is no theoretical reason why material velocity cannot be determined in such an NR area.

Basically, the procedure is to downward-continue both the theoretical downgoing wave and the observed upgoing wave. They are projected back down to the reflectors where their nearly constant ratio should represent the reflection coefficient as a function of offset. If they are projected downwards with an incorrect velocity, the ratio will be an oscillatory function of offset. The task, then, is to find the velocity which gives the best fit of the two waves. It does not matter whether the reflectors have any lateral continuity or not because the fitting is done for variable offset at a fixed midpoint at the reflector depth. When reflectors have no lateral continuity they may be called scatterers. An earth model with randomly located scatterers would produce migrated seismic data which was a random function of (moveout-corrected) time and midpoint but which was a constant function of offset.

It is easy to think of a good means to downward-continue the downgoing waves. From the shot point these waves expand spherically. For a homogeneous medium, we can just write down an analytic solution. For a moderately inhomogeneous medium, we can use the methods of earlier chapters. One problem is that the approximation $Q_{zz} \approx 0$ restricts validity to angles of about 15° from the vertical. This is easily improved by transforming from cartesian (x, z) coordinates to polar (r, θ) coordinates. The approximation $Q_{rr} \approx 0$ requires rays to stay within 15° of a radius line. Obviously a "stratified media coordinate frame" could be designed to handle even stronger velocity inhomogeneity of that type.

The problem which is more difficult is to find a good coordinate system for the upcoming waves. It took me two years to come up with a practical solution. A hint is provided by observing why, for the downgoing wave, the polar system is preferable to the cartesian system. For a quasi-spherical wave $Q_{\theta\theta}$ will be nearly 0, whereas Q_{xx} gets large quickly unless you are directly under the source. Because we deal with equations like $Q_{zt} = Q_{xx}$ or $Q_{rt} = Q_{\theta\theta}/r^2$, this means that Q_r will generally be small; whereas Q_z is small only on the z axis directly under the source. Consequently, the approximation $Q_{rr} \approx 0$ is much better than $Q_{zz} \approx 0$. Our observation is that the advantage of the (r, θ) coordinates is that the downgoing wave $D(r, \theta)$ is nearly independent of the lateral θ coordinate. What we need is a coordinate frame in which the upcoming wave U is nearly independent of the lateral coordinate. Experienced geophysicists will immediately recognize that normal moveout-corrected data fill this requirement. Normal moveout (NMO) correction is a compression of the time axis on far offset seismograms intended to make the far offset waves arrive at the same (NMO corrected) time as the vertically incident waves. Thus, the partial derivative of the wave field Q with respect to shot-geophone offset at a fixed NMO corrected time should be small.

The closer our data come to those from flat horizontal reflectors in an earth

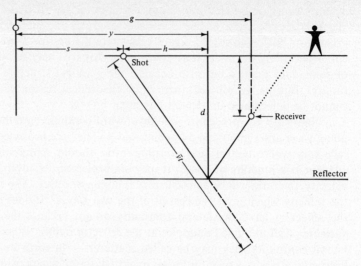

FIGURE 11-14
Geometry for normal-moveout correction of downward-continued data.

of known velocity the smaller the offset derivative will be. The purpose of a wave equation is to handle the departure from such an idealized situation.

This compression of the time axes of the far offset seismograms is really a coordinate change. The usual definition of NMO correction does not anticipate our desire to project our geophones deeply into the earth. As we project our geophones downward along a ray path we will retain the *surface midpoint y* and the *surface half-offset h = f/2* as lateral coordinates of the wave field. Lateral derivatives of idealized data should vanish.

Figure 11-14 shows the geometry for normal moveout correction of downward-continued data in homogeneous media of velocity $\bar{v}$. The transformation from interpretation variables to observation variables is

$$s(h, y, d, z) = y - h \tag{11-3-1a}$$

$$g(h, y, d, z) = y + \frac{(d - z)h}{d} \tag{11-3-1b}$$

$$t(h, y, d, z) = \frac{(d^2 + h^2)^{1/2}(2d - z)}{d\bar{v}} \tag{11-3-1c}$$

Either algebraic or geometric means yield the inverse transformation

$$d(s, g, t, z) = \tfrac{1}{2}\{[\bar{v}^2 t^2 - (g - s)^2]^{1/2} + z\} \tag{11-3-2a}$$

$$y(s, g, t, z) = \frac{1}{2}\left\{(g + s) + \frac{z(g - s)}{[\bar{v}^2 t^2 - (g - s)^2]^{1/2}}\right\} \tag{11-3-2b}$$

$$h(s, g, t, z) = \frac{1}{2}\left\{(g - s) + \frac{z(g - s)}{[\bar{v}^2 t^2 - (g - s)^2]^{1/2}}\right\} \tag{11-3-2c}$$

That (11-3-2) is indeed inverse to (11-3-1) is readily checked by substituting (11-3-1) into (11-3-2).

In a homogeneous medium of velocity $\tilde{v}$ we may write the solution for the downgoing wave as a delta function on an expanding circle

$$D(g, s, t, z) = \delta[(g - s)^2 + z^2 - \tilde{v}^2 t^2] \qquad (11\text{-}3\text{-}3)$$

The upcoming wave U will be computed in the (h, y, d, z) variables and we want to compare it to the downgoing wave D, expressed by (11-3-3) in (g, s, t, z) variables. We can convert D to (h, y, d, z) variables by substitution of (11-3-1) into (11-3-3); a meaningful simplification arises if we assume the medium velocity $\tilde{v}$ equals the moveout coordinate frame velocity $\bar{v}$. We get

$$D(h, y, d, z) = \delta[4d(z - d)] \qquad (11\text{-}3\text{-}4)$$

In this case, the downgoing wave turns out to be independent of the lateral coordinates h and y.

Now let us consider an earth model which contains only a single point scatterer located at (x_0, z_0). This scatterer is illuminated by a delta function source located at $(s, 0)$. Excluding horizontally propagating waves, we have for the upcoming wave $U(s, g, t, z)$ an infinitesimal distance above the scatterer

$$U(s, g, t, z_0 - 0) = \delta(g - x_0)\delta[\bar{v}^2 t^2 - (s - x_0)^2 - z_0^2] \qquad (11\text{-}3\text{-}5)$$

Substituting the transformation (11-3-1) at $z = d$ into (11-3-5), we obtain

$$U(h, y, d, z = d) = \delta(y - x_0)\delta[d^2 + h^2 - (y - h - x_0)^2 - z_0^2]$$

The existence of $\delta(y - x_0)$ allows us to set $y = x_0$ in the other delta function, getting

$$U(h, y, d, z = d) = \delta(y - x_0)\delta(d^2 - z_0^2)$$

We now see the central concept that the wave at the reflector in moveout corrected coordinates is indeed independent of the half-offset h. Obviously, the superposition of a random collection of point scatterers will create a migrated wave field which is random in y and d but still constant in the offset h. Indeed, the concept would also seem to be valid even if the scatterers were randomly distributed out of the plane of the section. In three-dimensional space it is only necessary to regard z as the radial distance from the traverse line.

The purpose of all this is to estimate velocity; but velocity is needed for the first step, namely the migration. Use of an erroneous velocity in the migration prevents total collapse to a delta function on the midpoint axis. This causes some destructive interference between adjoining midpoints representing some information loss for a random scatterer model but it is of no consequence in a layered earth model (where even the migration itself is unnecessary).

Stephen M. Doherty [Ref. 37] made a calculation to illustrate these concepts. Figure 11-15 shows an earth model. Figure 11-16 shows surface data and downward-continued data for the model.

From the point of view of velocity determination, it is immaterial what coordinate frame is used to downward-continue the observed waveforms. However,

$z = 2.5$

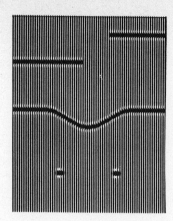

$z = 3.5$

FIGURE 11-15
An earth model used to illustrate velocity analysis with downward-continued data.

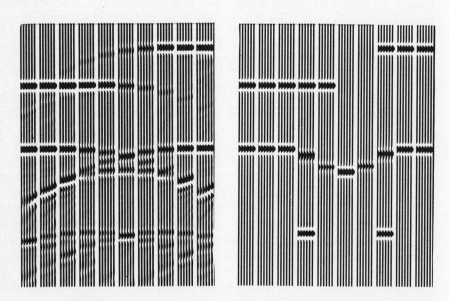

FIGURE 11-16
Surface data and downward-continued data for the model of Fig. 11-15. The coordinates are designed to display three-dimensional data (y, h, d) on a two-dimensional page. The vertical axis, as usual, is the d coordinate. For the horizontal axis the h coordinate has been sampled at six values of h which are displayed together in groups (common midpoint gathers). There are ten of these gathers spaced along the y axis. Within each group $h = 0$ is on the left and h_{max}, corresponding to about 45° rays, is on the right. The left-hand frame shows the surface data and the right-hand frame shows the data down at the reflectors. At the reflectors we see horizontal alignment of waveforms indicating that the data are independent of offset h.

it is convenient to downward-continue these waves in the NMO coordinate frame. This proceeds in a fashion similar to that in our earlier work. To simplify the algebra, first note that (11-3-2b) and (11-3-2c) imply that

$$\frac{\partial y}{\partial(g, t, z)} = \frac{\partial h}{\partial(g, t, z)} \qquad (11\text{-}3\text{-}6)$$

The wave equation

$$\left(\partial_{gg} + \partial_{zz} - \frac{\partial_{tt}}{\tilde{v}^2}\right)P = 0 \qquad (11\text{-}3\text{-}7)$$

in NMO coordinates will take the form

$$\left[(d_g \partial_d + y_g \partial_y + h_g \partial_h)^2 + (\partial_z + d_z \partial_d + y_z \partial_y + h_z \partial_h)^2 \right.$$
$$\left. - \frac{(d_t \partial_d + y_t \partial_y + h_t \partial_h)^2}{\tilde{v}^2}\right]Q = 0 \qquad (11\text{-}3\text{-}8)$$

As before, when we square these partial differential operators we will take the coefficients to be constant. This is the high frequency assumption that the wave field changes more rapidly than the coordinate frame. Before we compute all the required derivatives we define a simplifying combination b where

$$b = [\bar{v}^2 t^2 - (g - s)^2]^{1/2} \qquad (11\text{-}3\text{-}9)$$

The required derivatives are computed, recalling (11-3-6) to be

$$\begin{bmatrix} d_g & d_z & d_t \\ y_g & y_z & y_t \\ h_g & h_z & h_t \end{bmatrix} = \frac{1}{2}\begin{bmatrix} \dfrac{-(g-s)}{b} & 1 & \dfrac{\bar{v}^2 t}{b} \\ 1 + \dfrac{z\bar{v}^2 t^2}{b^3} & \dfrac{g-s}{b} & \dfrac{-(g-s)\bar{v}^2 tz}{b^3} \\ 1 + \dfrac{z\bar{v}^2 t^2}{b^3} & \dfrac{g-s}{b} & \dfrac{-(g-s)\bar{v}^2 tz}{b^3} \end{bmatrix} \qquad (11\text{-}3\text{-}10)$$

First we quickly discover that if moveout-correction velocity $\bar{v}$ equals media velocity $\tilde{v}$, say $\tilde{v} = \bar{v} = v$, then three of the terms in (11-3-8) vanish identically. By direct substitution, the reader may verify that

$$\left(d_g{}^2 + d_z{}^2 - \frac{d_t{}^2}{v^2}\right)Q_{dd} = 0 \qquad (11\text{-}3\text{-}11a)$$

$$2\left(d_g y_g + d_z y_z - \frac{d_t y_t}{v^2}\right)Q_{dy} = 0 \qquad (11\text{-}3\text{-}11b)$$

$$2\left(d_g h_g + d_z h_z - \frac{d_t h_t}{v^2}\right)Q_{dh} = 0 \qquad (11\text{-}3\text{-}11c)$$

Next, we obtain the three cross terms with ∂_z.

$$2y_z \, Q_{yz} = \frac{g-s}{bQ_{yz}} = \frac{h}{dQ_{yz}} \quad (11\text{-}3\text{-}12a)$$

$$2h_z \, Q_{hz} = \frac{g-s}{bQ_{hz}} = \frac{h}{dQ_{hz}} \quad (11\text{-}3\text{-}12b)$$

$$2d_z \, Q_{dz} = Q_{dz} \quad (11\text{-}3\text{-}12c)$$

From (11-3-6) we realize that the coefficients of Q_{yy}, Q_{hh}, and $2Q_{yh}$ are identical. Through a considerable amount of algebraic reduction we obtain

$$\left(y_g^2 + y_z^2 - \frac{y_t^2}{v^2}\right)(\partial_y + \partial_h)^2 Q = -\left(\frac{d}{2d-z}\right)^2 \left(\frac{1+h^2}{d^2}\right)(\partial_y + \partial_h)^2 Q \quad (11\text{-}3\text{-}13)$$

As usual, we make the Fresnel-like approximation by dropping the Q_{zz} term. In cartesian geometry, this limits accurate treatment of rays to within a cone of about $15°$ of the vertical. In the NMO geometry, this would seem to be more like a $15°$ limitation on structural dips. Of course, the higher-accuracy techniques can always be used where required. Gathering (11-3-11) to (11-3-13) together, we obtain the basic result

$$\left[\partial_d + \frac{h}{d}(\partial_y + \partial_h)\right]\partial_z Q = \left(\frac{d}{2d-z}\right)^2 \left(1 + \frac{h^2}{d^2}\right)(\partial_y + \partial_h)^2 Q \quad (11\text{-}3\text{-}14)$$

Equation (11-3-14) may be used for downward continuation of moveout-corrected unstacked sections for velocity determination.

It seems worthwhile to inspect (11-3-14) in some special cases. At the surface for zero offset Q_h vanishes by symmetry. At a point scatterer we saw that Q was independent of h. For idealized data from layered reflectors Q is a function of d only. In a wide variety of practical situations it turns out to be reasonable to simplify (11-3-14) with $Q_d \gg Q_y \gg Q_h$. This leaves us with

$$Q_{dz} = -\left(\frac{d}{2d-z}\right)^2 \left(1 + \frac{h^2}{d^2}\right)Q_{yy} \quad (11\text{-}3\text{-}15)$$

It seems natural to wonder about the variable coefficient $d/(2d-z)$ in comparison to the earlier equations with constant coefficients. We can now show that with regard to migration that there is no practical difference. Define a new variable

$$z' = \frac{zd}{2d-z} \quad (11\text{-}3\text{-}16)$$

Note that at the surface $z = 0$ we have z' equal to zero and at the reflectors $z = d$ we have $z' = d$. Thinking of $Q(z, d) = Q'(z', d)$ we find

$$Q_z = z_z' \partial_{z'} \, Q'$$

$$Q_d = (\partial_d + z_d' \partial_{z'})Q'$$

With these, the left-hand side of (11-3-15) becomes

$$Q_{dz} = (\partial_d - z_d' \partial_{z'} \cdot)z_z' \partial_{z'} \, Q$$

In a Fresnel-like approximation, we drop $\partial_{z'z'}$, obtaining

$$Q_{dz} = z_z' \, Q_{dz'} = \frac{2d^2}{(2d - z)^2} \, Q_{dz'}$$

which reduces (11-3-15) to

$$Q_{dz'} = -\frac{1}{2}\left(1 + \frac{h^2}{d^2}\right)Q_{yy} \quad (11\text{-}3\text{-}17)$$

To justify the factor of 2 which was asserted in Sec. 11-2, we may make another transformation from d to a two-way travel-time coordinate t' given by

$$t' = \frac{2d}{v}$$

which gives

$$Q_{t'z'} = -\frac{v}{4}\left[1 + \left(\frac{2h}{vt'}\right)^2\right]Q_{yy} \quad (11\text{-}3\text{-}18)$$

Of course (11-3-18) must be integrated from $z' = 0$ to $z' = t'v'/2$. A convenient rescaling of the depth axis is in terms of two-way travel time t'' where

$$t'' = \frac{2z'}{v}$$

This leads to the equation

$$Q_{t't''} = -\frac{v^2}{8}\left[1 + \left(\frac{2h}{vt'}\right)^2\right]Q_{yy} \quad (11\text{-}3\text{-}19)$$

in which t' is the two-way travel time and t'' is the two-way travel-time depth axis which is integrated from the surface $t'' = 0$ to the reflectors at $t'' = t'$.

Strictly speaking, (11-3-19) should be applied separately to data of each offset h before the data are summed over offset (stacked). For reasons of economy, the data are often stacked before migration with (11-3-19). In such a compromise, h in (11-3-19) is taken as zero or some average value of $2h/vt'$ is used.

So far, we have shown that downward-continued, moveout-corrected seismograms will be independent of offset if downward continued with the correct velocity. What we have not seen is how to estimate the velocity error from the downward-continued data. For this we must recognize another important term which has been omitted from the entire analysis. We saw this term in earlier studies of propagation in inhomogeneous media. We must carry through the distinction between media velocity $\bar{v}(x, z)$ and NMO velocity $\bar{v}$ [generalizable to $\bar{v}(z)$] which was abandoned for the sake of the simplifications beginning at (11-3-11). Recalling that for small departures from layered models, $Q_d \gg Q_y \gg Q_h$, we see that the first of the three terms in (11-3-11) will be the most important. Making the distinction between the two velocities, (11-3-11a) now introduces the significant term

$$\left(d_g{}^2 + d_z{}^2 - \frac{d_t{}^2}{\tilde{v}^2}\right) Q_{dd} \neq 0 \quad (11\text{-}3\text{-}20)$$

$$\frac{\bar{v}^2 t^2}{\bar{v}^2 t^2 - (g-s)^2}\left(1 - \frac{\bar{v}^2}{\tilde{v}^2}\right) Q_{dd} \neq 0$$

$$\left(1 + \frac{h^2}{d^2}\right)\left(1 - \frac{\bar{v}^2}{\tilde{v}^2}\right) Q_{dd} \neq 0 \quad (11\text{-}3\text{-}21)$$

Thus, with this new term but the other approximations, (11-3-15) becomes

$$Q_{dz} = -\left(\frac{d}{2d-z}\right)^2\left(1 + \frac{h^2}{d^2}\right) Q_{yy} + \left(1 + \frac{h^2}{d^2}\right)\left(1 + \frac{\bar{v}^2}{\tilde{v}^2}\right) Q_{dd} \quad (11\text{-}3\text{-}22)$$

Numerically, we can consider solving (11-3-22) by a splitting method where the solution is projected downward by alternate use of the two equations

$$Q_{dz} = -\left(\frac{d}{2d-z}\right)^2\left(1 + \frac{h^2}{d^2}\right) Q_{yy} \quad (11\text{-}3\text{-}23a)$$

$$Q_z = \left(1 + \frac{h^2}{d^2}\right)\left(1 - \frac{\bar{v}^2}{\tilde{v}^2}\right) Q_d \quad (11\text{-}3\text{-}23b)$$

Equation (11-3-23a) may be called the "diffraction" part and (11-3-23b) may be called the "thin-lens" part. The effect of (11-3-23b) is that as Q is projected in the z direction, each seismogram [a seismogram is a function of (moveout-corrected) time d at a fixed half-offset h and midpoint y] undergoes a steady time shift (d shift). The amount of the shift increases with the velocity error according to $1 - \bar{v}^2/\tilde{v}^2$ and it increases with offset according to $1 + h^2/d^2$. Thus, the effect of (11-3-23b) is to change the curvature of the data with half-offset h. However, (11-3-23a) contains Q_{yy} but it does not contain Q_h or Q_{hh}. This means that the operations of (11-3-23a) and (11-3-23b) commute. Thus, we can project all the way down to the reflectors with (11-3-23a) and then use (11-3-23b). It is significant that the hard part of the job, namely (11-3-23a), depends on the frame velocity $\bar{v}$, not the material velocity $\tilde{v}$. This means that we can rather economically test various media velocities $\tilde{v}$.

Before we can consider the task of selecting our best estimate of the media velocity $\tilde{v}$, we must consider matching the upcoming wave U to some reflection coefficient c multiplied by some downgoing wave D. The matching of these waves can be done in the field-recording coordinates, but we prefer to do the matching in the NMO coordinate system. First, let us get an expression for the downgoing wave in NMO coordinates. Insert (11-3-1) into (11-3-3) to obtain

$$D(h, y, d, z) = \delta\left[4d(z-d) + \left(1 - \frac{\tilde{v}^2}{\bar{v}^2}\right)\left(1 + \frac{h^2}{d^2}\right)(2d - z)^2\right] \quad (11\text{-}3\text{-}24)$$

At present, we are not trying to preserve slow magnitude variations [spherical spreading was omitted from (11-3-3)], so we can divide through the argument of the delta function by $-4d$. Since we are interested in small amounts of variation of $\tilde{v}$ from $\bar{v}$ the delta function will vanish very near to $z = d$. Thus, to a good approximation we can substitute z for d in the coefficient of $(1 - \tilde{v}^2/\bar{v}^2)$, obtaining

$$D(h, y, d, z) = \delta\left[d - z - \left(1 - \frac{\tilde{v}^2}{\bar{v}^2}\right)\left(1 + \frac{h^2}{z^2}\right)\frac{z}{4}\right]$$

$$= \delta(d - z - s) \quad (11\text{-}3\text{-}25)$$

where we have defined a time (d) shift function

$$s(h^2, z, \tilde{v}/\bar{v}) = \left(1 - \frac{\tilde{v}^2}{\bar{v}^2}\right)\left(1 + \frac{h^2}{z^2}\right)\frac{z}{4} \quad (11\text{-}3\text{-}26)$$

Now let us return to the task of matching the up- and downgoing wave. We might hope to determine a reflection coefficient, along with some angular dependence, in the form of a power series, for example:

$$c = c_0 + \frac{c_1 h}{z} + \frac{c_2 h^2}{z^2} + \cdots \quad (11\text{-}3\text{-}27)$$

To simplify the sequel we will estimate only the constant term c_0 by the minimization

$$\min_c \sum_h \sum_d [U(y, z, h, d) - c(y, z)D(y, z, h, d)]^2 \quad (11\text{-}3\text{-}28)$$

The solution is obviously

$$c(y, z) = \frac{\sum_h \sum_d UD}{\sum_h \sum_d D^2} \quad (11\text{-}3\text{-}29)$$

Because D vanishes almost everywhere, we can gain insight by replacing the double sum by a single sum, specifically for the numerator

$$\text{Numerator} = \sum_h \sum_d U(y, z, h, d)\, \delta(d - z - s)$$

$$= \sum_h U[y, z, h, d = z + s(h^2, z, \tilde{v}/\bar{v})] \quad (11\text{-}3\text{-}30)$$

Letting N denote the number of terms in the offset sum, we get for (11-3-29)

$$c(y, z, \tilde{v}) = \frac{1}{N}\sum_h U \quad (11\text{-}3\text{-}31)$$

Finally, we come to the part of determining the velocity $\tilde{v}$ which provides the best minimum of $U - cD$. For this, a computer scan over $\tilde{v}$ may be used to find the minimum

$$\min_{\tilde{v}} \sum_h \sum_d (U - cD)^2 = \min_{\tilde{v}} \sum_h (U - c)^2$$

$$= \min_{\tilde{v}} \sum_h^N \left(U - \frac{1}{N} \sum_h^N U \right)^2$$

$$= \min_{\tilde{v}} \left(\sum_h U^2 \right) - \frac{1}{N} \left(\sum U \right)^2 \geq 0 \quad (11\text{-}3\text{-}32)$$

In practice it is found that rather than minimize the sum squared minus the squared sum it is preferable to maximize the negative logarithm or the semblance ratio

$$\text{Semblance} = \frac{\left(\sum U \right)^2}{N \sum U^2} \leq 1 \quad (11\text{-}3\text{-}33)$$

The ratio has the advantage of being insensitive to the magnitude of the wave U and lends itself well to displays over a wide range of conditions.

11-4   MULTIPLE REFLECTIONS

Accurate modeling of multiple reflections makes it possible to subtract theoretical multiples from field data, thereby uncovering the more informative, primary reflections. We will first review a simplified layered model for multiple reflections and then modify it for application to a two-dimensionally inhomogeneous earth. We plan to solve both forward and inverse problems. The forward problem is given the two-dimensional spatial distribution of reflection coefficients to find the reflections including diffracted multiples with peglegs. The inverse problem is to deduce the two-dimensional spatial distribution of reflection coefficients from the waves.

The basic idea that we use for the inverse problem is that "reflectors exist at points in the earth where the first arrival of a downgoing wave is time-coincident with an upcoming wave." As a practical matter, we try to choose reflection coefficients which ensure that the upcoming wave vanishes before the onset of the downgoing wave. In Chap. 8 we learned that, for a layered medium, the Z transform polynomial for the downgoing wave $D(Z)$ is minimum-phase. This means that the inverse of $D(Z)$, namely $1/D(Z)$, can be expanded into positive powers of Z and the lead coefficient will be $1/d_0$. Consequently, in the nth layer the reflection coefficient at the bottom of the layer is the coefficient of the lowest power of Z in the expansion for $U^n(Z)/D^n(Z)$. Our plan is to show that $U^n(Z)/D^n(Z)$ is observable at the surface $n = 0$, and then show how U^{n+1}/D^{n+1} can be easily computed from U^n/D^n. First of all, pressure P and vertical velocity W must be taken as known at the surface for all time.

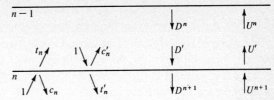

FIGURE 11-17
The defining conventions for waves in layers being scattered at the nth interface.

From Sec. 9-3 and equations (9-3-12a) and (9-3-12b), we have

$$\begin{bmatrix} U \\ D \end{bmatrix} = \tfrac{1}{2} \begin{bmatrix} 1 & -\dfrac{1}{Y} \\ 1 & \dfrac{1}{Y} \end{bmatrix} \begin{bmatrix} P \\ W \end{bmatrix} \qquad (11\text{-}4\text{-}1)$$

and

$$\begin{bmatrix} P \\ W \end{bmatrix} = \begin{bmatrix} 1 & 1 \\ -Y & Y \end{bmatrix} \begin{bmatrix} U \\ D \end{bmatrix} \qquad (11\text{-}4\text{-}2)$$

In (11-4-1) and (11-4-2) we think of the surface $n = 0$ pressure $P^0(Z) = p_0$ as a delta function at time $t = 0$, which is an instantaneous violation of the surface boundary condition $P^0 = 0$ in order to introduce energy into the medium. This causes a delta function behavior in W^0 at $t = 0$; and, later on, the reflectors in the earth cause W^0 to be nonvanishing while P^0 is vanishing. To apply our present method, we will require $U^0 = \tfrac{1}{2}(P - YW)$ which vanishes until the first echo and then becomes the negative of the observed returning waveform W and $D^0 = \tfrac{1}{2}(P + YW)$ which is a delta function at $t = 0$ followed by the observed waveform.

Figure 11-17 will recall the conventions of Chap. 8 for scattering at an interface. For propagation across the layer, we have the obvious delays which become multiplication by the half-unit delay operator $Z^{1/2} = \exp(i\omega \, \Delta t/2) = \exp(i\omega \, \Delta z/2v)$.

$$D' = Z^{1/2} D^n \qquad (11\text{-}4\text{-}3)$$

$$U^n = Z^{1/2} U' \qquad (11\text{-}4\text{-}4)$$

We have the scattering of downgoing waves into upgoing waves at the interface

$$U' = t_n U^{n+1} + c_n' D' \qquad (11\text{-}4\text{-}5)$$

Now for the theory to be perfectly general, as in Chap. 8, we would write an equation converse to (11-4-5) which states that downgoing waves are augmented by scattering from upgoing waves. However, to achieve simplicity in our first exposure to diffracted multiple reflections, we will assume that downgoing waves are generated from upcoming waves only at the surface $n = 0$. Everywhere else in the earth we will suppose that downgoing waves go along their merry way without any contributions from upcoming waves. With this presumption, we write

$$D^{n+1} = t_n' D' \qquad (11\text{-}4\text{-}6)$$

As a practical matter, the assumption we have built into (11-4-6) is not a bad one for marine data where the free-surface reflector is by far the strongest reflector. Although a large range of reflection coefficients is found in practice, a common case would be for a sea-floor coefficient of .1 and subsurface coefficients of .01. With land data, it is often important to account for a lot of absorption and scattering in the near-surface layers.

Anyway, eliminating the D' and U' variables from (11-4-3) to (11-4-6), we obtain

$$D^{n+1} = t'Z^{1/2}D^n \tag{11-4-7}$$

$$U^{n+1} = \frac{U' - c_n'D'}{t_n}$$

$$= \frac{Z^{-1/2}U^n + c_nZ^{1/2}D^n}{t_n} \tag{11-4-8}$$

Dividing (11-4-8) by (11-4-7), we achieve a simple relationship for the downward continuation of U/D. It is

$$\frac{U^{n+1}}{D^{n+1}} = \frac{Z^{-1}U^n/D^n + c_n}{t_n't_n} \tag{11-4-9}$$

To understand this simple recurrence relationship we must recall that the coefficient of Z^0 in U^n/D^n vanishes (see Fig. 11-17) and that the coefficient of Z in U^n/D^n is $c_n' = -c_n$. Thus, the Z^{-1} multiplier in (11-4-9) shifts U^n/D^n to have $-c_n$ as the coefficient of Z^0 and then the c_n in (11-4-9) extinguishes it. Clearly, the downward continuation is so simple that the results can almost be seen as the coefficients of the surface ratio $U^0(Z)/D^0(Z)$. The only complicating factor is the divisor $t_n't_n$ which gives a monotonically increasing scale factor of $1/\prod_{k=1}^n (1 - c_k^2)$ to the coefficient of Z^n in $U^0(Z)/D^0(Z)$.

Now let us see how we can generalize these layered-model ideas to a two-dimensionally inhomogeneous earth. The big difference is that upward and downward continuation of waveforms can no longer be accomplished simply by the delay operator Z. We could use (10-5-16) for upward and downward continuation; however, it seems more natural and certainly more economical to use (10-5-17a) and (10-5-18) which are

$$D'_{z't'} = \frac{\bar{v}}{2} D'_{x'x'} \tag{11-4-10}$$

$$U''_{z''t''} = -\frac{\bar{v}}{2} U''_{x''x''} - c'(z'')\partial_{t''} D'\left(x'', z'', t'' - \frac{2z''}{v''}\right) \tag{11-4-11}$$

Equations (11-4-10) and (11-4-11) may be compared with (11-4-7) and (11-4-8). It is apparent that transmission coefficients are not accounted for in (11-4-10) and (11-4-11). That is the sort of thing which is usually unimportant in the analysis of field data, but which can be recovered by a more exhaustive derivation if necessary.

First, we will consider the forward problem. It is easier to understand in a finite difference notation than in the differential equation system (11-4-10) and (11-4-11). Let us define a matrix T which is a tridiagonal matrix because of the second-difference operator δ_{xx}

$$T = -\left(\frac{\bar{v}\,\Delta z\,\Delta t}{8\,\Delta x^2}\right)\delta_{xx} \quad (11\text{-}4\text{-}12)$$

Then (11-4-10) with the boundary condition that D vanishes before $t' = 0$ can be written in tabular form

$$
\begin{array}{|c|c|}
\hline
T+I & T-I \\
\hline
T-I & (T+I) \\
\hline
\end{array}
\;*\;
$$

y				z
0	0	0	0	0
$d_0{}^0$	$d_0{}^1$	$d_0{}^2$	$d_0{}^3$	$d_0{}^4$
$d_1{}^0$	$d_1{}^1$	$d_1{}^2$	$d_1{}^3$	
$d_2{}^0$	$d_2{}^1$	$d_2{}^2$		
$d_3{}^0$	$d_3{}^1$			
$d_4{}^0$				

$$= 0 \quad (11\text{-}4\text{-}13)$$

(with t' labeling the rows)

where $*$ denotes convolution in the (z, t') plane. This table can be filled in by the knowledge of the zeros in the top row and the surface observations in the left column. We will do the forward problem by illustrating a typical step. Suppose both U and D are known for all z before $t = 4$ but not at or after $t = 4$. The typical step we will show is how to get U and D at $t = 4$. That means we know everything entered in the D table (11-4-13) except for $d_4{}^0$. The first step is to take D from the t' coordinate in (11-4-13) to the t'' coordinate. If $\Delta z = \bar{v}\,\Delta t/2$, this is achieved by downshifting successive columns in (11-4-13) to

				z
0	0	0	0	0
$d_0{}^0$	0	0	0	0
$d_1{}^0$	$d_0{}^1$	0	0	0
$d_2{}^0$	$d_1{}^1$	$d_0{}^2$	0	0
$d_3{}^0$	$d_2{}^1$	$d_1{}^2$	$d_0{}^3$	0
$d_4{}^0$	$d_3{}^1$	$d_2{}^2$	$d_1{}^3$	$d_0{}^4$

$$(11\text{-}4\text{-}14)$$

(with t'' labeling the rows)

The meaning of the t'' coordinate is that if reflection occurs, all elements on a given row of constant t'' can contribute to the upcoming wave received at a surface arrival time t''. Equation (11-4-11) may be rearranged to

$$\left(\frac{\bar{v}}{2}\partial_{xx} + \partial_{zt}\right)(-U) = c'(x, z)\partial_t D \quad (11\text{-}4\text{-}15)$$

which can be expressed in tabular form as

$T-I$	$T+I$		0	0	0	0	0
$(T+I)$	$T-I$	$*$	$d_1{}^0$	$-u_1{}^1$	0	0	0
			$d_2{}^0$	$-u_2{}^1$	$-u_2{}^2$	0	0
			$d_3{}^0$	$-u_3{}^1$	$-u_3{}^2$	$-u_3{}^3$	0

$$= \frac{\Delta z}{2}$$

-1	-1
1	1

$*$

$d_0{}^0$	0	0	0	0
$d_1{}^0$	$d_0{}^1$	0	0	0
$d_2{}^0$	$d_1{}^1$	$d_0{}^2$	0	0
$d_3{}^0$	$d_2{}^1$	$d_1{}^2$	$d_0{}^3$	0
	$d_3{}^1$	$d_2{}^2$	$d_1{}^3$	$d_0{}^4$

$(11\text{-}4\text{-}16)$

$*$

0	c_1'	c_2'	c_3'	c_4'

In the tabular equation (11-4-16) the unknown elements have been left blank. On the right-hand side the reflection coefficients are to be convolved upward into the downgoing wave to create the sources for the upgoing wave. The c_0 entry in the reflection coefficient row is set at zero because we will prescribe the boundary condition $d_t{}^0 = -u_t{}^0$ separately. Thus, we see the right-hand side of (11-4-16) is completely known, and the unknowns in the left-hand side may be filled in from right to left, thereby computing the upcoming wave table at $t'' = 4$ for all z. This gives $d_4{}^0$, enabling us to go back and fill in another row in the downgoing wave table which enables us to fill another row in the upgoing wave table, ad infinitum.

The inverse calculation proceeds in a similar fashion. Suppose $d_t{}^0$ is known for all t and we wish to calculate c_1', c_2', etc., in a recursive fashion. It is sufficient to show how to compute c_1'. Suppose we skim off the two left-hand columns of (11-4-16). We get

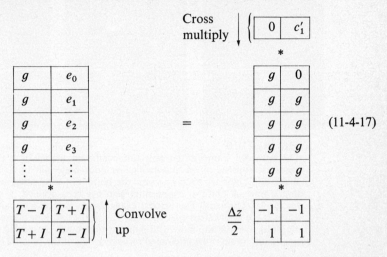

$$(11\text{-}4\text{-}17)$$

In (11-4-17) all the boxes filled by g on the left are given. The boxes with g on the right are readily computable as before. If c_1' were known, it would be a straightforward task to compute successively $\ldots, e_3, e_2, e_1, e_0$. We would be compelled to initialize the computation with an approximation such as $e_N = 0$ for some large N. If the correct value of c_1' had been used, then we should find that e_0 vanishes. Since we do not know what value of c_1' to use, we try $c_1' = +1$ obtaining e_0^+ and we try $c_1' = -1$ obtaining e_0^-. The correct value of c_1' is the appropriately weighted linear combination

$$0 = \alpha e_0^+ + \beta e_0^- \qquad (11\text{-}4\text{-}18)$$

where

$$1 = \alpha + \beta \qquad (11\text{-}4\text{-}19)$$
$$c_1' = \alpha - \beta \qquad (11\text{-}4\text{-}20)$$

which inverts to

$$2\alpha = 1 + c_1' \qquad (11\text{-}4\text{-}21)$$
$$2\beta = 1 - c_1' \qquad (11\text{-}4\text{-}22)$$

reducing (11-4-11) to

$$0 = (1 + c_1')e_0^+ + (1 - c_1')e_0^-$$

or

$$c_1' = \frac{e_0^+ + e_0^-}{e_0^+ - e_0^-} \qquad (11\text{-}4\text{-}23)$$

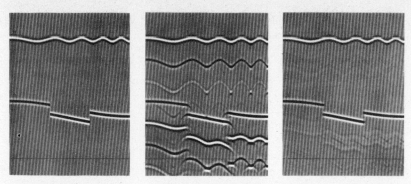

FIGURE 11-18
The left-hand frame is the reflection-coefficient model of a two-dimensional earth. It consists of an undulating sea floor underlain by a faulted, dipping structure. The horizontal line near the bottom of the frames is the one-second timing line. A uniform exponential gain has been applied to all three frames for display purposes. The vertical exaggeration is 5 on all frames. The center frame is the synthesized time section. Below the sea floor are seen the sea-floor multiples. Below the faulted, dipping structure are the pegleg multiples. The right-hand frame is the attempted reconstruction of the original model.

Thus, the idea which we have used to compute c_1' is the principle that the upgoing wave at depth z_1 (which is e_0) must vanish before the onset of a downgoing wave at z_1.

A number of practical difficulties must be overcome before realistic behavior can be expected of either the one-dimensional- or the two-dimensional algorithms which were presented. First of all is the problem of the source spreading out in three dimensions. This calls for a spherical spreading correction. Luckily, since the main idea involves a *ratio* between up- and downgoing waves, the true amplitude does not seem to be very important. In fact, the calculation may be shown to be invariant under exponential gain readjustment. One real problem which does arise is that the amplitude of the initial shot delta function cannot be properly measured and it is required for the calculation. In fact, there is a shot *waveform* to be accounted for which may also have a slight timing error due to residual moveout. Luckily, the shot waveform can be calibrated in deep water where the sea floor is flat. Say at such a place the returned wave is scaled to zero everywhere except around the sea-floor primary reflection; call this waveform P. Also, another waveform M is created by weighting the returned wave to 0 everywhere except around the first multiple. Ideally, P convolved onto itself should give M; the departure allows estimation of shot amplitude, timing, and waveform.

A fact to realize is that strong reflective interfaces may not fall precisely on a sampled data point. This means that $(4, 0)$, $(3, 1)$, $(2, 2)$, $(1, 3)$, and $(0, 4)$ must all be regarded as successive delays of the same impulsive waveform. The problem that their spectra differ at high frequencies is the same type of problem as differential equations departing from difference equations for wavelengths shorter than about ten sample points. There is no need to model the high frequencies accurately; just be sure to sample the data densely enough. The real problem with high frequencies is keeping them from causing instability. For example, in the calculation of $U^0/D^0 = -\hat{W}(Z)/[1 + \hat{W}(Z)]$ it is first necessary to estimate the $\hat{W}$ which would be recorded if the shot P/I had been an ideal delta function. Naturally, the high frequencies in $\hat{W}$ are rather meaningless. The only reason you care about them is that they should not be such as to make $1 + \hat{W}$ nonminimum-phase which would prevent the division. In this case, the high frequencies can be filtered out from $\hat{W}$ and the divisor will tend toward a positive real function. Don C. Riley [Ref. 38] has established that these various practical difficulties can be overcome and Fig. 11-18 shows an example of one of his calculations.

REFERENCES

[1] FLINN, EDWARD A., ENDERS A. ROBINSON, and SVEN TREITEL (Eds.): The MIT Geophysical Analysis Group (GAG) Reports, *Geophysics*, vol. 32, no. 3, 1967.

[2] HASKELL, N. A.: The Dispersion of Surface Waves on Multilayered Media, *Bull. Seismal Soc. Am.*, vol. 43, pp. 17–34, 1953.

[3] CLAERBOUT, JON F.: Coarse Grid Calculations of Waves in Inhomogeneous Media with Application to Delineation of Complicated Seismic Structure, *Geophysics*, vol. 35, no. 3, 1970.

[4] CLAERBOUT, JON F.: Numerical Holography, *Acoustic Holography*, vol. 3, Plenum Press, 1971.

[5] CLAERBOUT, JON F.: Toward a Unified Theory of Reflector Mapping, *Geophysics*, vol. 36, no. 3, 1971.

[6] CLAERBOUT, JON F., and A. G. JOHNSON: Extrapolation of Time Dependent Waveforms along Their Path of Propagation, *Geophys. J. R. Astron. Soc.*, vol. 26, nos. 1–4, pp. 285–295, 1971.

[7] LANDERS, T., and JON F. CLAERBOUT: Numerical Calculations of Elastic Waves in Laterally Inhomogeneous Media, *J. Geophys. Res.*, vol. 77, no. 8, pp. 1476–1482, 1972.

[8] CLAERBOUT, JON F., and S. M. DOHERTY: Downward Continuation of Moveout Corrected Seismograms, *Geophysics*, vol. 37, no. 5, pp. 741–768, 1972.

[9] *IEEE Trans. Audio Electrostat.*, Special Issue on fast Fourier transform, June, 1967.

[10] WOLD, HERMAN: "Stationary Time Series," Almquist and Wiksell, Stockholm, 1938.

[11] ROBINSON, ENDERS A.: "Random Wavelets and Cybernetic Systems," Griffin, London, 1962.

[12] GRENANDER, ULF, and GABOR SZEGO: "Toeplitz Forms and Their Applications," University of California Press, Berkeley and Los Angeles, 1958.

[13] LEVINSON, NORMAN: The Wiener RMS (Root Mean Square) Error Criterion in Filter Design and Prediction, Appendix B *in* NORBERT WIENER: "Extrapolation, Interpolation and Smoothing of Stationary Time Series," Technology Press of the Massachusetts Institute of Technology, Cambridge, 1947.

[14] ATKINSON, F. V.: "Discrete and Continuous Boundary Problems," Academic Press, New York, 1964.

[15] WHITTLE, P.: "Prediction and Regulation," English Universities Press, Ltd., London, 1963.

[16] KOLMOGOROV, A.: Sur l'interpolation et l'extrapolation des suites stationnaires, *C. R. Acad. Sci.* (Paris), vol. 208, pp. 2043–2045, 1939.

[17] GABOR, D.: Theory of Communication, *J. IEEE*, vol. 93, no. 1, pp. 429–441, 1946.

[18] MOSTELLER, F., and R. E. K. ROURKE: "Sturdy Statistics," Addison Wesley Publishing Co., Inc., Reading, Mass., 1973.

[19] CLAERBOUT, J. F.: Spectral Factorization of Multiple Time Series, *Biometrica*, vol. 53, nos. 1 and 2, pp. 264–6, 1966.

[20] SENETA, EUGENE: "Non-Negative Matrices," John Wiley and Sons, Inc., New York, 1973.

[21] GOLUB, G.: Numerical Methods for Solving Linear Least Squares Problems, *Numerische Mathematik*, vol. 7, pp. 206–216, 1965.

[22] SINGLETON, R. S.: Algorithm 347 Sort, *Comm. ACM*, vol. 12, 1969.

[23] HOARE, C. A. R.: Quicksort, *Comput. J.*, vol. 5, pp. 10–15, 1962.

[24] PLACKETT, R. L.: Studies in the History of Probability and Statistics, Chapt. XXIX, *in* The Discovery of the Method of Least Squares: *Biometrica*, vol. 59, no. 2, pp. 239–251, 1972.

[25] CLAERBOUT, JON F., and FRANCIS MUIR: Robust Modeling with Erratic Data, *Geophysics*, vol. 38, no. 5, pp. 826–844, 1973.

[26] CLAERBOUT, JON F., and E. A. ROBINSON: The Error in Least-Squares Inverse Filtering, *Geophysics*, vol. 29, no. 1, pp. 118–120, 1964.

[27] BURG, JOHN PARKER: The Relationship Between Maximum Entropy Spectra and Maximum Likelihood Spectra, *Geophysics*, vol. 37, no. 2, pp. 375–376, 1972.

[28] WIDROW, B., P. E. MANTEY, J. J. GRIFFITHS, and B. B. GOODE: Adaptive Antenna Systems, *Proc. IEEE*, vol. 55, pp. 2143–2159, 1967.

[29] LEVIN, M. J.: Maximum Likelihood Array Processing, MIT Lincoln Laboratory Technical Report, 1964.

[30] GOUPILLAUD, P.: An Approach to Inverse Filtering of Near Surface Layer Effects from Seismic Records, *Geophysics*, vol. 26, no. 6, pp. 754–760, 1961.

[31] KUNETZ, G.: Generalization des opérateurs d'antirésonance a un nombre quelconque de réflecteurs, *Geophys. Prosp.*, vol. 12, pp. 283–289, 1964.

[32] WHITE, J. E.: Use of Reciprocity Theorem for Computation of Low-Frequency Radiation Patterns, *Geophysics*, vol. 25, no. 3, pp. 613–624, 1960.

[33] RICHTMYER, ROBERT D., and K. W. MORTON: "Difference Methods for Initial Value Problems," John Wiley and Sons, Inc., New York, 1967.

[34] MITCHELL, A. R.: "Computational Methods in Partial Differential Equations," John Wiley and Sons, Inc., New York, 1969.

[35] GOODMAN, J. W.: "Introduction to Fourier Optics," McGraw-Hill Book Company, New York, 1968.

[36] TREITEL, SVEN, J. L. SHANKS, and C. W. FRASIER: Some Aspects of Fan Filtering, *Geophysics*, vol. 32, no. 5, pp. 789–800, 1967.

[37] DOHERTY, STEPHEN M.: "Structure Independent Seismic Velocity Estimation," doctoral dissertation, Geophysics Department, Stanford University, Stanford, California, 1975.

[38] RILEY, DON CLINTON: "Wave Equation Synthesis and Inversion of Diffracted Multiple Seismic Reflections," doctoral dissertation, Geophysics Department, Stanford University, Stanford, California, 1975.